Topological Charge of Optical Vortices

This book is devoted to the consideration of unusual laser beams – vortex or singular beams. It contains many numerical examples, which clearly show how the phase of optical vortices changes during propagation in free space, and that the topological charge is preserved.

Topological Charge of Optical Vortices shows that the topological charge of an optical vortex is equal to the number of screw dislocations or the number of phase singularities in the beam cross-section. A single approach is used for the entire book: based on M. Berry's formula. It is shown that phase singularities during beam propagation can be displaced to infinity at a speed greater than the speed of light. The uniqueness of the book is that the calculation of the topological charge for scalar light fields is extended to vector fields and is used to calculate the Poincare–Hopf singularity index for vector fields with inhomogeneous linear polarization with V-points and for the singularity index of vector fields with inhomogeneous elliptical polarization with C-points and C- lines.

The book is written for opticians, and graduate students interested in an interesting section of optics – singular optics. It will also be of interest to scientists and researchers who are interested in modern optics. In order to understand the content of the book, it is enough to know paraxial optics (Fourier optics) and be able to calculate integrals.

Topological Charge of Optical Vortices

Victor V. Kotlyar, Alexey A. Kovalev,
and Anton G. Nalimov

CRC Press
Taylor & Francis Group
Boca Raton London New York

CRC Press is an imprint of the
Taylor & Francis Group, an **informa** business

The authors thank Alexey Kovalev for providing the cover image.

MATLAB® is a trademark of The MathWorks, Inc. and is used with permission. The MathWorks does not war-
rant the accuracy of the text or exercises in this book. This book's use or discussion of MATLAB® software or
related products does not constitute endorsement or sponsorship by The MathWorks of a particular pedagogical
approach or particular use of the MATLAB® software.

First edition published 2023
by CRC Press
6000 Broken Sound Parkway NW, Suite 300, Boca Raton, FL 33487-2742

and by CRC Press
4 Park Square, Milton Park, Abingdon, Oxon, OX14 4RN

CRC Press is an imprint of Taylor & Francis Group, LLC

© 2023 Victor V. Kotlyar, Alexey A. Kovalev, and Anton G. Nalimov

ISBN: 978-1-032-34553-6 (HB)
ISBN: 978-1-032-35309-8 (PB)
ISBN: 978-1-003-32630-4 (EB)

DOI: 10.1201/9781003326304

Typeset in Times
by Deanta Global Publishing Services, Chennai, India

Contents

Preface

The book is devoted to the consideration of unusual laser beams – vortex or singular beams. They are called vortex beams because their energy flow rotates in a spiral around the optical axis when the beam propagates along this axis. And they are called singular beams because in the cross-section of such beams there are isolated points with zero intensity, in which the phase is not defined. These points are singular (unusual points). Vortex laser beams are described by two integral (averaged) characteristics: orbital angular momentum (OAM) and topological charge (TC). The OAM of the beam shows with what moment of force the light will act on a microparticle placed at the focus of the vortex laser beam. And the TC of the beam shows how many complete jumps by 2π the phase will acquire in the cross-section of the beam when it is traversed along a closed contour that encompasses the entire beam. If the OAM of the beam is preserved during the propagation of a scalar paraxial laser beam, then the TC is also preserved if we take into account the phase singularities located at the periphery in the beam cross-section (at infinity). The OAM normalized to the beam power can be either an integer or a fractional number. And the TC of the beam (except for the initial plane and at infinity) is always an integer. In the book, the TC of the axial superposition of Laguerre–Gauss (LG) vortex beams is calculated. It turned out to be equal to the maximal TC of the beams in the superposition. Therefore, if the amplitude of the light field is expanded according to the basis of the LG beams in an infinite series, then the TC of such a light field is infinite. The book contains examples of vortex beams with infinite TC. Note that the OAM of the beam is determined by the amplitude and phase, and TC is determined only by the phase of the beam. Therefore, there are light beams in which the TC is zero, and the OAM is nonzero (for example, elliptical Gaussian beams). And vice versa, there are beams in which the OAM is equal to zero, and the TC is nonzero. The book shows that the phase singularities of optical vortices can "go" to infinity (and "come" from infinity) with a speed greater than the speed of light in a vacuum. Examples of light beams with a half-integer TC are also given, but the fractional part of the TC is at infinity ("hidden" phase). The book shows that the TC of the superposition of two parallel LG beams depends on which beam is on the left and which is on the right. And if we rearrange the LG beams in places (left to right, and right to left), then the TC of such a superposition will change to 1. The book will be of interest to everyone who is interested in optics.

A sufficient number of monographs are devoted to the study of the OAM of vortex beams, and there are no monographs on the study of TC. This book is precisely devoted to the study of the topological charge of vortex laser beams.

MATLAB ® is a registered trademark of The MathWorks, Inc. For product information, please contact:

The MathWorks, Inc.
3 Apple Hill Drive
Natick, MA 01760-2098 USA
Tel: 508 647 7000
Fax: 508-647-7001
E-mail: info@mathworks.com
Web: www.mathworks.com

Acknowledgments

This book contains research results financially supported by the Russian Science Foundation grant No. 18-19-00595.

The authors are grateful to PhD A.P. Porfirev and Prof. S. Rasouli for carrying out experiments on the propagation of beams with optical vortices.

Authors

Victor V. Kotlyar is Head of the Laboratory at Image Processing Systems Institute of the Russian Academy of Science, a branch of the Federal Scientific Research Center "Crystallography and Photonics", and Professor of Computer Science at Samara National Research University, Russia. He earned his MS, PhD, and DrSc degrees in Physics and Mathematics from Samara State University (1979), Saratov State University (1988), and Moscow Central Design Institute of Unique Instrumentation, the Russian Academy of Sciences (1992). He is a SPIE- and OSA-member. He is co-author of 400 scientific papers, 7 books, and 7 inventions. His current interests are diffractive optics, gradient optics, nanophotonics, and optical vortices.

Alexey A. Kovalev graduated in 2002 from Samara National Research University, Russia, majoring in Applied Mathematics. He earned his PhD in Physics and Maths in 2012. He is senior researcher of Laser Measurements at the Image Processing Systems Institute of the Russian Academy of Science, a branch of the Federal Scientific Research Center "Crystallography and Photonics". He is a co-author of more than 270 scientific papers. His research interests are mathematical diffraction theory, photonic crystal devices, and optical vortices.

Anton G. Nalimov graduated from Samara State Aerospace University, Russia, in February 2003. He entered postgraduate study in 2003 with a focus on the specialty 05.13.18 "Mathematical Modeling and Program Complexes". He finished it in 2006 with the specialty 01.04.05 "Optics". Nalimov works in the Technical Cybernetics department at Samara National Research University as an associate professor, and also works as a scientist in the Image Processing Systems Institute of the Russian Academy of Science, a branch of the Federal Scientific Research Center "Crystallography and Photonics" in Samara. He is a PhD candidate in Physics and Mathematics, co-author of 200 papers and 3 inventions.

Introduction

Laser optical vortices are light fields that have singularity points in the phase distribution, i.e. points where the phase is undetermined. Optical vortices also have screw dislocations in their wavefront. In the intensity distribution of optical vortices, there are isolated points of zero intensity. Topological charge (TC) is one of the main characteristics of optical vortices. This is an integer number equal to the number of phase jumps by 2π along an infinite-radius circle in the beam cross-section. The TC is positive if the phase increases counterclockwise, and TC is negative otherwise. This definition of the topological charge was given by M.V. Berry and it shows that the cross-section of an optical vortex can contain both a finite and an infinite number of singularity points (local optical vortices), that can reside in the periphery of the laser beam in areas with almost zero intensity. Such peripheral points of phase singularity cannot be detected experimentally, but they contribute to the TC and cannot be neglected. Optical vortices can have only an integer TC or an indefinite TC. An optical vortex can have a fractional TC only in the initial plane, since, in this plane, arbitrary TC can be given. However, on propagation in free space, the initial fractional TC generates an infinite number of local optical vortices with opposite TCs, which are located at different distances from the optical axis of the beam. Therefore, the TC of such a beam is indefinite, since there are different numbers of phase jumps by 2π on transverse circles with different radii. The TC is conserved on propagation in free space, similarly to the orbital angular momentum of vortex beams. There are works where the authors demonstrated that the TC of the combined beams and of the beams with an initial fractional TC is not conserved on propagation. However, these works did not take into account local optical vortices located in the beam periphery, since these vortices cannot be detected experimentally or in simulation within the paraxial limits. Such peripheral optical vortices can be detected by nonparaxial simulation by using the Rayleigh–Sommerfeld integrals.

In this book, topological charges are obtained for a superposition (coaxial and noncoaxial) of the Laguerre–Gaussian and Bessel–Gaussian beams, for asymmetric beams, and the Hermite–Gaussian vortex beams. The TC evolution is shown for two combined Laguerre–Gaussian beams with different waists radii. It is shown how the TC is generated in optical vortices with an initial fractional topological charge. It is demonstrated that the TC is conserved on propagation in space, as well as after passing through an arbitrary amplitude mask, and is resistant to random phase distortions.

1 Topological Charge of Superposition

Conservation of Topological Charge

1.1 TOPOLOGICAL CHARGE AND ASYMPTOTIC PHASE INVARIANTS OF VORTEX LASER BEAMS

There are several well-known non-diffracting and propagation-invariant light fields. The most prominent examples in 3D space are the Bessel beams [1], parabolic beams [2], Mathieu beams [3], as well as Laguerre–Gaussian and Hermite–Gaussian paraxial modes [4]. In 2D space, there are also the Airy and Weber beams [5,6]. A thorough review of propagation-invariant fields can be found in [7,8]. Besides propagating in free space, the interaction of such beams with matter is also studied, including non-linear processes [9]. Potential applications of such beams are, for example, wireless communications and optical interconnections. In addition to the beams that preserve their shape on propagation, there are several properties of the beam cross-section which are also propagation-invariant. These properties can be used as indicators that can help the receiver to identify the incoming signal beam. For instance, well-known indicators are the orbital angular momentum (OAM) and the topological charge (TC) of vortex beams. Many works were dedicated to the conservation of these properties, either on propagation in atmospheric turbulence [10] or after amplitude distortions [11]. Many of these works were about determining the OAM [12,13] or TC [14,15] of an optical signal beam.

These two indicators are often used interchangeably since for conventional rotationally symmetric optical vortices both OAM and TC give the same value. However, the nature of these indicators is quite different physically. While OAM is an integral property of a light field transverse intensity and phase distributions which are calculated by integration over the whole transverse plane [16,17,18], TC is a purely phase property which is calculated by integration of phase angular derivative over an infinite-radius circle [19] or closed curve. Thus, TC can be treated as an asymptotic phase property. Propagation invariance of the OAM can be easily proven mathematically since the propagation operator (Fresnel transform) is unitary. Conservation of both the OAM and of the spin angular momentum on free-space propagation was proven in [20] (Section 4 "Eigenoperator description of laser beams"). The invariance of the TC cannot be proven in this way. This can be proven intuitively since it

DOI: 10.1201/9781003326304-1

is known that phase singularities can disappear only as the result of the annihilation of two singularities of opposite topological charge [21]. Thus, on propagation, *TC* should not change its value. However, there is a well-known work by M.S. Soskin et al., where a superposition of Gaussian and Laguerre–Gaussian beams with different-waist radii can change the *TC* [22]. This seems to contradict the idea of *TC* conservation. Recently, we revisited this problem, studying the *TC* change on the propagation of two different-waist LG beams [23] and showed that, when nearing the *TC*-change plane, certain vortices move away from the optical axis to infinity. Thus, the summary *TC* of all the vortices, including those in infinity, remains. Therefore, two questions arise: can the *TC* conservation be proven mathematically, and are there some other propagation-invariant asymptotic phase invariants of light fields?

In this section, to prove that *TC* is conserved upon propagation, we introduce a huge-ring approximation, which is similar to the paraxial approximation, but, on the contrary, the distance from the optical axis is much larger than the propagation distance. Using this approximation, we prove that *TC* value does not change from one transverse plane to another. In addition, we show that another asymptotic phase propagation-invariants can be constructed similarly to the *TC*.

1.1.1 ORBITAL ANGULAR MOMENTUM AND TOPOLOGICAL CHARGE

If a light field propagates along the optical axis z and has the complex amplitude $E(r, \varphi, z)$, where (r, φ, z) are the cylindrical coordinates, then its normalized OAM (OAM J_z divided by beam power W) in a transverse plane reads as [16,17,18]:

$$\frac{J_z}{W} = \frac{\mathrm{Im} \int_0^\infty \int_0^{2\pi} E^*(r,\varphi,z) \frac{\partial E(r,\varphi,z)}{\partial \varphi} r dr d\varphi}{\int_0^\infty \int_0^{2\pi} E^*(r,\varphi,z) E(r,\varphi,z) r dr d\varphi}, \tag{1.1}$$

with Im being the imaginary part of a complex number, while *TC* μ is defined as the integral over an infinite-radius circle [19]:

$$\mu = \frac{1}{2\pi} \lim_{r \to \infty} \int_0^{2\pi} \frac{\partial}{\partial \varphi} \left[\arg E(r,\varphi,z) \right] d\varphi. \tag{1.2}$$

1.1.2 PROPAGATION OF A LIGHT FIELD IN FREE SPACE AND CONSERVATION OF ITS ORBITAL ANGULAR MOMENTUM

The complex amplitude of a monochromatic light field in homogeneous medium obeys the Helmholtz equation, which in the cylindrical coordinates reads as:

$$\frac{\partial^2 E}{\partial r^2} + \frac{1}{r} \frac{\partial E}{\partial r} + \frac{1}{r^2} \frac{\partial^2 E}{\partial \varphi^2} + \frac{\partial^2 E}{\partial z^2} + k^2 E = 0, \tag{1.3}$$

where $k = 2\pi / \lambda$ is the module of wavevector for light with the wavelength of λ. For paraxial propagation, the Helmholtz equation reduces to:

$$2ik\frac{\partial E}{\partial z} + \frac{\partial^2 E}{\partial r^2} + \frac{1}{r}\frac{\partial E}{\partial r} + \frac{1}{r^2}\frac{\partial^2 E}{\partial \varphi^2} = 0. \tag{1.4}$$

It is well-known that if E is a solution of Equation (1.4) then the complex amplitude $E(r, \varphi, z)$ in a transverse plane is related to that in the initial plane $(z = 0)$ by the Fresnel transform [24]:

$$E(r,\varphi,z) = \frac{-ik}{2\pi z}\exp\left(ikz + \frac{ikr^2}{2z}\right)$$

$$\times \int_0^\infty \int_0^{2\pi} E(\rho,\theta,0)\exp\left[\frac{ik\rho^2}{2z} - i\frac{k}{z}r\rho\cos(\theta - \varphi)\right]\rho d\rho d\theta. \tag{1.5}$$

It can be shown the dot product of two functions is equal to the dot product of their Fresnel transforms. In addition, if E is a solution of Equation (1.4) then it is obvious that the functions E^* and $\partial E/\partial \varphi$ are also solutions of Equation (1.4). Therefore, both numerator and denominator in Equation (1.1) are conserved on propagation and thus the normalized OAM is conserved too. The detained proof of the OAM conservation can be found in [20] (Section 4 "Eigenoperator description of laser beams").

As to nonparaxial free-space propagation, the OAM should be analyzed as a vectorial quantity. Its z-component reads as [25]:

$$J_z = \text{Im}\sum_{i=x,y,z}\int_0^\infty\int_0^{2\pi} E_i^*(r,\varphi,z)\frac{\partial E_i(r,\varphi,z)}{\partial \varphi}rdrd\varphi, \tag{1.6}$$

$$E(x,y,z) = \int_{-\infty}^\infty\int_{-\infty}^\infty A(\alpha,\beta)\exp\left[ik\left(\alpha x + \beta y + z\sqrt{1-\alpha^2-\beta^2}\right)\right]d\alpha d\beta, \tag{1.7}$$

with (x, y) being the Cartesian coordinates in a transverse plane and (α, β) being the Cartesian coordinates in the Fourier plane (cosines of the angles defining the directions of plane waves), $A(\alpha, \beta)$ being the angular spectrum of plane waves.

1.1.3 Conservation of the Topological Charge

Unfortunately, the conservation of TC (Equation (1.2)) cannot be proven so easily. According to the TC definition in Equation (1.2), the field should be analyzed in its periphery, at an infinite distance r from the optical axis. Thus, a paraxial approximation is inappropriate here. Therefore, we introduce another approximation here, quite opposite to the paraxial. Generally, without the paraxial limits, if a light field propagates along the z-axis, its complex amplitudes in two transverse planes (source plane and observation plane) are related by the Rayleigh–Sommerfeld integral [26]:

$$E(r,\varphi,z) = \frac{-1}{2\pi} \int_0^\infty \int_0^{2\pi} E(\rho,\theta,0) \frac{\partial}{\partial z}\left[\frac{\exp(ikL)}{L}\right] \rho\, d\rho\, d\theta, \qquad (1.8)$$

where L is the distance between a point in the source plane $(\rho,\theta,0)$ and a point in the observation plane (r,φ,z):

$$L = \left[z^2 + r^2 + \rho^2 - 2r\rho\cos(\theta-\varphi)\right]^{1/2}. \qquad (1.9)$$

The complex amplitude given by Equation (1.8) is an exact solution of the Helmholtz equation and describes a light field without paraxial approximation. The Fresnel transform in Equation (1.5) can be obtained from Equation (1.8) for a case when the propagation distance is large compared to the transverse coordinates (paraxial propagation) (Figure 1.1(a)) and therefore:

$$L \approx z + \frac{r^2 + \rho^2 - 2r\rho\cos(\theta-\varphi)}{2z}. \qquad (1.10)$$

To calculate the topological charge, we now suppose that, on the contrary, r is much greater than z and ρ (Figure 1.1(b)). Thus, the distance L is given by:

$$L \approx r + \frac{z^2 + \rho^2}{2r} - \rho\cos(\theta-\varphi). \qquad (1.11)$$

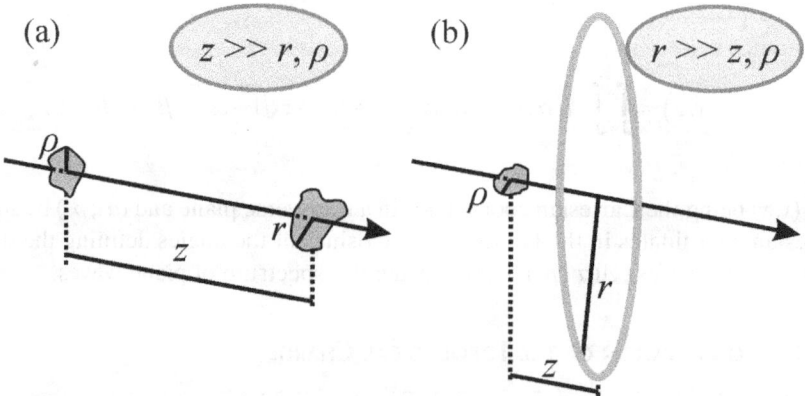

FIGURE 1.1   (a) Paraxial approximation: propagation distance z is much larger than transverse coordinates ρ and r in the input and output planes respectively, (b) huge-ring approximation: transverse coordinate r in the output plane is much larger than the propagation distance z and the transverse coordinate ρ in the input plane.

We call here this expression a huge-ring approximation. Using it, we obtain the complex amplitude on a very large radius ring in a transverse plane. The Rayleigh–Sommerfeld integral can be rewritten as:

$$E(r,\varphi,z) = \frac{-z}{2\pi} \int\limits_{0}^{\infty} \int\limits_{0}^{2\pi} E(\rho,\theta,0) \left(\frac{ik}{L^2} - \frac{1}{L^3} \right) \exp(ikL) \rho d\rho d\theta. \tag{1.12}$$

Similarly to the derivation of the Fresnel transform, we use Equation (1.11) for the exponent in the integrand, while the expression $(ik/L^2 - 1/L^3)$ we write approximately as ik/r^2:

$$E(r,\varphi,z) = \frac{-ikz}{2\pi r^2} \exp\left(ikr + ik\frac{z^2}{2r} \right)$$

$$\times \int\limits_{0}^{\infty} \int\limits_{0}^{2\pi} E(\rho,\theta,0) \exp\left[\frac{ik\rho^2}{2r} - ik\rho\cos(\theta - \varphi) \right] \rho d\rho d\theta. \tag{1.13}$$

Equation (1.13) is the main equation in this section and it allows us to estimate the complex amplitude on a circle of a very large radius, just as needed to calculate TC. It is seen in Equation (1.13) that the integral is independent of z and depends only on the angular polar coordinate φ. On the contrary, the multipliers before the integrals are φ-independent. Therefore, the φ-derivative of the field phase $\partial(\arg E)/\partial \varphi$ is independent of z:

$$\frac{\partial}{\partial z} \left[\lim_{r \to \infty} \frac{\partial}{\partial \varphi} \arg E(r,\varphi,z) \right] = 0. \tag{1.14}$$

Consequently, z-independent are any quantities obtained from $\partial(\arg E)/\partial \varphi$ at large radii r. The most prominent example is the topological charge of Equation (1.2). Thus, TC conservation is just a partial case following Equation (1.14), and below we consider some other partial cases.

1.1.4 Asymptotic Phase Invariants of Vortex Laser Beams

Since the field should be continuous at $\varphi = 0$ and $\varphi = 2\pi$, $\arg E$ should change by an integer number of 2π, thus forcing TC to be an integer number. Even if in the initial plane it was fractional, then, on propagation, it becomes an integer [19]. However, Equation (1.14) indicates that, by using the function $\partial(\arg E)/\partial \varphi$ at large radii r, other propagation-invariant quantities may also be constructed, e.g.:

$$\mu_g = \frac{1}{2\pi} \lim_{r \to \infty} \int\limits_{0}^{2\pi} g(\varphi) \frac{\partial}{\partial \varphi} \left[\arg E(r,\varphi,z) \right] d\varphi \tag{1.15}$$

with $g(\varphi)$ being an arbitrary function. To confirm our theory, we made some numerical experiments. For constructing the invariants, we choose two functions: $g_1(\varphi) = \cos\varphi$ and $g_2(\varphi) = \text{rect}(\varphi / \pi)$ (i.e. $g_2(\varphi) = 1$ at $-\pi/2 \leq \varphi \leq \pi/2$ and $g_2(\varphi) = 0$ otherwise). So, we test whether or not the following values:

$$\mu_1 = \frac{1}{2\pi} \lim_{r \to \infty} \int_0^{2\pi} \cos\varphi \frac{\partial}{\partial\varphi} \left[\arg E(r,\varphi,z) \right] d\varphi, \qquad (1.16)$$

$$\mu_2 = \frac{1}{2\pi} \lim_{r \to \infty} \int_{-\pi/2}^{\pi/2} \frac{\partial}{\partial\varphi} \left[\arg E(r,\varphi,z) \right] d\varphi \qquad (1.17)$$

are asymptotic phase invariants. We note that, in general, the functions $g(\varphi)$ and the argument of the complex amplitude ($\arg E$) in Equation (1.15) are not analytic functions (for instance, $g_2(\varphi)$ in Equation (1.17) is not analytic). Therefore, the integral within Equation (1.15) cannot be evaluated by transforming to a contour integral in the complex plane and by using the theory of residues.

1.1.5 NUMERICAL SIMULATION

The theory is obtained for the invariant quantities, computed over an infinite-radius circle. This is impractical, but gives an idea that these quantities can also conserve in other, realistic, conditions. Below, we consider two paraxial light beams and choose feasible simulation parameters quite opposite to the huge-ring approximation, i.e. when the circle radius is much smaller than the propagation distance. For a test beam, we take a superposition of two Gaussian beams with optical vortices. In the initial plane ($z = 0$), such a beam has the following complex amplitude:

$$E(r,\varphi,0) = \exp\left(-\frac{r^2}{w^2} \right) \left[A_1 \exp(in_1\varphi) + A_2 \exp(in_2\varphi) \right], \qquad (1.18)$$

where w is the Gaussian beam waist radius, n_1 and n_2 are the topological charges of the vortices, A_1 and A_2 are the superposition coefficients.

Figure 1.2 illustrates the intensity and phase distributions of the test beam (Equation 1.18) in several transverse planes for the following parameters: wavelength $\lambda = 633$ nm, waist radius $w = 1$ mm, topological charges $n_1 = 10$ and $n_2 = 5$, superposition coefficients $A_1 = 1$ and $A_2 = 0.5$, propagation distances $z = 0$ (initial plane), $z = z_0/2$ ($z_0 = kw^2/2$ is the Rayleigh range), and $z = 2z_0$ (far field). The calculation area is $-R \leq x, y \leq R$ with (x, y) being the Cartesian coordinates and R being the half-size of the area: $R = 5$ mm for $z = 0$, $R = 10$ mm for $z = z_0/2$, and $R = 20$ mm for $z = 2z_0$. Dashed lines on the phase distributions show the circles along which we calculate TC (Equation (1.2)) and the invariants shown in Equations (1.16) and (1.17) (all circles are of radius $0.8R$). Figure 1.2(a, d) is obtained by Equation (1.18), while Figure 1.2(b, c, e, f) is obtained by using an expression for the Fresnel diffraction of a Gaussian

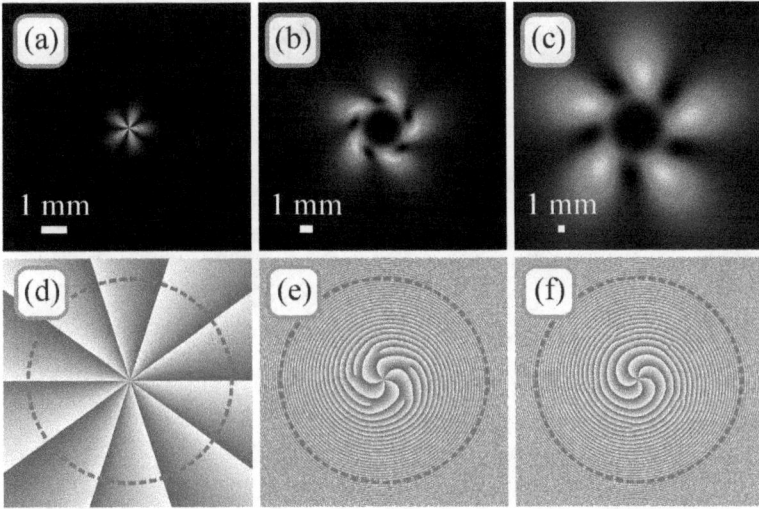

FIGURE 1.2 Intensity (a–c) and phase (d–f) distributions of a superposition of two Gaussian vortices (Equation (1.18)) in several transverse planes for the following parameters: wavelength $\lambda = 633$ nm, waist radius $w = 1$ mm, topological charges $n_1 = 10$ and $n_2 = 5$, superposition coefficients $A_1 = 1$ and $A_2 = 0.5$, propagation distances $z = 0$ (a, d), $z = z0/2$ (b,e), and $z = 2z0$ (c,f), calculation area $-R \leq x, y \leq R$ with $R = 5$ mm (a, d), $R = 10$ mm (b, e), and $R = 20$ mm (c, f).Dashed circles on the phase distributions are those along which TC (Equation (1.2)) and invariants from Equations (1.16) and (1.17) are calculated.

optical vortex [27,28]. Calculation of TC by Equation (1.2) yields nearly the same values for all three propagation distances: $\mu = 9.9999$ at $z = 0$, $\mu = 9.9695$ at $z = z_0/2$, and at $z = 2z_0$ (theoretical value of μ is 10 [13]). Calculation by Equation (1.16) yields the value $\mu_1 = -0.0015$ for all three distances z. Calculation by Equation (1.17) yields the value $\mu_2 = 4.8514$ at $z = 0$, $\mu_2 = 4.9702$ at $z = z_0/2$, and $\mu_2 = 4.9654$ at $z = 2z_0$ (it is obvious that the theoretical value of μ_2 should be $10/2 = 5$). As another test beam, we consider a Gaussian beam with multiple vortices located on a circle of a radius r_0. Such a beam can be obtained as a coaxial superposition of a Laguerre–Gaussian vortex beam and of a Gaussian beam [29]. The complex amplitude of this beam reads as:

$$E(r,\varphi,z) = \frac{1}{q}\exp\left(-\frac{r^2}{qw^2}\right)\left(\frac{r^m e^{im\varphi}}{q^m} - r_0^m\right), \qquad (1.19)$$

where w is the Gaussian beam waist radius, m is the number of optical vortices with unit topological charge, $q = 1 + iz/z_0$.

Figure 1.3 depicts the intensity and phase distributions of the test beam from Equation (1.19) in several transverse planes for the following parameters: wavelength $\lambda = 633$ nm, waist radius $w = 1$ mm, number of vortices $m = 3$, the radius of the circle with vortices $r_0 = 0.7w$, propagation distances $z = 0$, $z = z_0/2$, and $z = 2z_0$, calculation area $-R \leq x, y \leq R$ with $R = 5$ mm for $z = 0$, $R = 10$ mm for $z = z_0/2$, and $R = 20$ mm

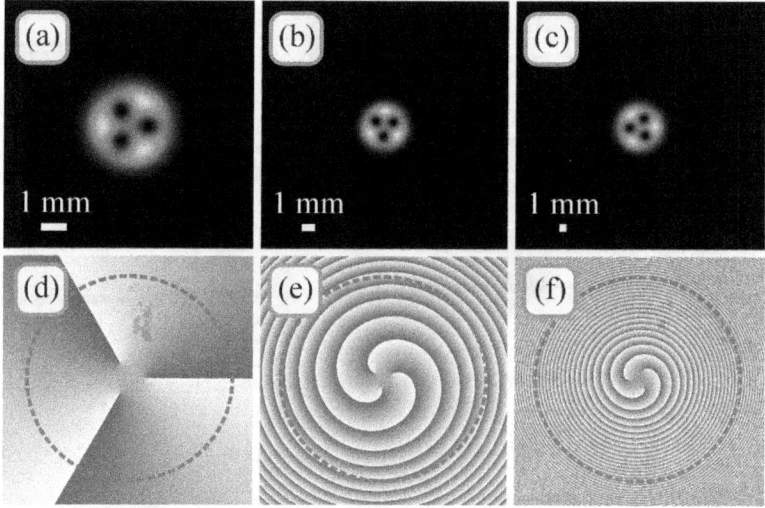

FIGURE 1.3 Intensity (a–c) and phase (d–f) distributions of a Gaussian beam with several vortices (Equation (1.19)) in several transverse planes for the following parameters: wavelength $\lambda = 633$ nm, waist radius $w = 1$ mm, number of vortices $m = 3$, radius of the circle with vortices $r_0 = 0.7w$, propagation distances $z = 0$ (a,d), $z = z_0/2$ (b, e), and $z = 2z_0$ (c, f), calculation area $-R \leq x, y \leq R$ with R = 5 mm (a, d), R = 10 mm (b, e), and R = 20 mm (c, f). Dashed circles on the phase distributions are those along which TC (Equation (1.2)) and invariants from Equations (1.16) and (1.17) are calculated.

for $z = 2z_0$. Dashed lines on the phase distributions show the circles along which we calculate TC (Equation (1.2)) and the invariants in Equations (1.16) and (1.17) (all circles are of radius 0.8R). All patterns in Figure 1.3 are obtained by Equation (1.19).

Calculation of TC by Equation (1.2) yields $\mu = 3.0000$ at $z = 0$, $\mu = 2.9996$ at $z = z_0/2$, and $\mu = 2.9941$ at $z = 2z_0$ (theoretical value of μ is 3). Calculation by Equation (1.16) yields the value $\mu_1 \approx 0$ ($\mu_1 \sim 10^{-4}$) for all three distances z. Calculation by Eq. (1.17) yields the value $\mu_2 = 1.4980$ at $z = 0$, $\mu_2 = 1.4995$ at $z = z_0/2$, and $\mu_2 = 1.4970$ at $z = 2z_0$ (i.e. $\mu_2 \approx 1.5$ in all transverse planes).

Thus, we note that despite the above theory proving conservation of the TC and other asymptotic phase invariants, when they are calculated over an infinite-radius circle and when the light field propagates by a finite distance, the simulation; however, demonstrates that, in practice, for some specific light fields these quantities can be conserved even when calculated over circles comparable to the beam transverse sizes, and much smaller than the propagation distance. This gives a potential for using these quantities for identifying incoming signals in optical wireless communications. It is hardly possible to estimate the necessary circle radius (R relative to z) for an arbitrary beam, but in all parts of Figures 1.2 and 1.3, this radius is about several times the effective beam width.

In conclusion, we have suggested in this section an alternative way to prove the conservation of the topological charge of a light field on propagation [30]. Our proof is based on a so-called huge-ring approximation of the Huygens–Fresnel principle, which is opposite to the paraxial approximation and which we suggested here for the observation point on an infinite-radius ring. It turned out, that, in addition to the topological charge, phase distribution in areas far from the optical axis allows obtaining other quantities that are also propagation-invariant and the number of these invariants is theoretically infinite. In a simulation, we suggested two such invariants and tested them on two paraxial light fields. Of course, of practical interest are the invariants that conserve the rings of finite radii, and in the simulation we also used finite-radius rings. However, there may exist light fields, for which these invariants fail to conserve. Thus, construction of the propagation-invariants by using the condition (Equation 1.15), investigating their applicability to various-kind light fields, and developing the methods for measuring these invariants, is yet to be studied. The results of this work can find applications in optical data transmission. In general cases, for correct measuring of the TC of non-symmetric incoming optical signal, phase distribution should be obtained, for example, by the Shack–Hartmann wavefront sensor, as was studied experimentally in [31,32]. Identifying an incoming beam by using the invariants, like the partial TC (Equation 1.17), allows measuring the wavefront in a smaller area (rather than over the whole circle in the beam periphery).

1.2 TOPOLOGICAL CHARGE OF A LINEAR COMBINATION OF OPTICAL VORTICES: TOPOLOGICAL COMPETITION

Laser optical vortices (OV) are a particular type of laser beam that carries an orbital angular momentum (OAM) [33]. The OAM associated with paraxial, nonparaxial, and vector beams has been amply studied, as can be seen from works published in 2019 alone [34,35,36,37,13,38,39]. Well-known examples of laser OV are presented by Laguerre–Gauss modes [4], Bessel [1], Bessel–Gauss beams [40], Hypergeometric [41], and Circular [42] beams. The listed radially symmetric beams carry the same OAM normalized to the beam power, which is equal to the beams' integer TC, n. Non-axially symmetric OVs have also been described and known to carry different OAM, for which a variety of formulae have been deduced [43,44]. As well as carrying OAM, optical vortices are also characterized by the topological charge TC, which was defined in [19]. The literature dealing with the calculation of TC of composite OVs is very scarce. For example, in [10], TC was shown to conserve in a medium-turbulence atmosphere over a distance of several kilometers, while in [45], TC variations were numerically studied of vortex soliton in a non-linear medium. Sometimes authors identify TC and OAM, moreover, sometimes they assert that TC can be changed via diffraction by simple dielectric obstacles. On the other hand, in the articles [46,11] it was shown that the sectorial aperture can significantly change the OAM while the TC remains constant and equal to the initial value. Thus, it became necessary to provide some clarification on this issue.

In this section, we focused on some noteworthy examples of the TC behavior in optical vortex arrays that show a cautious approach to calculating TC. In particular, we demonstrate that TC of an optical vortex is conserved in spite of amplitude distortion and shift of the OV center across the carrier beam. It is also shown that in a linear superposition of simple OVs whose amplitude is given by $A(r)\exp(in\varphi)$ (where (r, φ) are the polar coordinates in the beam cross-section) the constituent beams enter a "competition": TC of the resulting beam is defined by both magnitude and sign of the constituent vortex, $+n$, $-n$, as well as being dependent on the amplitude of weight coefficients of the linear combination.

1.2.1 TC OF AN OV AFTER PASSING AN AMPLITUDE MASK

Below, we analyze changes in TC resulting from "cut-off" of a sector-shaped portion from an optical vortex. OVs with a "cut-off" sector have been discussed in detail by Volyar et al. [46]. This work has given an impetus to study the topic of TC conservation following different types of distortions and transformation of an OV. The definition of TC of an OV (and an arbitrary paraxial light field) was given by Berry [19] and J. Nye [21]. For an arbitrary light field with complex amplitude $E(r, \varphi)$, where (r, φ) are the polar coordinates, can be written in the form [19]:

$$TC = \frac{\lim_{r \to \infty} \frac{1}{2\pi} \int_0^{2\pi} d\varphi \frac{\partial}{\partial \varphi} \arg E(r,\varphi) = \frac{1}{2\pi} \lim_{r \to \infty} \text{Im} \int_0^{2\pi} d\varphi \frac{\partial E(r,\varphi)/\partial \varphi}{E(r,\varphi)}. \quad (1.20)$$

This means that the TC monochromatic beam is specified as the total number of optical vortices in the transverse cross-section of a light stream taking into account their signs. Let us write the complex amplitude $E_n(r, \varphi)$ with a cut-off sector as:

$$E(r,\varphi) = A(r)\exp(in\varphi) f(\varphi), \quad (1.21)$$

where the sector function reads as:

$$f(\varphi) = \begin{cases} 1, -\alpha < \varphi < \alpha, \\ \delta \lll, \text{otherwise.} \end{cases} \quad (1.22)$$

Substituting Equation (1.21) into Equation (1.20) yields:

$$TC = \frac{\lim_{r \to \infty} \frac{\text{Im}}{2\pi} \int_0^{2\pi} d\varphi \frac{inE(r,\varphi) + A(r)e^{in\varphi} \frac{\partial f(\varphi)}{\partial \varphi}}{E(r,\varphi)}$$

$$= \frac{\text{Im}}{2\pi} \lim_{r \to \infty} \int_0^{2\pi} d\varphi \left(in + \frac{\partial f(\varphi)}{\partial \varphi} \frac{1}{f(\varphi)} \right) = n. \quad (1.23)$$

The final equality in Equation (1.23) reflects the fact that the second term within the brackets is real. We may infer that if the aperture is only φ-angle dependent, TC of an OV remains unchanged. Although, if strictly $\delta=0$ in Equation (1.22), then instead of Equation (1.23) it should be written that $TC = \alpha n/\pi$. But fractional TC can only be in the initial plane. The proof of Equation (1.23) can easily be repeated for an arbitrarily-shaped amplitude filter (Equation 1.22), which is defined by both angle φ and radius r:

$$f(r,\varphi) = \begin{cases} 1, (r,\varphi) \in \Omega, \\ \\ \delta \lll 1, (r,\varphi) \notin \Omega, \end{cases} \tag{1.24}$$

where Ω is the diaphragm cut-off area. Then, instead of Equation (1.23), we obtain a similar relation:

$$TC = \frac{\text{Im}}{2\pi} \lim_{r \to \infty} \int_0^{2\pi} d\varphi \left(in + \frac{\partial f(r,\varphi)}{\partial \varphi} \frac{1}{f(r,\varphi)} \right) = n. \tag{1.25}$$

A weak transmission ($\delta \ll 1$) was introduced in Equations (1.22) and (1.24) in the region where the diaphragm should not transmit light in order to avoid the 0/0 uncertainty in the division of $E(r,\varphi)$ in Equations (1.23) and (1.25) on $E(r,\varphi)$. We note that the derivative $\partial\,(n\varphi\,)/\partial\,\varphi$ is equal to n only if the angle φ can be arbitrary ($0 < \varphi < 2\pi$). This means that instead of conditions in Equations (1.22) and (1.24) it is sufficient that a closed curve existed around the singular point (OV center) with nonzero field amplitude on this curve. A simulation (Figure 1.4) confirms this requirement. An indirect confirmation of the conservation of TC of an optical vortex with a cut-off

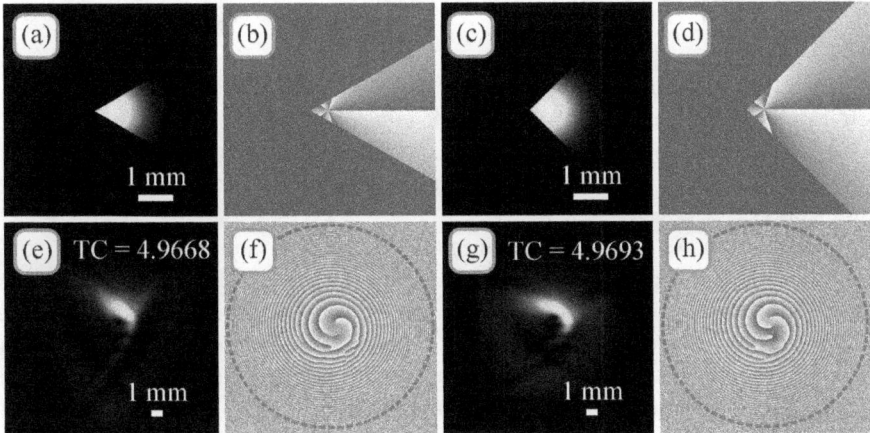

FIGURE 1.4 Distributions of intensity (a,c,e,g) and phase (b, d, f, h) of a Gaussian optical vortex bounded by a sector-shape diaphragm in the initial plane $z = 0$ (a–d) and after propagation in free space (e–h) for two different angles of the sector aperture $\alpha = \pi/6$ (a, b, e, f) and $\alpha = \pi/4$ (c, d, g, h). Dashed rings (f, h) show the circle over which the TC was calculated. White text (e, g) shows the TC.

sector Equations (1.23) and (1.25) is that the OAM of such a beam is equal to the topological charge. Indeed, the OAM normalized to the beam power:

$$J_z = \text{Im} \frac{1}{2\pi} \int\limits_0^\infty \int\limits_0^{2\pi} \bar{E}(r,\varphi,z) \left(\frac{\partial E(r,\varphi,z)}{d\varphi} \right) r dr d\varphi$$

$$= \text{Im} \frac{1}{2\pi} \int\limits_0^\infty \int\limits_{-\alpha}^{\alpha} A(r) e^{in\varphi} \left(in\bar{A}(r) e^{-in\varphi} \right) r dr d\varphi = \frac{\alpha n}{\pi} \int\limits_0^\infty |A(r)|^2 r dr, \qquad (1.26)$$

$$\frac{J_z}{W} = n, \quad W = \frac{\alpha}{\pi} \int\limits_0^\infty |A(r)|^2 r dr.$$

It is seen from Equation (1.25) that multiplying the complex amplitude (Equation 1.21) of the OV by any real function does not change TC of the original OV, as a real function does not change the complex amplitude argument in Equation (1.20). Optically speaking, the multiplication of the light field amplitude by a real function is equivalent to the passage of light through a thin amplitude mask (with the above conditions taken into account).

When the spiral phase plate is bounded by a sector aperture (Equation 1.21), a Gaussian beam after passing has the following complex amplitude in the Cartesian coordinates:

$$E(x,y,0) = \exp\left[-\frac{x^2+y^2}{w^2} + in \arg(x+iy) \right] \text{rect}\left\{ \frac{\arg\left[(x-x_0) + i(y-y_0) \right]}{2\alpha} \right\}, \quad (1.27)$$

where w is the Gaussian beam waist radius, n is the TC of the spiral phase plate, α from Equation (1.22) is the half-angle of the sector aperture, (x_0, y_0) is the shift vector of the sector aperture (the singular point should be inside the sector), rect(x) = 1 at $|x| \leq 1/2$ and rect(x) = 0 at $|x| > 1/2$, and arg(x) is meant as the principal value of the argument (i.e. $-\pi < \arg(x) \leq \pi$).

Figure 1.4 shows distributions of intensity (a,c,e,g) and phase (b,d,f,h) of a Gaussian optical vortex bounded by a sector-shape diaphragm in the initial plane $z = 0$ (a–d) and after propagation in free space (e–h) for two different angles of the sector aperture $\alpha = \pi/6$ (a,b,e,f) and $\alpha = \pi/4$ (c,d,g,h). Distributions in the initial plane are obtained by Equation (1.27), while distributions at a distance are obtained by the Fresnel transform implemented numerically in MATLAB® as a convolution using the fast Fourier transform. The following parameters are used in the calculations: wavelength $\lambda = 532$ nm, Gaussian beam waist radius $w = 1$ mm, TC of the spiral phase plate $n = 5$, vector of the sector diaphragm shift $(x_0, y_0) = (-0.5, 0)$ mm, propagation distance $z = z_0/2$ ($z_0 = kw^2/2$ is the Rayleigh range), calculation area $-R \leq x, y \leq R$ ($R = 12.5$ mm, although Figure 1.4 shows smaller areas), number of pixels $N = 4096$. The obtained values are 4.9668 for $\alpha = \pi/6$ and 4.9693 for $\alpha = \pi/4$.

1.2.2   TC OF AN OFF-AXIS OPTICAL VORTEX

In this subsection, we analyze how TC changes upon an off-axis shift of the OV center from the optical axis of a radially symmetric beam with amplitude $A(r)$. Can an OV be shifted by an arbitrary vector (r_0, φ_0). Then, instead of Equation (1.21), the complex amplitude $E_n(r, \varphi)$ takes the form:

$$E_n(r,\varphi) = \left(\frac{re^{i\varphi} - r_0 e^{i\varphi_0}}{w} \right)^n A(r). \tag{1.28}$$

Substituting Equation (1.28) into Equation (1.20) yields:

$$TC = \lim_{r \to \infty} \frac{\mathrm{Im}}{2\pi} \int_0^{2\pi} d\varphi \, \frac{inre^{i\varphi}}{re^{i\varphi} - r_0 e^{i\varphi_0}} = \frac{1}{2\pi} \mathrm{Im} \lim_{r \to \infty} \int_0^{2\pi} d\varphi \, \frac{inre^{i\varphi}}{re^{i\varphi} - r_0 e^{i\varphi_0}} = n. \tag{1.29}$$

The final equality in Equation (1.29) stems from the fact that for large radii ($r \gg r_0$), only the first term is retained in the denominator. It is seen from Equation (1.29) that an off-axis shift of the OV center relative to a radially symmetric beam (e.g. a Gaussian beam) does not lead to a change in TC. In the meantime, for a beam with an off-axis phase singularity center, the normalized OAM is lower than TC of the whole beam, with the former decreasing with increasing shift magnitude r_0 [47,48].

Figure 1.5 shows the distribution of the intensity and phase of a Gaussian beam with an off-axis optical vortex in the initial plane and after propagation in space for different displacements of the vortex from the optical axis. The complex amplitude in the initial plane is $E_n(x,y) = \left[\left(re^{i\varphi} - r_0 e^{i\varphi_0} \right)/w \right]^n \exp\left[-\left(x^2 + y^2 \right)/w^2 \right]$, where w is the waist radius of the Gaussian beam, n and (r_0, φ_0) are the topological charge of the optical vortex, and the vector (in polar coordinates) of its displacement from the optical axis. The complex amplitude after propagation in space is calculated using a numerical Fresnel transform realized in the form of convolution using the fast Fourier transform. The following calculation parameters were used: $w = 1$ mm, $n = 7$, $\varphi_0 = 0$, $r_0 = w_0/4$ (Figure 1.5(a, b)), $r_0 = w_0/2$ (Figure 1.5(c, d)), $r_0 = 2w_0$ (Figure 1.5(e, f)), space distance $z = z_0/2$, computational domain $-R \leq x, y \leq R$ ($R = 5$ mm). The TC in the initial plane, calculated numerically by the formula in Equation (1.20) (along a ring of radius $0.8R$), is 6.9997 for $r_0 = w_0/4$ and $r_0 = w_0/2$, 6.9995 for $r_0 = 2w_0$, i.e. in all cases about 7. At a distance of TC, it turned out to be 6.9989, 6.9989, and 6.9986, respectively.

An interesting case occurs when an optical vortex is bounded by a diaphragm in the initial plane and therefore it is impossible to use the limit $r \to \infty$ like in Equation (1.29). For example, if a spiral phase plate (SPP) is bounded by a circular diaphragm with a radius R and is shifted horizontally from the optical axis by a distance x_0, a plane wave after passing such SPP acquires the following complex amplitude:

$$E(r,\varphi) = \mathrm{circ}\left(\frac{r}{R} \right) \exp\left[in \arctan\left(\frac{r\sin\varphi}{r\cos\varphi - x_0} \right) \right], \tag{1.30}$$

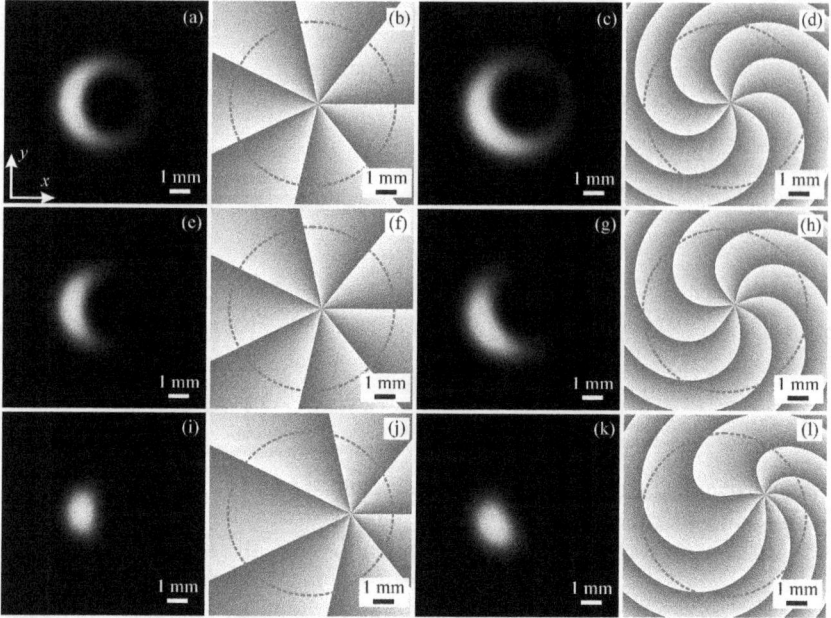

FIGURE 1.5 Distributions of intensity (a, c, e, g, i, k) and phase (b, d, f, h, j, l) of a Gaussian beam with an off-axis optical vortex in the initial plane (a, b, e, f, i, j) and after propagation in space (c, d, g, h, k, l) for different lateral displacements of the vortex from the optical axis. Calculation parameters: waist radius $w = 1$ mm, TC is $n = 7$, displacement $r_0 = w_0/4$ (a–d), $r_0 = w_0/2$ (e–h), $r_0 = 2w_0$ (i–l); $\varphi_0 = 0$ in all figures, the propagation distance in space is $z = z_0/2$ (z_0 is the Rayleigh distance). Thedashed rings on the phase distributions denote the radius of the ring by which the TC was calculated by Equation(1.20).

where $circ(r/R) = 1$ for $r \leq R$ and $circ(r/R) = 0$ for $r > R$. Topological charge (Equation (1.20)) of the initial vortex field from Equation (1.30) is given by:

$$TC = \frac{n}{2\pi} \int_0^{2\pi} \frac{r^2 - rx_0\cos\varphi}{R^2 + x_0^2 - 2rx_0\cos\varphi} d\varphi = \begin{cases} n, & x_0 < R, \\ n/2, & x_0 = R. \end{cases} \tag{1.31}$$

Equation (1.31) illustrates that shifting the SPP center conserves TC which equals n if the SPP center is within the diaphragm.

If the SPP center is on the diaphragm edge, TC decreases two times immediately. Equation (1.31) is consistent with the condition from the previous section which states that there should not be zero amplitude around the center of singularity. Interestingly, the OAM of the beam from Equation (1.30) decreases continuously till zero when the shift distance x_0 increases from 0 to R:

$$\frac{J_z}{W} = n\left(1 - \frac{x_0^2}{R^2}\right). \tag{1.32}$$

From Equation (1.32), if $x_0 = R$, the beam's OAM is zero.

1.2.3 TC OF AN OPTICAL VORTEX WITH MULTI-CENTER OPTICAL SINGULARITIES

Below, we analyze a laser Gaussian beam with m embedded simple ($TC = +1$) phase singularities distributed uniformly on a circle of radius a, i.e., at points defined by the Cartesian coordinates:

$$\begin{cases} x = a\cos\varphi_p, \\ \\ y = a\sin\varphi_p, \end{cases} \tag{1.33}$$

where $\varphi_p = 2\pi p/m$, $p = 0, \ldots, m-1$. The complex amplitude of such an OV at an arbitrary distance from the waist can be shown to be given by:

$$E(r,\varphi,z) = \frac{1}{\sigma}\left(\frac{\sqrt{2}}{w_0}\right)^m \exp\left(-\frac{r^2}{\sigma w_0^2}\right)\left(\frac{r^m e^{im\varphi}}{\sigma^m} - a^m\right), \tag{1.34}$$

where $\sigma = 1 + iz/z_0$ and $z_0 = kw_0^2/2$ is the Rayleigh range (k is the wave number). Substituting Equation (1.34) into Equation (1.20) yields:

$$TC = \frac{1}{2\pi}\lim_{r\to\infty}\text{Im}\left\{\int_0^{2\pi}\frac{im\sigma^{-m}r^m e^{im\varphi}}{\sigma^{-m}r^m e^{im\varphi} - a^m}\,d\varphi\right\} = m. \tag{1.35}$$

Because at $r \to \infty$, the term a^m in the denominator is negligibly small, TC of the beam Equation (1.34) turns out to be independent on the distance z passed and the radius a of the circle of the OV centers, instead, being equal to the number of simple OVs in the beam. This result can be extended onto an arbitrary case of m OV centers with multiplicity m_p are found at points (r_p, φ_p), where $p = 1, 2, \ldots m$ and the carrier amplitude $A(r)$ is axially symmetric. Such a complex OV is given by the complex amplitude [49,50]:

$$E_m(r,\varphi,z=0) = A(r)\prod_{p=1}^{m}\left(re^{i\varphi} - r_p e^{i\varphi_p}\right)^{m_p}. \tag{1.36}$$

Substituting (1.36) into (1.20) yields:

$$TC = \frac{1}{2\pi}\lim_{r\to\infty}\text{Im}\left\{\int_0^{2\pi} ire^{i\varphi}\sum_{p=1}^{m}\frac{m_p}{re^{i\varphi} - r_p e^{i\varphi_p}}\,d\varphi\right\} = \sum_{p=1}^{m}m_p. \tag{1.37}$$

Equation (1.37) suggests that in a beam with axially symmetric amplitude and several degenerate simple OVs of Equation (1.36), with their centers located at arbitrary points over the beam cross-section, TC equals the sum of multiplicity (degeneracy) values of all constituent vortices.

1.2.4   TC OF AN ON-AXIS COMBINATION OF OPTICAL VORTICES

Here, we discuss a light field whose complex amplitude is described by a linear combination of a finite number of Laguerre–Gaussian (LG) modes with the numbers $(n, 0)$:

$$E_{N,-M}(r, \varphi, z = 0) = \exp\left(-\frac{r^2}{w^2}\right) \sum_{n=-M}^{N} C_n \left(\frac{r}{w}\right)^{|n|} e^{in\varphi}. \tag{1.38}$$

Substituting Equation (1.38) into Equation (1.20) yields a relation for TC:

$$TC = \frac{1}{2\pi} \lim_{r \to \infty} \text{Im} \left\{ \int_0^{2\pi} i \frac{\displaystyle\sum_{n=-M}^{N} nC_n \left(\frac{r}{w}\right)^{|n|} e^{in\varphi}}{\displaystyle\sum_{n=-M}^{N} C_n \left(\frac{r}{w}\right)^{|n|} e^{in\varphi}} \, d\varphi \right\}. \tag{1.39}$$

Following a limiting passage $r \to \infty$ under the integral sign in Equation (1.39), the numerator and denominator each retain just one highest-power term under the sum sign. If $M > N$, then TC of the beam in Equation (1.38) is $TC = -M$, if $M < N$, then TC in Equation (1.38) equals $TC = N$. Finally, if $M = N$, instead of Equation (1.39), we obtain:

$$TC = \frac{1}{2\pi} \text{Im} \left\{ \int_0^{2\pi} iN \frac{\left(C_N e^{iN\varphi} - C_{-N} e^{-iN\varphi}\right)}{\left(C_N e^{iN\varphi} + C_{-N} e^{-iN\varphi}\right)} \, d\varphi \right\}. \tag{1.40}$$

Thus, we can infer that if in a linear combination of a finite number of LG modes with different TC, the absolute value of the maximum positive TC is larger than the maximum negative TC, the TC of the entire beam equals the positive $TC = N$. If the opposite is the case, the resulting TC equals the negative $TC = -M$. Finally, in the next section we show that for $M = N$, the integral in Equation (1.40) can be taken analytically and, based on Equation (1.42), $TC = N$ if $|C_N| > |C_{-N}|$ or $TC = -N$ if $|C_N| < |C_{-N}|$. When $|C_N| = |C_{-N}|$, TC of the entire beam equals zero.

1.2.5   TC OF THE SUM OF TWO OPTICAL VORTICES

Now, let us analyze a simple but rather interesting case that produces an unexpected result. Assume a light field with a complex amplitude in the initial plane that describes an axial superposition of two Gaussian OVs with different TC and different amplitudes:

$$E(r, \varphi) = \left(ae^{in\varphi} + be^{im\varphi}\right) e^{-r^2/w^2}, \tag{1.41}$$

where w is the Gaussian beam waist radius, n and m are integer topological charges of the OVs, a and b are weight coefficients in the OV superposition, which are generally complex. Substituting Equation (1.41) into Equation (1.20) yields a relation for TC:

$$TC = \frac{1}{2\pi} \lim_{r \to \infty} \text{Im} \left\{ \int_0^{2\pi} \frac{\partial E(r,\varphi)/\partial\varphi}{E(r,\varphi)} d\varphi \right\} = \frac{1}{2\pi} \text{Re} \left\{ \int_0^{2\pi} \frac{nae^{in\varphi} + mbe^{im\varphi}}{ae^{in\varphi} + be^{im\varphi}} d\varphi \right\}. \quad (1.42)$$

The integral in the right-hand side of Equation (1.42) can be reduced to a sum of two integrals:

$$TC = \frac{1}{2\pi} \int_0^{2\pi} \left(\frac{n+m}{2} + \frac{n-m}{2} \frac{|a|^2 - |b|^2}{|a|^2 + |b|^2 + 2|a||b|\cos t} \right) dt. \quad (1.43)$$

With the first integral in Equation (1.43) being trivial, the second integral can be rearranged as:

$$TC = \frac{n+m}{2} + \frac{1}{2\pi} \frac{n-m}{2} \frac{|a|^2 - |b|^2}{|a|^2 + |b|^2} \int_0^{2\pi} \frac{dt}{1 + \dfrac{2|a||b|}{|a|^2 + |b|^2}\cos t}. \quad (1.44)$$

With the coefficient before the cosine function being not larger than unity, this is a reference integral (expression 3.613.1 in [51]):

$$\int_0^\pi \frac{\cos(nx)dx}{1 + a\cos x} = \frac{\pi}{\sqrt{1-a^2}} \left(\frac{\sqrt{1-a^2} - 1}{a} \right)^n \quad [a^2 < 1, n \geq 0]. \quad (1.45)$$

In the case the integration interval is from zero to 2π, rather than being to π, the expression needs to be multiplied by two. Then, Equation (1.42) takes the form:

$$TC = \frac{n+m}{2} + \frac{n-m}{2} \frac{|a|^2 - |b|^2}{||a|^2 - |b|^2|}. \quad (1.46)$$

For completeness sake, the normalized OAM of the beam in Equation (1.41) can be given in the form:

$$OAM = \frac{na^2 + mb^2}{a^2 + b^2}. \quad (1.47)$$

From Equation (1.46) it follows that if $|a| > |b|$, then $TC = n$ and if $|a| < |b|$, then $TC = m$. If $m = n$, as can be expected, we obtain that $TC = n$. Thus, TC of the resulting beam in Equation (1.41) equals that of the constituent OV with the larger amplitude. At $|a| = |b|$, there occurs degeneracy (photon entanglement), with Equation (1.46) becoming invalid due to uncertainty 0/0. Because of this, at $|a| = |b|$, the field in Equation (1.41) needs to be rearranged to the form:

$$E(r,\varphi) = |a| \left(e^{in\varphi + i \arg a} + e^{im\varphi + i \arg b} \right) e^{-r^2/w^2}$$

$$= 2|a| \cos \left(\frac{n\varphi - m\varphi + \arg a - \arg b}{2} \right) \tag{1.48}$$

$$\times \exp \left(-\frac{r^2}{w^2} + i \frac{n\varphi + m\varphi + \arg a + \arg b}{2} \right).$$

Substituting Equation (1.48) into Equation (1.20) yields

$$TC = \lim_{r \to \infty} \frac{1}{2\pi} \int_0^{2\pi} \frac{\partial}{\partial \varphi} \left(\frac{n\varphi + m\varphi + \arg a + \arg b}{2} \right) d\varphi = \frac{n+m}{2}. \tag{1.49}$$

As we can infer from Equation (1.49), the superposition of two same-amplitude OVs, with one TC being even and the other odd, produces an OV with a fractional (semi-integer) TC. It should be noted that it is only in the initial plane that TC of a beam can be fractional, whereas during propagation TC needs to be an integer for the amplitude to be continuous. It is worth noting that OAM in Equation (1.47) equals TC in Equations (1.46) and (1.49) only when either $a = 0$, or $b = 0$, or $a = b$. At the same time, if the beam is degenerate ($a = b$), the content of the constituent angular harmonics of the beam cannot be derived from the known TC. For instance, all the below-listed beams have the same TC and OAM, which is equal to 4:

$$E_1(r,\varphi) = \left(e^{i\varphi} + e^{i7\varphi} \right) e^{-r^2/w^2},$$

$$E_2(r,\varphi) = \left(e^{i2\varphi} + e^{i6\varphi} \right) e^{-r^2/w^2},$$

$$E_3(r,\varphi) = \left(e^{i3\varphi} + e^{i5\varphi} \right) e^{-r^2/w^2}, \tag{1.50}$$

$$E_4(r,\varphi) = e^{i4\varphi} e^{-r^2/w^2}.$$

Figure 1.6(a, b) shows the intensity and phase of the superposition of two Gaussian vortices in the initial plane for the following calculation parameters: waist radius $w = 1$ mm, topological charges $n = 12$ and $m = 7$, weight coefficients are unit modulo, but with a random phase: $a = e^{2.9616\,i}$, $b = e^{0.2247\,i}$, computational domain $-R \leq x, y \leq R$ ($R = 1$ mm). The TC calculated numerically by formula (1.20) is 9.4708, i.e. approximately $(12 + 7)/2$. Shown in Figure 1.6 (c, d) are the intensity and phase of the same superposition, but at the Fresnel distance (for the wavelength $\lambda = 532$ nm) and in a wider calculated region ($R = 10$ mm). The TC calculated numerically by Equation (1.20) is 11.8167, that is, about 12. In both cases, the TC was calculated by integration over a

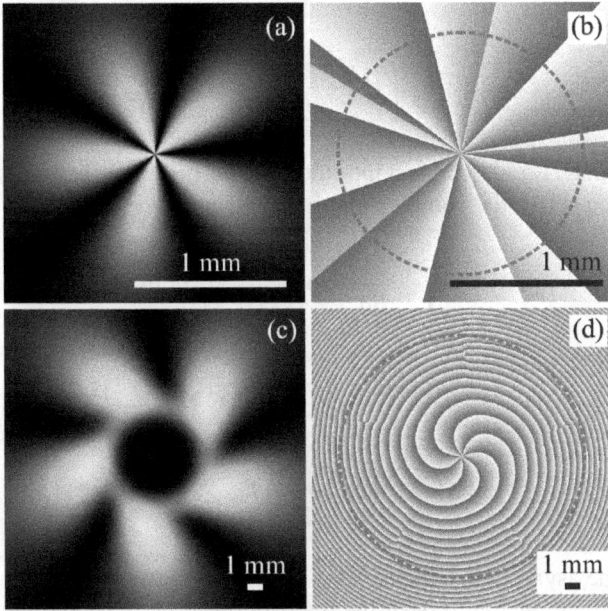

FIGURE 1.6 The intensity (a, c) and phase (b, d) of the axial superposition of two Gaussian OVs with *TC* 12 and 7, but with the same weight amplitudes (in Equation (1.41)) in the initial plane (a, c) and at the Rayleigh distance (c, d). Dashed rings on the phase distributions denote the radius of the ring by which the topological charge was calculated by the Equation (1.20).

ring of radius 0.8*R*. This example corresponds to the situation described by amplitude in Equation (1.46), and when the amplitude moduli are equal $|a| = |b|$ for two vortices, the *TC* of the entire beam will be equal to the larger of the two *TC*, i.e. 12.

1.2.6 TOPOLOGICAL CHARGE IN AN ARBITRARY PLANE

In this subsection, we shall demonstrate that a combination of two same-amplitude ($a = b$) Gaussian OVs of Equation (1.41) with different *TC* produces an OV with half-integer *TC* of Equation (1.49) in the initial plane, generating an OV with integer *TC* as it propagates. Actually, if in the initial plane there is a Gaussian OV:

$$E(r,\varphi) = e^{-r^2/w^2 + in\varphi}, \tag{1.51}$$

following the propagation through an ABCD-system, its complex amplitude is given by:

$$E_z(\rho,\theta) = (-i)^{n+1} \sqrt{\frac{\pi}{2}} \frac{z_0}{Bq_1} \exp\left(\frac{ikD\rho^2}{2B} + in\theta\right)$$

$$\times \sqrt{\xi} \exp(-\xi)\left[I_{\frac{n-1}{2}}(\xi) - I_{\frac{n+1}{2}}(\xi)\right], \tag{1.52}$$

where:

$$\xi = \left(\frac{z_0}{B}\right)^2 \left(\frac{\rho}{w}\right)^2 \left(\frac{1}{2q_1}\right), \quad q_1 = 1 - i\frac{A}{B}z_0. \tag{1.53}$$

Since in Equation (1.52), $I_\nu(x)$ is a modified Bessel function, then for the superposition in Equation (1.41), the complex amplitude is:

$$E_z(\rho,\theta) = -i\sqrt{\frac{\pi}{2}}\frac{z_0}{Bq_1}\exp\left(\frac{ikD\rho^2}{2B}\right)\sqrt{\xi}\exp(-\xi)\times$$

$$\times\left\{a(-i)^n\exp(in\theta)\left[I_{\frac{n-1}{2}}(\xi) - I_{\frac{n+1}{2}}(\xi)\right]\right. \tag{1.54}$$

$$\left. + b(-i)^m\exp(im\theta)\left[I_{\frac{m-1}{2}}(\xi) - I_{\frac{m+1}{2}}(\xi)\right]\right\}.$$

Retaining in the asymptotic expansion of the modified Bessel function just first two terms yields a relationship to describe the difference of two modified Bessel functions of adjacent orders at large values of the argument:

$$I_{\frac{n-1}{2}}(\xi) - I_{\frac{n+1}{2}}(\xi) \sim \frac{e^\xi}{\sqrt{2\pi\xi}}\left\{\left[1 - \frac{4\left(\frac{n-1}{2}\right)^2 - 1}{8\xi}\right] - \left[1 - \frac{4\left(\frac{n+1}{2}\right)^2 - 1}{8\xi}\right]\right\} = \frac{ne^\xi}{2\xi\sqrt{2\pi\xi}}.$$

$$\tag{1.55}$$

Then, at large values of ρ, Equation (1.54) takes the form:

$$E_z(\rho,\theta) = -i\sqrt{\frac{\pi}{2}}\frac{z_0}{Bq_1}\exp\left(\frac{ikD\rho^2}{2B}\right)\sqrt{\xi}\exp(-\xi)$$

$$\times\left[a(-i)^n\exp(in\theta)\frac{ne^\xi}{2\xi\sqrt{2\pi\xi}} + b(-i)^m\exp(im\theta)\frac{me^\xi}{2\xi\sqrt{2\pi\xi}}\right] \tag{1.56}$$

$$= \frac{-iz_0}{4Bq_1\xi}\exp\left(\frac{ikD\rho^2}{2B}\right)\left[an(-i)^n\exp(in\theta) + bm(-i)^m\exp(im\theta)\right].$$

In view of Equations (1.42) and (1.46), Equation (1.56) suggests that at $|a| = |b|$, in the initial plane $TC = (m+n)/2$. At the same time, in any other plane, modules of the coefficients in front of $e^{in\theta}$ and $e^{im\theta}$ are proportional to $|n|$ and $|m|$, being no more equal to each other (at $n \neq m$), hence, according to Equation (1.46), $TC = \max(n, m)$.

Note, however, that if in Equation (1.56) $|an| = |bm|$, we again find ourselves in the situation of degeneracy, because in view of Equation (1.46) and at $z > 0$, TC of two OVs of Equation (1.41) equals the arithmetic mean of Equation (1.49): $TC = (n + m)/2$. This situation can be addressed as follows: with the equality $|an| = |bm|$ meaning that $|a| \neq |b|$, Equation (1.46) suggests that the total TC of the field in the initial plane equals that of the OV with the larger amplitude (respectively, $|a|$ or $|b|$). In the meantime, the integer TC in the initial plane conserves upon propagation.

1.2.7 TOPOLOGICAL CHARGE FOR AN OPTICAL VORTEX WITH AN INITIAL FRACTIONAL CHARGE

For an OV with fractional $TC = \mu$ (μ is an arbitrary real number), a relation to describe the corresponding fractional TC has been derived [52]. The mutual transformations between beams with fractional-order and integer-order vortices were considered in detail in [53]. An OV with fractional TC, which is possible only in the initial plane, can be decomposed in terms of OVs with integer TC n (μ is an arbitrary real number) as follows:

$$E_\mu(r,\varphi,z) = \exp(-i\mu\varphi)\Psi(r,z) = \frac{e^{i\pi\mu}\sin\pi\mu}{\pi}\Psi(r,z)\sum_{n=-\infty}^{\infty}\frac{e^{in\varphi}}{\mu-n}. \qquad (1.57)$$

In Equation (1.57), the function $\Psi(r, z)$ is real. Substituting the right-hand side of Equation (1.57) into a general relation for OAM:

$$J_z = \mathrm{Im}\int_0^\infty\int_0^{2\pi}\bar{E}(r,\varphi,z)\left(\frac{\partial E(r,\varphi,z)}{\partial\varphi}\right)rdrd\varphi \qquad (1.58)$$

yields:

$$J_z = W\frac{\sin^2(\pi\mu)}{\pi^2}\sum_{n=-\infty}^{\infty}\frac{n}{(\mu-n)^2}, \qquad (1.59)$$

where W is the energy (power) of the beam:

$$W = \int_0^\infty\int_0^{2\pi}E(r,\varphi,z)\bar{E}(r,\varphi,z)rdrd\varphi. \qquad (1.60)$$

The series in the right-hand side of Equation (1.59) can be reduced to a reference series [51]:

$$\sum_{n=1}^{\infty}\frac{n^2}{\left(n^2\pm a^2\right)^2} = \frac{\pi}{4a}\left[\pm\begin{Bmatrix}\coth\pi a\\\cot\pi a\end{Bmatrix}\mp a\begin{Bmatrix}\mathrm{cosech}^2\,\pi a\\\mathrm{cosec}^2\,\pi a\end{Bmatrix}\right], \qquad (1.61)$$

using which, the final relation for the normalized OAM of the field in Equation (1.59) is rearranged to:

$$\frac{J_z}{W} = \mu - \frac{\sin 2\pi\mu}{2\pi}. \tag{1.62}$$

From Equation (1.62) it follows that OAM equals $TC = \mu$ only if μ is integer and half-integer. This conclusion is in agreement with Equations (1.53) and (1.55) for the linear combination composed of two angular harmonics.

We obtain the expression for the TC of the optical vortex in the Fresnel diffraction zone for the initial field with a fractional topological charge from Equation (1.57), but for definiteness we choose the amplitude function in the form of a Gaussian one. Then instead of Equation (1.57) we get:

$$E_\mu(r,\varphi,z=0) = \exp(-i\mu\varphi - \left(\frac{r}{w}\right)^2) = \frac{e^{i\pi\mu}\sin\pi\mu}{\pi}\sum_{n=-\infty}^{\infty}\frac{e^{in\varphi-r^2/w^2}}{\mu-n}. \tag{1.63}$$

In view of Equation (1.52), the amplitude of the optical vortex in Equation (1.63) for any z will be equal to ($B = z$, $A = D = 1$):

$$E_2(\rho,\theta) = \frac{1}{\sqrt{2\pi}}\left(\frac{-iz_0}{q_1 z}\right)\exp\left(\frac{ik\rho^2}{2z} + i\pi\mu\right)\sin(\pi\mu)\sqrt{x}\exp(-x)$$

$$\times \sum_{m=-\infty}^{\infty}(-i)^m(\operatorname{sgn} m)^{|m|}\frac{\exp(im\theta)}{\mu-m}\left[I_{\frac{|m|-1}{2}}(x) - I_{\frac{|m|+1}{2}}(x)\right] \tag{1.64}$$

We substitute Equation (1.64) in Equation (1.20) and, when passing to the limit in Equation (1.20), we take into account the asymptotic behavior in Equation (1.36), then we obtain the expression for calculating the TC of the optical vortex (1.63):

$$TC = \frac{\operatorname{Re}}{2\pi}\left\{\int_0^{2\pi}\frac{\displaystyle\sum_{n=-\infty}^{\infty}\frac{(-i)^{|n|}n|n|e^{in\varphi}}{\mu-n}}{\displaystyle\sum_{n=-\infty}^{\infty}\frac{(-i)^{|n|}|n|e^{in\varphi}}{\mu-n}}d\varphi\right\}. \tag{1.65}$$

Equation (1.65) is remarkable in that the answer is known, which was numerically obtained in [19], but has not yet been obtained analytically. Calculation (1.65) can be called the Berry problem [19]. The right-hand side of Equation (1.65) should give only whole TC, closest to μ :

$$TC = \sum_{n=-\infty}^{\infty}n\operatorname{rect}(\mu-n), \quad \operatorname{rect}(x) = \begin{cases}1, |x| \leq 1/2, \\ 0, |x| > 1/2\end{cases} \tag{1.66}$$

From a comparison of Equations (1.65) and (1.66), it can be said that the *TC* in Equation (1.65) is equal to the *TC* of the angular harmonic in the series in the numerator and the denominators for which the weight coefficient is greater in absolute value. This is also consistent with the results for a complete linear combination of LG modes in Equation (1.38) and for the sum of two angular harmonics in Equation (1.41).

The *TC* of an optical vortex can be measured using a cylindrical lens by the method described in [15]. Figure 1.7 shows the intensity distributions at a double focal length from a cylindrical lens for optical vortices with an initial fractional *TC* in Equation (1.57). It can be seen that on the line at an angle of −45 degrees in the center of the picture are two zeros (two dark lines) (Figure 1.7a) for $\mu < 2.5$ and three zeros (three dark lines) for $\mu > 2.5$ (Figure 1.7b, c, d). As shown in Figure 1.7, for arbitrary initial fractional *TC* between 2 and 2.5, *TC* of the optical vortex equals 2, and for arbitrary initial fractional *TC* (Equation (1.57)) higher than 2.5 and lower than 3, *TC* of the beam equals 3. The experiment in Figure 1.7 confirms the numerical result in Equation (1.66).

1.2.8 Topological Charge of an Elliptic Optical Vortex Embedded in a Gaussian Beam

Let us analyze a simple example of an OV with introduced phase distortion by making it ellipse-shaped. While for a conventional OV the complex amplitude in the initial plane is given by:

$$E(r,\varphi) = A(r)\exp(in\varphi), \tag{1.67}$$

for an elliptic vortex imbedded, say, into a Gaussian beam (or any other radially symmetric beam) it takes the form:

$$E_e(x,y) = A(\sqrt{x^2+y^2})(x+i\alpha y)^n$$
$$= A(\sqrt{x^2+y^2})(x^2+\alpha^2 y^2)^{n/2}\exp\left(in\arctan\left(\frac{\alpha y}{x}\right)\right). \tag{1.68}$$

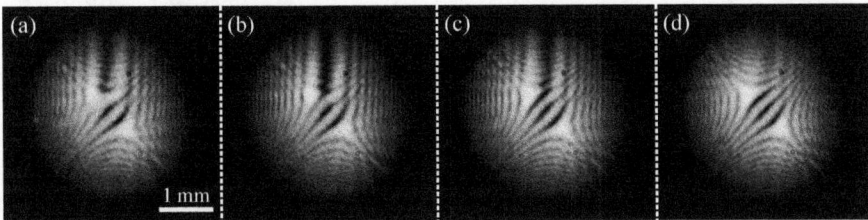

FIGURE 1.7 Intensity distributions measured at a distance $z = 200$ mm (at a double focal length from a cylindrical lens) from a spiral phase plate with fractional order μ : (a) 2.3, (b) 2.5, (c) 2.7, (g) 2.9. The sizes of the images are 4,000 by 4,000 microns.

Substituting Equation (1.68) into Equation (1.20) yields:

$$TC = \frac{1}{2\pi} \int\limits_0^{2\pi} d\varphi \frac{\partial}{\partial \varphi} \arg E_e(r,\varphi) = \frac{1}{2\pi} \int\limits_0^{2\pi} d\varphi \frac{\partial}{\partial \varphi} \left(n \arctan \left(\alpha \tan \varphi \right) \right)$$

(1.69)

$$= \left(\frac{n\alpha}{2\pi} \right) \int\limits_0^{2\pi} \frac{d\varphi}{\cos^2 \varphi + \alpha^2 \sin^2 \varphi} = n.$$

Note that a result similar to Equation (1.69), but only for $n = 1$, was previously obtained in [54]. From Equation (1.69) it follows that the fact that an optical vortex or SPP is ellipse-shaped does not change the TC of the original simple OV in Equation (1.67). At any degree of ellipticity (any α), an elliptic OV has $TC = n$. In the meantime, OAM of an elliptic OV is always lower than n, being equal to:

$$\frac{J_z}{W} = \frac{nP_{n-1}(y)}{P_n(y)} < n,$$

(1.70)

where $y = (1 + \alpha^2)/(2\alpha) > 1$ and $P_n(y)$ is Legendre polynomial. Figure 1.8 shows the intensity and phase distributions of a Gaussian beam with an elliptical vortex in the initial plane and after propagation in space for different ellipticities. The complex amplitude in the initial plane is $E_e(x,y) = \exp\left[-\left(x^2 + y^2 \right)/w^2 \right] (x + i\alpha y)^n$, where w is the waist radius of the Gaussian beam, n and α are the topological charge and ellipticity of the optical vortex, respectively. The following calculation parameters were used: $w = 1$ mm, $n = 7$, $\alpha = 1.1$ (Figure 1.8(a,b,c,d)), $\alpha = 1.5$ (Figure 1.8(e,f,g,h)), $\alpha = 3$ (Figure 1.8(i,j,r,l)), the distance of propagation in space $z = z_0/2$, the computational domain is $-R \le x, y \le R$ ($R = 5$ mm). The TC in the initial plane, calculated numerically by Equation (1.20) (along a ring of radius 0.8R), is 6.9997 at $\alpha = 1.1$, 6.9996 at $\alpha = 1.5$, 6.9987 at $\alpha = 3$, that is, in all cases about 7. TC turned out to be 6.9989, 6.9988, and 6.9979, respectively. That is, it is also approximately equal to 7.

Summing up, it has been theoretically shown [55] that OVs conserve the integer TC when passing through an arbitrary aperture or shifted from the optical axis of an arbitrary axisymmetric carrier beam. If the beam contains a finite number of off-axis optical vortices with different-value same-sign TC, the total TC of the resulting beam has been shown to be equal to the sum of all constituent TCs. In this case, there is no topological competition, because it takes place as a result of on-axis superposition of OVs. By way of illustration, if an on-axis superposition is composed of a finite number of Laguerre–Gaussian modes $(n, 0)$, the resulting TC equals that of the constituent mode with the highest TC (including sign). If the highest positive and negative TCs of the constituent modes are equal in magnitude, the "winning" TC is the one with the larger absolute value of the weight coefficient. If the constituent modes have the same weight coefficients, the resulting TC equals zero. If the beam is composed of two on-axis different-amplitude Gaussian vortices with different TC, the resulting TC equals that of the constituent vortex with the larger absolute

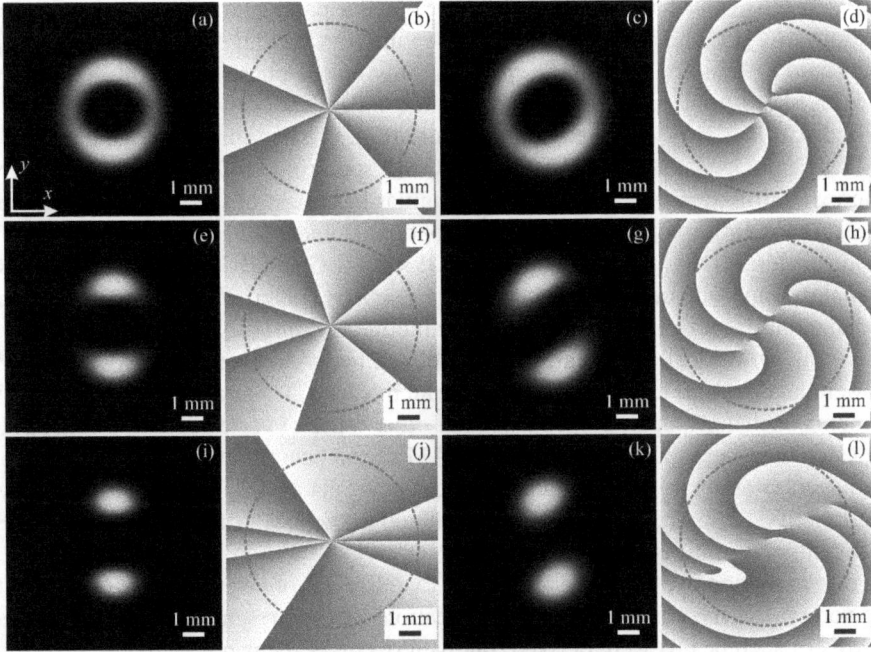

FIGURE 1.8 Distributions of intensity (a, c, e, g, i, k) and phase (b, d, f, h, j, l) of a Gaussian beam with an elliptical vortex in the initial plane (a, b, e, f, i, j) and after propagation in space (c, d, g, h, k, l) for different ellipticities. Dashed rings on the phase distributions denote the radius of the ring by which the *TC* was calculated by the Equation (1.20).

value of the weight coefficient amplitude, irrespective of the correlation between the individual TCs. If the constituent beams have equal weight coefficients, there occurs degeneracy, with the resulting *TC* being equal to the mean arithmetic of the constituent Gaussian OVs. If in the superposition of two Gaussian OVs, one *TC* is odd and the other is even, the resulting *TC* in the initial plane is half-integer. As the beam propagates, degeneracy is eliminated, with the resulting *TC* becoming equal to the larger (positive) integer constituent *TC*. This effect has been given the name "topological competition of optical vortices". Theoretical predictions have been corroborated by numerical simulation and experiments.

1.3 TOPOLOGICAL CHARGE OF ASYMMETRIC OPTICAL VORTICES

Presently, laser vortex beams [56], or optical vortices (OV), have been actively studied because they have found uses in many optical applications. For instance, OVs are utilized in quantum information science [57], cryptography [58], wireless communication systems [59], data transmission in optical fibers [60], second-harmonic generation [61], short-pulse interferometry [62], and probing of turbulent media

[63]. Vortex beams are characterized by two major parameters, namely, topological charge (*TC*) [19] and orbital angular momentum (OAM) [33], which describe different aspects of an OV. With *TC* depending only on the phase of a light field, OAM is both phase- and amplitude- (intensity) dependent. *TC* can be measured using a cylindrical lens [15] or a triangular aperture [64]. For measuring OAM, a cylindrical lens can also be utilized [12,13]. The OAM spectrum of OVs, which defines the energy contribution in each constituent angular harmonic of the laser beam, can be measured with a multi-order diffractive optical element [65] or based on intensity moments [66,11]. For radially symmetric OVs (e.g. LG and BG beams [4,40]), whose complex amplitude can be given by $E(r,\varphi,z) = A(r,z)\exp(in\varphi)$, where $A(r,z)$ is the radial component of the beam's complex amplitude, n is *TC* of the beam, and (r, φ, z) are the cylindrical coordinates, *TC* is defined by OAM normalized to the beam's power and equals n. It is worth noting that an integer *TC* of a radially symmetric OV remains unchanged upon propagation. For other types of vortex beams, *TC* needs to be calculated individually. Meanwhile OAM of the beam remains unchanged upon propagation and can be calculated in the source plane, for *TC* this is not always the case. For instance, *TC* of a combined beam composed of two LG modes with different-waist radii is not conserved [22].

In this section, we derive relationships to define *TC* of certain radially asymmetric vortex laser beams. In previous publications of the present authors, normalized OAMs of such beams were derived, but patterns of *TC* behavior were not analyzed. Below, we derive relationships to describe *TC* of asymmetric LG, BG, and Kummer beams [44,67,68], superposition of two HG modes [69], and a vortex HG beam [70]. We note that the considered asymmetrical beams are obtained in different ways. Asymmetric LG and Kummer beams are obtained from the conventional symmetric LG and Kummer beams by a transverse complex shift in the Cartesian coordinates. Asymmetric BG beams are obtained by a hybrid technique: complex shift is applied only to the Bessel function, whereas the Gaussian function remains unshifted. Vortex HG beams are derived from the conventional HG beam by an astigmatic transform using a cylindrical lens. And another vortex beam is a superposition of two HG modes with complex weight coefficients. Therefore, we calculated *TC* for each type of beam separately.

1.3.1   TC of an Asymmetric LG Beam

Upon free-space propagation of an asymmetric LG (aLG) beam, at a distance z its complex amplitude is given by [44]:

$$E\left(x,y,z\right) = \frac{w\left(0\right)}{w\left(z\right)}\left[\frac{\sqrt{2}}{w\left(z\right)}\right]^{|l|}\left[\left(x-x_0\right)+i\theta(l)\left(y-y_0\right)\right]^{|l|}$$

$$\times L_p^{|l|}\left[\frac{2\rho^2}{w^2\left(z\right)}\right]\exp\left[-\frac{\rho^2}{w^2\left(z\right)}+\frac{ik\rho^2}{2R\left(z\right)}-i\left(|l|+2p+1\right)\zeta\left(z\right)\right],$$

(1.71)

where:

$$\rho^2 = \left(x - x_0\right)^2 + \left(y - y_0\right)^2,$$

$$w\left(z\right) = w\sqrt{1 + \left(\frac{z}{z_R}\right)^2},$$

$$R\left(z\right) = z\left[1 + \left(\frac{z_R}{z}\right)^2\right],$$ (1.72)

$$\zeta\left(z\right) = \arctan\left(\frac{z}{z_R}\right),$$

where $\theta(l) = \{1, l \geq 0; -1, l < 0\}$, (x, y, z) and (r, φ, z) are the Cartesian and cylindrical coordinates, (x_0, y_0) are the complex coordinates of the off-axis shift of the LG beam, w is the Gaussian beam waist radius, l is TC of the optical vortex, $L_p^l(x)$ is the associated Laguerre polynomial, $z_R = kw^2/2$ is the Rayleigh range, and $k = 2\pi /\lambda$ is the wavenumber of light of wavelength λ . The transverse intensity of the beam is not radially symmetric, unlike conventional LG beams [4]. If (x_0, y_0) are real, beam in Equation (1.71) becomes a conventional off-axis LG mode.

Below, we discuss the TCs of various OVs derived in this work in relation to their OAMs derived in the previous studies. In doing so, we shall make use of formulae for calculating the OAM of paraxial laser beams and beam power [44]:

$$J_z = \operatorname{Im} \iint_{\mathbb{R}^2} E^* \left(x \frac{\partial E}{\partial y} - y \frac{\partial E}{\partial x} \right) dxdy,$$ (1.73)

$$W = \iint_{\mathbb{R}^2} E^* E \, dxdy.$$ (1.74)

For an an LG beam, OAM normalized to power is given by [44]:

$$\frac{J_z}{W} = l + \frac{2\operatorname{Im}\left(x_0^* y_0\right)}{w^2}\left[\frac{L_p^1\left(\dfrac{Q^2}{2w^2}\right)}{L_p\left(\dfrac{Q^2}{2w^2}\right)} + \frac{L_{p+l}^1\left(\dfrac{Q^2}{2w^2}\right)}{L_{p+l}\left(\dfrac{Q^2}{2w^2}\right)} - 1\right].$$ (1.75)

where:

$$Q = 2i\sqrt{\left(\operatorname{Im} x_0\right)^2 + \left(\operatorname{Im} y_0\right)^2}.$$ (1.76)

Unlike *TC*, an increase or decrease in OAM is fully determined by the sign of the quantity $\text{Im}\left(x_0^* y_0\right)$, because the relation in square brackets in Equation (1.75) is always larger than or equal to the unit.

Below, we calculate the *TC* of an aLG beam of Equation (1.71), using Berry's formula [19]:

$$TC = \lim_{r\to\infty} \frac{1}{2\pi} \int_0^{2\pi} d\varphi \frac{\partial}{\partial\varphi} \arg E(r,\varphi) = \frac{1}{2\pi} \lim_{r\to\infty} \text{Im} \int_0^{2\pi} d\varphi \frac{\partial E(r,\varphi)/\partial\varphi}{E(r,\varphi)}. \quad (1.77)$$

We note that the standard definition of the *TC* is the number of 2π phase changes on a closed loop. Unfortunately, this definition is not constructive, and does not allow analytical calculation of the *TC* of optical vortices. It is a great merit of Berry that he proposed a more constructive *TC* in Equation (1.77), which we use in this paper. Both of these definitions lead to the same result.

Assuming ($l > 0$) that the complex shift in Equation (1.72) is given by $x_0 = aw$, $y_0 = iaw$, the term $[(x-x_0)+i(y-y_0)]^l$ in Equation (1.71) takes a simple form $r^l e^{il\varphi}$, with the variable ρ^2 in Equation (1.72) taking the following form: $\rho^2 = (x-x_0)^2 + (y-y_0)^2 = r^2 - 2awre^{i\varphi}$, where a is a dimensionless constant whose magnitude defines the asymmetry of the beam. With due regard to the above considerations, the derivative with respect to the angle φ of the function in Equation (1.71) takes the form:

$$\frac{\partial E(r,\varphi,z)}{\partial\varphi} = ilE(r,\varphi,z)$$

$$-\left[-\frac{1}{w^2(z)}+\frac{ik}{2R(z)}\right]\left(2iawre^{i\varphi}\right)E(r,\varphi,z) \quad (1.78)$$

$$-\frac{4iawre^{i\varphi}}{w^2(z)}\frac{1}{L_p^{|l|}(\xi)}\frac{d}{d\xi}L_p^{|l|}(\xi)E(r,\varphi,z),$$

where $\xi = \dfrac{2\rho^2}{w^2(z)}$

Substituting Equation (1.78) into Equation (1.77) yields ($w=w(0)$):

$$TC = \frac{1}{2\pi}\lim_{r\to\infty}\text{Im}\int_0^{2\pi}d\varphi\left\{il-\left[-\frac{1}{w^2(z)}+\frac{ik}{2R(z)}\right]\left(2iarwe^{i\varphi}\right)-\frac{i4awre^{i\varphi}}{w^2(z)}\frac{1}{L_m^{|l|}(\xi)}\frac{\partial L_m^{|l|}(\xi)}{\partial\xi}\right\}$$

$$= l + \lim_{r\to\infty}\text{Re}\left[\left(-\frac{2}{w^2(z)}+\frac{ik}{R(z)}\right)raw\int_0^{2\pi}d\varphi e^{i\varphi}\right]-\lim_{r\to\infty}\text{Re}\int_0^{2\pi}d\varphi\frac{2r|l|awe^{i\varphi}}{r\left(r-2awe^{i\varphi}\right)}=l.$$

$$(1.79)$$

In deriving the third term in Equation (1.79), in the limiting case of $r \to \infty$, we utilized the asymptotic formula for the associated Laguerre polynomials: $[L_p^{|l|}(x)]^{-1} \partial L_p^{|l|}(x)/\partial x \approx |l|/x$. Thus, at $r \to \infty$, the third term in Equation (1.79) equals zero, because the radial variable r has the first degree in the numerator and second degree in the denominator. The second term in Equation (1.79) is also zero but for a different reason. Here, notwithstanding the presence of the radial variable r, which tends to infinity, there is also an integral function $\exp(i\varphi)$ integrated from 0 to 2π, which equals zero. From Equation (1.79), TC of an aLG mode is seen to equal l. Thus, for a conventional LG mode, a complex shift of coordinates causes changes in its form Equation (1.71) and OAM (Equation (1.75)), with its TC (Equation (1.79)) remaining unchanged. We also note that for the aLG beam of Equation (1.71), TC remains equal to l upon propagation because Equation (1.79) is valid for any z.

In this subsection, we denoted the topological charge as l, which is commonly used for LG beams. In the next subsections, we denote TC as n.

1.3.2   TC OF AN ASYMMETRIC BG BEAM

The complex amplitude of a BG beam [40] in the source plane $z = 0$ can be given by:

$$E_n\left(r,\varphi,z=0\right) = \exp\left(-\frac{r^2}{\omega_0^2} + in\varphi\right) J_n\left(\alpha r\right), \qquad (1.80)$$

where $\alpha = k \sin\theta_0 = (2\pi/\lambda)\sin\theta_0$ is a scaling factor, $k = 2\pi/\lambda$ is the wavenumber of light of wavelength λ, θ_0 is the angle of a conical wave that forms a Bessel beam. In any other plane z, the complex amplitude in Equation (1.80) takes the form:

$$E_n(r,\varphi,z) = q^{-1}(z)\exp\left(ikz - \frac{i\alpha^2 z}{2kq(z)}\right)\exp\left(-\frac{r^2}{\omega_0^2 q(z)} + in\varphi\right) J_n\left[\frac{\alpha r}{q(z)}\right], \quad (1.81)$$

where $q(z) = 1 + iz/z_0$, $z_0 = k\omega_0^2/2$ is the Rayleigh range, ω_0 is the radius waist of the Gaussian beam, $J_n(x)$ is the Bessel function of the first kind and n-th order. For an asymmetric Bessel–Gaussian (aBG) beam, the complex amplitude takes the form [67]:

$$E_n\left(r,\varphi,z;c\right) = \frac{1}{q(z)}\exp\left(ikz - \frac{i\alpha^2 z}{2kq(z)} - \frac{r^2}{q(z)\omega_0^2} + in\varphi\right)$$

$$\times\left[\frac{\alpha r}{\alpha r - 2cq(z)\exp(i\varphi)}\right]^{n/2} J_n\left\{\frac{1}{q(z)}\sqrt{\alpha r\left[\alpha r - 2cq(z)\exp(i\varphi)\right]}\right\}.$$

$$(1.82)$$

In Equation (1.82), c is a dimensionless constant that defines the asymmetry of the aBG beam. Unlike the aLG beam of Equation (1.70), the aBG beam of Equation (1.82) is derived not through a complex shift of coordinates but through the superposition of conventional Bessel beams, which is mathematically expressed using a reference series [67]:

$$\sum_{k=0}^{\infty} \frac{t^k}{k!} J_{k+v}(x) = x^{v/2} \left(x - 2t\right)^{-v/2} J_v\left(\sqrt{x^2 - 2tx}\right).$$ (1.83)

It is possible to derive a relationship for OAM carried by the aBG beam, normalized to power [67]:

$$\frac{J_z}{W} = n + \left[\sum_{p=0}^{\infty} \frac{c^{2p} p I_{n+p}(y)}{(p!)^2}\right] \cdot \left[\sum_{p=0}^{\infty} \frac{c^{2p} I_{n+p}(y)}{(p!)^2}\right]^{-1},$$ (1.84)

where $y = \alpha^2 \omega_0^2 / 4$. We did not manage to further simplify Equation (1.84). From Equation (1.84), the OAM of the aBG beams is seen to be larger than n, because all constituent terms of the series in Equation (1.84) are positive. Thus, with increasing parameter c, asymmetry of the aBG beam increases and OAM also increases near linearly. Growth of the OAM in Equation (1.84) follows partly from the series in Equation (1.83), where all Bessel functions have an order higher than that of the Bessel function in the aBG beam from Equation (1.82). The physical reason of the OAM growth with the asymmetry degree is that the "center of mass" [71] of the aBG beam shifts from the singularity center located in the origin [67]. *TC* of the aBG beam in Equation (1.82) can be found using Equation (1.77). First, write down the derivative of function in Equation (1.82) with respect to the azimuthal angle (assuming $J_n(x) \neq 0$, to avoid dividing by zero):

$$\frac{\partial E_n(r, \varphi, z; c)}{\partial \varphi} = in E_n(r, \varphi, z; c)$$

$$+ \frac{incq(z)e^{i\varphi}}{\left(\alpha r - 2cq(z)\exp(i\varphi)\right)} E_n(r, \varphi, z; c)$$ (1.85)

$$- \frac{ic(\alpha r)^{1/2} e^{i\varphi} J_n^{-1}(x)}{\left(\alpha r - 2cq(z)\exp(i\varphi)\right)^{1/2}} \frac{\partial J_n(x)}{\partial x} E_n(r, \varphi, z; c),$$

where $x = q^{-1}(z)\left\{\alpha r \left[\alpha r - 2cq(z)\exp(i\varphi)\right]\right\}^{1/2}$. Then, Equation (1.77) takes the form:

$$TC = \frac{1}{2\pi} \lim_{r \to \infty} \mathrm{Re} \int_0^{2\pi} d\varphi \left\{ n + \frac{ncq(z)e^{i\varphi}}{\left[\alpha r - 2cq(z)e^{i\varphi}\right]} - \frac{c(\alpha r)^{1/2} e^{i\varphi}}{\left[\alpha r - 2cq(z)e^{i\varphi}\right]^{1/2}} \frac{1}{J_n(x)} \frac{\partial J_n(x)}{\partial x} \right\}$$

$$= n + \lim_{r \to \infty} \mathrm{Re} \left[\frac{c J_n^{-1}(x)}{2\pi} \frac{\partial J_n(x)}{\partial x} \int_0^{2\pi} d\varphi e^{i\varphi} \right] = n.$$

(1.86)

In Equation (1.86), the second term of the integrand tends to zero at $r \to \infty$, because in the denominator the radial variable r has the first degree. Although for a different reason, the third term is also rearranged to a zero-valued term because it contains a ratio of the derivative of a Bessel function to a Bessel function. At $r \to \infty$, the said ratio can take any value since, at a large argument, the asymptotic functions of a Bessel function and the derivative thereof are given by:

$$J_n(x \gg 1) \approx \sqrt{\frac{2}{\pi x}} \cos\left(x - \frac{n\pi}{2} - \frac{\pi}{4}\right),$$

$$\frac{dJ_n}{dx}(x \gg 1) \approx -\sqrt{\frac{2}{\pi x}} \sin\left(x - \frac{n\pi}{2} - \frac{\pi}{4}\right),$$

(1.87)

with their ratio given by a tangent. The third term of Equation (1.86) also contains an integral of $\exp(i\varphi)$ taken with respect to the angle φ from 0 to 2π, which equals zero, making the entire term equal zero. From Equation (1.86) it follows that at any asymmetry degree (any c) and any distance from the source plane (any z) of an aBG beam, TC equals n. Meanwhile, the OAM of the aBG beam increases with increasing asymmetry degree, Equation (1.84).

1.3.3 TC OF AN ASYMMETRIC KUMMER BEAM

Asymmetric Kummer beams (aK beams) that represent exact solutions of a paraxial Helmholtz equation have been reported [68]. The complex amplitude of an aK beam is deduced by shifting a conventional Hypergeometric or Kummer beam [68] into a complex coordinate plane ($x \to x - aw$, $y \to y - iaw$, where a is a dimensionless real number) at any z:

$$E_s(r,\varphi,z) = \frac{(-i)^{n+1}}{n!} \Gamma\left(\frac{m+n+2+i\gamma}{2}\right)\left(\frac{z_0}{zq(z)}\right) q^{-(m+i\gamma)/2}(z)$$

$$\times \left(\frac{kwr}{2z\sqrt{q(z)}}\right)^n \exp\left(in\varphi + \frac{iks^2}{2z}\right) {}_1F_1\left(\frac{m+n+2+i\gamma}{2}, n+1, -\xi\right),$$

(1.88)

where $s^2 = r(r - 2awe^{i\varphi})$, $\xi = [kws/(2zq^{1/2}(z))]^2$, and ${}_1F_1(a,b,z)$ is a Kummer function (degenerate hypergeometric function). In the source plane ($z = 0$), the complex amplitude in Equation (1.88) of a shifted Kummer beam takes the form:

$$E_s(r,\varphi,z=0) = \frac{r^n e^{in\varphi}}{w^n}\left(\frac{s}{w}\right)^{m-n+i\gamma} \exp\left(-\frac{s^2}{w^2}\right),$$

(1.89)

The derivative of (1.89) with respect to φ is:

$$\frac{\partial E}{\partial \varphi} = inE - (m-n+i\gamma)\frac{irawe^{i\varphi}}{s^2}E + \frac{2irae^{i\varphi}}{w}E.$$

(1.90)

Substituting Equation (1.90) in Equation (1.77) yields:

$$TC = \frac{1}{2\pi} \lim_{r \to \infty} \text{Im} \int_0^{2\pi} d\varphi \left(in - \frac{i(m-n+i\gamma)rawe^{i\varphi}}{r(r-2awe^{i\varphi})} + \frac{2irae^{i\varphi}}{w} \right)$$

(1.91)

$$= n + \frac{ar}{\pi w} \int_0^{2\pi} \cos\varphi d\varphi = n.$$

The integrand in Equation (1.91) contains three terms, with the first equal to n and the second equal to zero at $r \to \infty$ (considering that the radial variable r has power two in the denominator and power one in the numerator). The third term also equals zero because the integral of the cosine function taken with respect to the angle on a period exactly equals zero, despite being multiplied by infinity at $r \to \infty$. Thus, in the source plane, an aK beam has TC equal to n. Using Equation (1.88), it can be shown in a similar way that at any z TC (Equation 1.88) equals n.

1.3.4   TC of an OV Composed of Two HG Modes

It has previously been demonstrated [69] that the superposition of two HG modes, $(n, n + 1)$ and $(n + 1, n)$, with a phase shift by $\pi/2$ generates an OV that carries OAM proportional to $-(n + 1)$. In this section, we obtain a relationship to define TC of such an OV. In the source plane $(z = 0)$, the superposition of two HG beams has the complex amplitude given by:

$$E(x,y,0) = \exp\left[-\frac{w^2}{2}(x^2 + y^2) \right]$$

(1.92)

$$\times \left[H_n(wx)H_{n+1}(wy) + i\gamma H_{n+1}(wx)H_n(wy) \right],$$

where $w = \sqrt{2}/w_0$, w_0 is the Gaussian beam waist radius, also assuming constant $\gamma = 1$. For light field in Equation (1.92), the normalized OAM for any integer n can be written as [69]:

$$\frac{J_z}{W} = -(n+1).$$

(1.93)

Owing to the fact that the sum of the numbers in both modes in Equation (1.92) is the same, the linear combination thereof in Equation (1.92) is also a mode, with both constituent modes having the same Gouy phase $((m+n+1)\arctan(z/z_0))$, which retains its form upon propagation, changing only in scale.

Even before one starts calculating TC from Equation (1.77), the answer can be predicted in advance. Actually, the first term in Equation (1.92) has n vertical zero lines and $(n + 1)$ horizontal zero lines. Meanwhile, the second term in Equation

(1.92), on the contrary, has $(n + 1)$ vertical zero lines (which do not coincide with those of the first term) and n horizontal zero lines. Hence, the intensity nulls in Equation (1.92) are found at intersection points of the horizontal and vertical zero lines. There are n^2 intersection points in the vicinity of zero $x + iy$ and $(n + 1)^2$ intersection points in the vicinity of zero $y + ix$. Therefore, for the Equation (1.92), TC is defined by $TC = n^2 - (n+1)^2 = -(2n+1)$. To determine TC of the beam by its complex amplitude from Equation (1.92), we express HG modes in Equation (1.92) via LG modes (expression (3.11) in [72]):

$$i^m H_n(\xi) H_m(\eta) = \sum_{k=0}^{[(n+m)/2]} {}'(-2)^k k! P_k^{(n-k,m-k)}(0)$$

$$\times \left[(\xi + i\eta)^{n+m-2k} + (-1)^m (\xi - i\eta)^{n+m-2k} \right] L_k^{n+m-2k} (\xi^2 + \eta^2),$$

(1.94)

where P are the Jacobi polynomials and the hatch near the sum means that for even numbers n and m the last term should be divided by 2. Then, products of the Hermite polynomials in Equation (1.92) read as:

$$H_n(wx) H_{n+1}(wy) = i^{-n-1} \sum_{k=0}^{n} (-2)^k k! P_k^{(n-k,n+1-k)}(0) w^{2n+1-2k}$$

$$\times L_k^{2n+1-2k} (w^2 x^2 + w^2 y^2) \left[(x+iy)^{2n+1-2k} + (-1)^{n+1} (x-iy)^{2n+1-2k} \right],$$

(1.95)

$$H_{n+1}(wx) H_n(wy) = i^{-n} \sum_{k=0}^{n} (-2)^k k! P_k^{(n+1-k,n-k)}(0) w^{2n+1-2k}$$

$$\times L_k^{2n+1-2k} (w^2 x^2 + w^2 y^2) \left[(x+iy)^{2n+1-2k} + (-1)^n (x-iy)^{2n+1-2k} \right].$$

(1.96)

Using these expressions, we can rewrite Equation (1.92) in the polar coordinates:

$$E(r,\varphi,0) = (-i)^{n+1} \exp\left(-\frac{w^2 r^2}{2} \right) \sum_{k=0}^{n} (-2)^k k! (wr)^{2n+1-2k} L_k^{2n+1-2k} (w^2 r^2)$$

(1.97)

$$\times \left[A_k^- e^{i(2n+1-2k)\varphi} - (-1)^n A_k^+ e^{-i(2n+1-2k)\varphi} \right].$$

where $A_k^{\pm} = P_k^{(n-k,n+1-k)}(0) \pm P_k^{(n+1-k,n-k)}(0)$.

Thus, we obtained a superposition of a finite number of LG modes. Such finite superposition has the TC equal to that of the LG mode with the highest (by modulus) order [55]. There are two such modes in Equation (1.97) (at $k = 0$), and therefore TC of such superposition equals either $2n+1$ or $-(2n+1)$, depending on which mode has the greater weight coefficient. It is known for the Jacobi polynomials that [51]:

$$P_k^{(\alpha,\beta)}(-z) = (-1)^k P_k^{(\beta,\alpha)}(z). \tag{1.98}$$

Therefore, $P_0^{(n,n+1)}(0) = P_0^{(n+1,n)}(0)$ and the weight coefficient A_0^- of the LG mode of the order $2n+1$ equals zero. Thus, TC of the superposition of two HG beams in Equation (1.92) is given by:

$$TC = -(2n+1). \tag{1.99}$$

To get the positive topological charge $TC = 2n+1$, one can swap the variables x and y in Equation (1.92).

1.3.5 TC of a Vortex HG Beam

It has been shown [70] that it is possible to superimpose several HG modes using a reference expression from [51]:

$$\sum_{k=0}^{n} \frac{n!\,t^k}{k!(n-k)!} H_k(x) H_{n-k}(y) = \left(1+t^2\right)^{n/2} H_n\left(\frac{tx+y}{\sqrt{1+t^2}}\right), \tag{1.100}$$

it is possible to generate a vortex HG beam whose complex amplitude is given by:

$$U_n(x,y,z) = i^n \exp\left(-\frac{x^2+y^2}{2}\right)\left(1-a^2\right)^{n/2} H_n\left(\frac{iax+y}{\sqrt{1-a^2}}\right). \tag{1.101}$$

Considering that in superposition in Equation (1.100) the sum of numbers of the constituent HG modes is constant, $k+(n-k)=n=\text{const}$, all constituent HG modes in Equation (1.100) have the same phase velocity (i.e. the same Gouy phases, $(n+m+1)\arctan(z/z_0)$), meaning that the whole beam in Equation (1.101) is a paraxial mode, which retains its transverse intensity structure upon propagation, only changing in scale and rotating. OAM of the beam from Equation (1.101) is given by [70]:

$$\frac{J_z}{I} = -\frac{2an}{1+a^2}. \tag{1.102}$$

The real parameter a in Equations (1.101) and (1.102) can be related to the rotation angle α of a cylindrical lens [15,72] that converts a conventional HG beam $(0, n)$ into a vortex HG beam in Equation (1.101):

$$\cos\alpha = \frac{a}{\sqrt{1+a^2}}, \quad \sin\alpha = \frac{1}{\sqrt{1+a^2}}. \tag{1.103}$$

Let us derive *TC* of vortex HG beam substituting Equation (1.101) into Equation (1.77), which gives:

$$TC = \frac{1}{2\pi} \lim_{r \to \infty} \mathrm{Im} \int_0^{2\pi} d\varphi \, \frac{\partial E(r,\varphi)/\partial \varphi}{E(r,\varphi)}$$

$$= \frac{n}{2\pi} \lim_{r \to \infty} \mathrm{Im} \int_0^{2\pi} d\varphi \, \frac{H_{n-1}(ir\cos\varphi\cos\alpha + r\sin\varphi\sin\alpha)(-ir\sin\varphi\cos\alpha + r\cos\varphi\sin\alpha)}{H_n(ir\cos\varphi\cos\alpha + r\sin\varphi\sin\alpha)}.$$

$$(1.104)$$

Putting $r \to \infty$ in (1.104) and replacing the Hermite polynomials with the highest-power monomials, $H_n(x \gg 1) \approx (2x)^n$, Equation (1.104) is rearranged to:

$$TC = \frac{1}{2\pi} \mathrm{Im} \int_0^{2\pi} d\varphi \, \frac{n(-i\sin\varphi\cos\alpha + \cos\varphi\sin\alpha)}{(i\cos\varphi\cos\alpha + \sin\varphi\sin\alpha)}$$

$$(1.105)$$

$$= \frac{-n}{2\pi} \int_0^{2\pi} d\varphi \, \frac{\tan\alpha}{\cos^2\varphi + \tan^2\alpha\sin^2\varphi} = -n.$$

In deriving (1.105), the reference integral was used [51]:

$$\int_0^{2\pi} \frac{d\varphi}{\cos^2\varphi + \gamma^2\sin^2\varphi} = \frac{2\pi}{\gamma}. \qquad (1.106)$$

From the comparison of relationships for OAM in Equation (1.102) and *TC* in Equation(1.105) of the vortex HG beam, *TC* is seen to depend neither on the parameter *a* nor the tilt angle α of the cylindrical lens (Equation (1.103)), with OAM being dependent of the parameter α. We note that to make the topological charge in Equation (1.105) of the vortex HG beam positive, one can swap the variables *x* and *y*. Because of this, both OAM and *TC* are equally important characteristics when describing OVs.

1.3.6 NUMERICAL SIMULATION

Figure 1.9 depicts intensity and phase patterns for an aBG beam of Equation (1.82) in three transverse planes: $z = 0$ (source plane, Figure 1.9(a,b)), $z = z_0$ (at the Rayleigh range, Figure 1.9(c,d)), and $z = 10z_0$ (far field, Figure 1.9(e,f)). In the source plane, alongside an OV of the eight order, a multitude of OVs of the first order can be observed. At the Rayleigh range, the only vortex can be found in the bottom left of Figure 1.9(d). In the far field, the only remaining vortex is the central OV of the

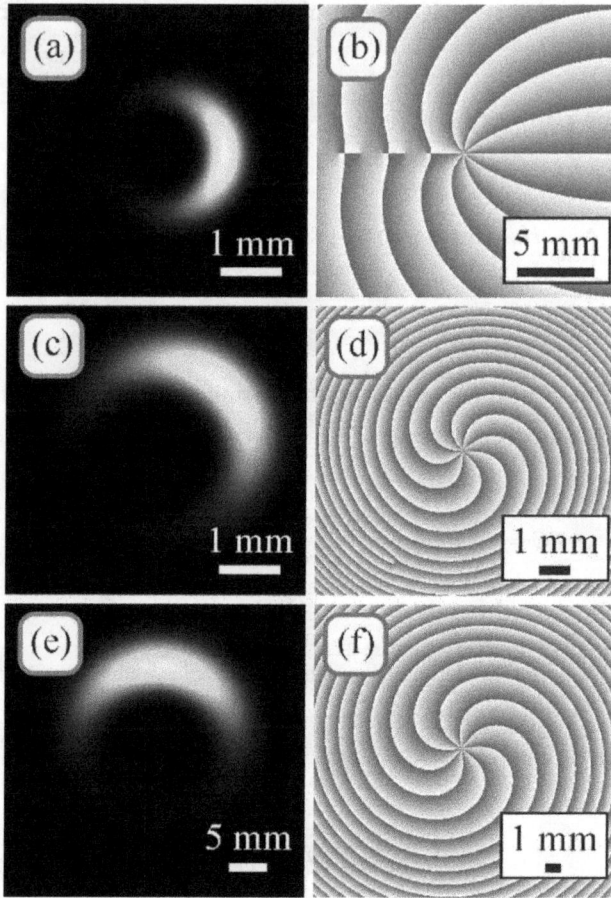

FIGURE 1.9 Patterns of intensity (a, c, e) and phase (b, d, f) for an aBG beam in three different planes. Simulation was conducted at a wavelength of $\lambda = 532$ n, beam radius $w_0 = 0.5$ mm, an OV of the eighth order $n = 8$, scaling factor $\alpha = 1/w_0$, asymmetry parameter $c = w_0/40$, propagation distance $z = 0$ (source plane) (a, b), $z = z_0$ (Rayleigh range) (c, d), and $z = 10z_0$ (far field) (e, f), for the calculation domain $-R \leq x, y \leq R$, where $R = 10$ mm ($z = 0$), $R = 10$ mm ($z = 0$), and $R = 20$ mm ($z = 10z_0$), and the number of pixels 2048 × 2048. TC was calculated along a circle of $x^2 + y^2 = R_1^2$, where R1 = 0.8R. The resulting values of TC are 11.9926 (a, b), 8.8837 (c, d), and 7.9393 (e, f).

eighth order. The numerical simulation gave the following values of TC: 11.9926 (at $z = 0$), 8.8837 (at $z = z_0$), and 7.9393 (at $z = 10z_0$).

Figure 1.10 depicts similar patterns of intensity and phase distributions of aLG beam (1.71) in three transverse planes: $z = 0$ (source plane, Figure 1.10 (a,b)), $z = z_0$ (at the Rayleigh range, Figure 1.10 (c,d)), and $z = 10z_0$ (far field, Figure 1.10 (e,f)). In the source plane, alongside an eighth-order OV, there are several first-order OVs. At the Rayleigh range, there are several such vortices, with three OVs clearly seen on

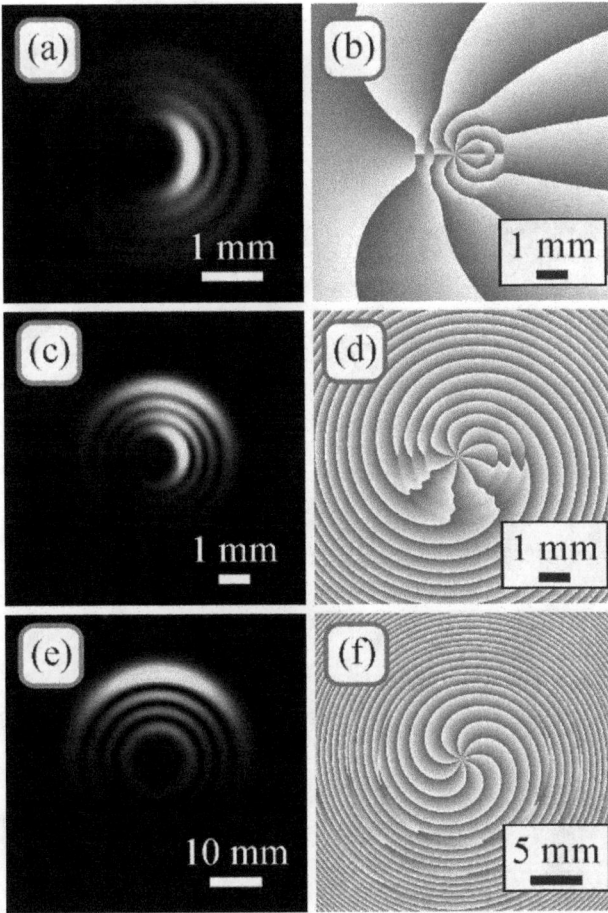

FIGURE 1.10 Patterns of intensity (a, c, e) and phase (b, d, f) for an aLG beam in three different planes. The simulation was conducted at a wavelength of $\lambda = 532$ nm, waist radius $w_0 = 0.5$ mm, OV order $n = 8$, mode radial index $m = 3$, shift vector $(x_0, y_0) = (0, iw_0/4)$, propagation distance $z = 0$ (source plane) (a, b), $z = z_0$ (Rayleigh range) (c, d), and $z = 10z_0$ (far field) (e, f), computation domain $-R \leq x, y \leq R$, where $R = 5$ mm ($z = 0$), $R = 5$ mm ($z = z_0$), $R = 30$ m ($z = 10z_0$), and the number of pixels 2048×2048. *TC* was calculated along a circle of radius $x^2 + y^2 = R1\ 2$, where $R1 = 0.8R$. The resulting values of *TC* are 7.9974 (a, b), 7.9925 (c, d), and 7.9226 (e, f).

the left of the center in Figure 1.10(d). In the far field, only an eighth-order central OV is seen. The numerically simulated values of *TC* are 7.9974 (at $z = 0$), 7.9925 (at $z = z_0$), and 7.9226 (at $z = 10z_0$).

Figure 1.11 depicts patterns of intensity and phase of an asymmetric Kummer function in Equation (1.88) in three transverse planes: $z = 0$ (source plane, Figure 1.11 (a,b)), $z = z_0$ (Rayleigh length, Figure 1.11 (c,d)), and $z = 10z_0$ (far field, Figure 1.11 (e,f)). It is seen that despite the asymmetric intensity pattern, in all the planes of

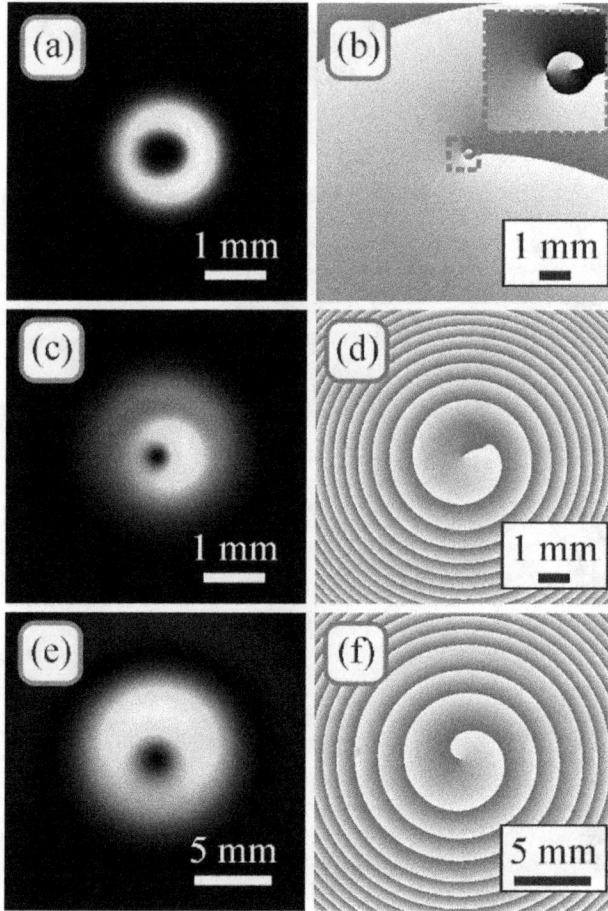

FIGURE 1.11 Patterns of intensity (a, c, e) and phase (b, d, f) for an asymmetric Kummer beam in three different planes. The simulation was conducted at a wavelength of $\lambda = 532$ nm, waist radius of the Gaussian beam, $w_0 = 0.5$ mm, OV order $n = 1$, parameters m and γ taken to be $m = 3$ and $\gamma = 0$, shift parameter $a = 0.2$, propagation distance $z = 0$ (source plane) (a, b), $z = z_0$ (Rayleigh length) (c, d), and $z = 10z_0$ (far field) (e, f), computation domain $-R \leq x, y \leq R$, where R = 5 mm ($z = 0$), R = 5 mm ($z = z_0$), R = 10 mm ($z = 10z_0$), in each plot, the number of pixels is 2048 × 2048. The inset in Figure 1.11 (b) depicts a magnified central fragment. *TC* was calculated along a circle of radius $x^2 + y^2 = R1\,2$, where R1 = 0.8R. The resulting values of *TC* are 0.9981 (a, b), 0.9992 (c, d), and 0.9999 (e, f).

interest there is a first-order OV at the center. The numerically calculated values of *TC* are 0.9981 (at $z = 0$), 0.9992 (at $z = z_0$), and 0.9999 (at $z = 10z_0$).

Figure 1.12 depicts patterns of intensity and phase for the superposition of two HG modes in Equation (1.92) in the source plane. Despite a great number of isolated intensity nulls, almost all of them are seen to compensate for each other (OVs of the plus and minus first order), with only a limited number of phase jumps remaining at

FIGURE 1.12 Patterns of intensity (a) and phase (b) for the sum of two HG modes in the source plane. The simulation was conducted at a wavelength of $\lambda = 532$ nm, waist radius of the Gaussian beam $w_0 = 0.5$ mm, OV order $n = 5$, computation domain $-R \leq x, y \leq R$, where $R = 5$ mm, the number of pixels 2048 × 2048. TC was calculated along a circle of radius $x^2 + y^2 = R_1^2$, where $R_1 = 0.8R$. The resulting TC is -10.9550.

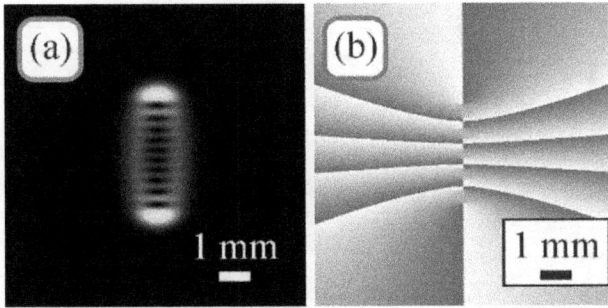

FIGURE 1.13 Patterns of intensity (a) and phase (b) for a vortex HG mode in the source plane. The simulation was conducted at a wavelength of $\lambda = 532$ nm, waist radius of the Gaussian beam $w_0 = 0.5$ mm, OV order $n = 10$, asymmetry parameter $a = 0.3$, computation domain $-R \leq x, y \leq R$, where $R = 5$ mm, and the number of pixels is 2048 × 2048. TC was numerically calculated along a circle $x^2 + y^2 = R_1^2$, where $R_1 = 0.8R$. The resulting TC value is -9.9993.

infinity. The numerically simulated value of TC equals -10.9550, corresponding to a theoretically predicted value of $-(2n + 1)$ at $n = 5$.

Figure 1.13 depicts patterns of intensity and phase for the vortex HG modes in Equation (1.101) in the source plane. On the vertical axis, there are ten isolated intensity nulls of the minus first order, the sum of which give TC equal to -10. The numerically simulated value of TC is -9.9993.

In this section, we have shown that if familiar radially symmetric laser beams (like LG and Kummer beams) are modified by shifting the coordinate system in a complex plane, both the form of the transverse intensity pattern and OAM are also changed depending on the degree of asymmetry, whereas TC of the asymmetric LG and Kummer beams remains unchanged and equals n [73]. It has also been shown that if a familiar radially symmetric BG beam with $TC = n$ undergoes a hybrid transformation in which the Gaussian beam remains unchanged while the Bessel beam is shifted into the complex plane, in the resulting asymmetric BG beam TC remains unchanged $TC = n$. The HG beam has been known to carry no TC, being topologically neutral. After passing through a cylindrical lens whose axis makes a certain angle with the Cartesian axes, the resulting vortex HG beam will carry OAM depending on the rotation angle of the cylindrical lens. However, the TC of such a beam has been found to be independent of the cylindrical lens rotation, being equal to n. A vortex HG beam is a finite superposition of conventional vortex-free HG beams. We have shown that a vortex beam can be generated by superimposing just two conventional HG beams with the numbers $(n, n + 1)$ and $(n + 1, n)$ and a $\pi/2$-phase shift. In the resulting vortex HG beam, TC equals $-(2n + 1)$, counting in terms of the number of intensity nulls. However, the straightforward use of Equation (1.77) for calculating TC of the field gives an incorrect result.

2 Evolution of an Optical Vortex with an Initial Fractional Topological Charge

2.1 CHANGE IN TC DURING PROPAGATION IN FREE SPACE AND STABILITY TO PHASE NOISE

Key characteristics of laser vortex beams [56] include topological charge (TC) and orbital angular momentum (OAM). M. Berry was the first to formulate the definition of TC of an optical vortex (OV) [19], whereas the notion of OAM was introduced into optics in [33]. While the total OAM of a paraxial light field is conserved upon free-space propagation, TC is sometimes conserved and sometimes not. TC is conserved upon propagation if the amplitude of the original light field can be given by $E(r, \varphi) = A(r)\exp(in\varphi)$, where $A(r)$ is the amplitude constituent that depends only on the radial variable r, φ is the azimuthal angle, and n is the integer TC of OV. Examples of such light fields include well-known Bessel–Gaussian and Laguerre–Gaussian beams. Examples of TC nonconservation upon OV propagation may be found in [22,74,75,52,76,77]. A simple superposition of a Gaussian beam and a Laguerre–Gaussian (LG) mode $(0, n)$ having different waist radii is discussed in [22]. Upon free-space propagation, total TC is changing due to different divergence of constituent beams. If in the original plane the waist radius of the Gaussian beam is larger than that of the LG mode, the total TC of superposition initially equals zero. Upon propagation, the difference between the waist radii decreases, with the waist radius of the LG mode at some point becoming larger than that of the Gaussian beam. From this point on, TC of superposition becomes equal to n. Another example of non-conserving TC is described in [74], where TC decreases with propagation distance after a circularly polarized beam passes through a slit with the shape of the Fermat spiral. It has been shown theoretically [19] and experimentally [75,52] that the original Gaussian beam with fractional TC has an integer TC in the near field, being equal to the nearest integer to the fractional number. However, it has been found that in the course of further propagation, TC of the beams undergoes other, previously unknown changes. For instance, it has been numerically and experimentally shown [76] that, as it propagates as far as the Fresnel zone, the original OV with fractional TC is converted to an OV with an integer TC equal to the nearest integer to the original fractional TC plus one. In a similar study conducted in [77], TC was actually measured in the Fourier plane (focal plane of a spherical lens), producing

DOI: 10.1201/9781003326304-2

results different from those reported in [76]. In [77], for a Gaussian OV $\exp(-r^2/w^2 + i\mu\,\varphi)$ with original fractional TC $\mu = (2k+1)+\varepsilon$ ($0.1 < |\varepsilon| < 1$), the far-field TC was shown to be $2k+1$. If the original fractional TC was $\mu = 2k+\varepsilon$ ($0.1 < |\varepsilon| < 1$), the far-field TC was also shown to be $2k+1$. The said studies have shown that, first, TC is not always conserved on propagation and, second, even and odd TCs show different stability towards changes. This conclusion can be drawn from the fact that regardless of the original fractional TC (close to an even integer or an odd integer) the far-field TC is always an odd integer. We note that TC can be measured using a triangular diaphragm [64,78] or a cylindrical lens [15].

In this section, we study theoretically and numerically the evolution of a Gaussian beam with original fractional TC. We show that there are only four evolution scenarios for an OV with original fractional TC. We also show that TC was measured in the near field in [19,75,52], in the Fresnel zone in [76], and in the far field in [77].

As a rule, an OV propagating in a turbulent atmosphere is identified by measuring its OAM [79,63]. However, due to minor "jitters" of both the entire laser beam and its constituent beams, there are continuous variations in OAM. For weak turbulence, OAM variations are small, otherwise, they are strong. When measuring TC, we should bear in mind that TC can only change discretely, while remaining integer. Hence, under weak turbulences, TC is not supposed to change at all. Analyzing the diffraction of an OV by a random phase screen, we show that TC remains unchanged until distortions of the random phase of the screen become essential.

2.1.1 THEORETICAL BACKGROUND

TC is calculated using a formula proposed by Berry [19]:

$$TC = \frac{1}{2\pi}\lim_{r\to\infty}\mathrm{Im}\int_0^{2\pi}d\varphi\,\frac{\partial E(r,\varphi)/\partial\varphi}{E(r,\varphi)}. \tag{2.1}$$

The amplitude of a Gaussian beam with the original fractional TC is given by [19]:

$$E(r,\varphi,z=0) = e^{-i\mu\varphi - r^2/w^2} = \frac{e^{i\pi\mu}\sin\pi\mu}{\pi}\sum_{n=-\infty}^{\infty}\frac{e^{in\varphi - r^2/w^2}}{\mu - n}. \tag{2.2}$$

In the Fresnel zone, the amplitude of field in Equation (2.2) is expressed as a difference of modified Bessel functions $I_\nu(x)$ [80]:

$$E(\rho,\theta,z) = \frac{\sin(\pi\mu)}{\sqrt{2\pi}}\left(\frac{-iz_0}{qz}\right)\exp\left(\frac{ik\rho^2}{2z} + i\pi\mu\right)\sqrt{x}e^{-x}$$
$$\times\sum_{m=-\infty}^{\infty}(-i)^{|m|}\frac{e^{im\theta}}{\mu - m}\left[I_{\frac{|m|-1}{2}}(x) - I_{\frac{|m|+1}{2}}(x)\right], \tag{2.3}$$

where $x = (z_0/z)^2(\rho/w)^2/(2q)$, $q = 1 - i(z_0/z)$. Polar coordinates (r, φ) in Equations (2.1) and (2.2) are used for the initial plane $(z = 0)$, while for the field in other planes $(z > 0)$ we use Equation (2.3) and from now on the polar coordinates (ρ, θ).

In the near field $z \ll z_0$, the parameters in Equation (2.3) take the form: $x = ik\rho^2/(2z)$, $q = -i(z_0/z)$. Considering that the modified and conventional Bessel functions are connected as $I_\nu(ix) = i^\nu J_\nu(x)$, in the near field, we obtain [19,15]:

$$E(\rho,\theta,z) = \frac{\sin(\pi\mu)}{\sqrt{2\pi}}\exp\left(\frac{ik\rho^2}{2z} + i\pi\mu\right)\sqrt{iy}\,e^{-iy}$$

$$\times \sum_{m=-\infty}^{\infty}(-i)^{|m|}\frac{e^{im\theta}}{\mu - m}i^{(|m|-1)/2}\left[J_{\frac{|m|-1}{2}}(y) - iJ_{\frac{|m|+1}{2}}(y)\right],$$

(2.4)

where $y = k\rho^2/(2z)$. In the limit, at $\rho \to \infty$, using the asymptotic form of Bessel function $J_\nu(y \gg 1) \approx [2/(\pi y)]^{1/2}\cos(y - \nu\pi/2 - \pi/4)$, we get:

$$E(\rho \to \infty, \theta, z) \approx \frac{\exp(i\pi\mu)\sin(\pi\mu)}{\pi}\sum_{m=-\infty}^{\infty}\frac{\exp(im\theta)}{\mu - m}.$$

(2.5)

Equations (2.5) and (2.2) are identical at $w \to \infty$. Substituting Equation (2.5) into Equation (2.1), we derive a relationship for calculating TC in the near field:

$$TC = \frac{1}{2\pi}\text{Re}\left\{\int_0^{2\pi}\left[\sum_{n=-\infty}^{\infty}\frac{ne^{in\varphi}}{\mu - n}\right]\cdot\left[\sum_{n=-\infty}^{\infty}\frac{e^{in\varphi}}{\mu - n}\right]^{-1}d\varphi\right\}.$$

(2.6)

Calculations with the aid of Equation (2.6) produce a well-known step function to describe TC jumps at $\mu = n + 1/2$ (Figure 2.1). Figure 2.1 depicts in which manner TC of field in Equation (2.2) depends, in the near field, on the original fractional TC μ,

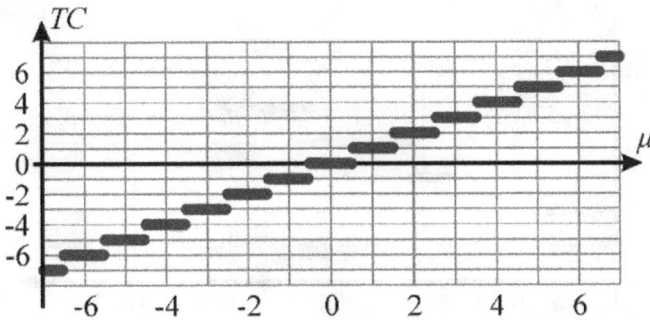

FIGURE 2.1 TC in the near field for a Gaussian beam with the original fractional TC, calculated using Equation (2.6).

with calculations based on Equation (2.6) conducted on the interval $-7 \leq \mu \leq 7$ with a 0.05 increment. It can be seen from Figure 2.1 that the TC jumps occur at half-integer values, i.e. at $\mu = n + 0.5$, which is in agreement with [19].

Next, we calculate the TC in the Fresnel zone for the field with the original fractional TC of Equation (2.2). Here, we can use the asymptotic form of the modified Bessel function at large argument values: $I_{(n-1)/2}(\xi) - I_{(n+1)/2}(\xi) \sim ne^{\xi} /[2\xi (2\pi \xi)^{1/2}]$ (it can be derived by using expression 9.7.1 in [81]). So, Equation (2.3) is reduced to:

$$E\left(\rho \to \infty,\theta,z\right) = \left(\frac{-iz}{z_0}\right)\left(\frac{w}{\rho}\right)^2 \sin\left(\pi\mu\right)$$

$$\times \exp\left(\frac{ik\rho^2}{2z} + i\pi\mu\right) \sum_{m=-\infty}^{\infty} (-i)^{|m|} \frac{\exp\left(im\theta\right)}{\mu - m}|m|. \qquad (2.7)$$

Substituting Equation (2.7) into (2.1), we obtain:

$$TC = \frac{\mathrm{Re}}{2\pi} \int_0^{2\pi} \left[\sum_{n=-\infty}^{\infty} \frac{(-i)^{|n|}n|n|e^{in\varphi}}{\mu - n}\right]\left[\sum_{n=-\infty}^{\infty} \frac{(-i)^{|n|} |n|e^{in\varphi}}{\mu - n}\right]^{-1} d\varphi \qquad (2.8)$$

Figure 2.2 depicts TC of a Gaussian beam with the original fractional TC of Equation (2.2) calculated for the Fresnel diffraction zone using Equation (2.8) for $5 \leq \mu \leq 10$ with a 0.05 step. From Figure 2.2, the TC jumps are seen to be found near every integer number, when $\mu \approx n + 0.1$, which is in agreement with [76]. However, according to [76] such a pattern was characteristic of the Fraunhofer diffraction zone, rather than the Fresnel zone. However, a similar study of TC of an OV in the far field (specifically, in the Fraunhofer diffraction zone or in the focus of a spherical Fourier lens) was conducted in [77], producing a different result.

In the far field (the focus of a Fourier lens), the amplitude of a light field takes the form:

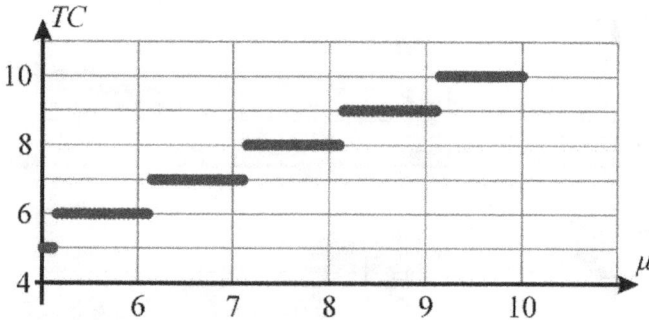

FIGURE 2.2 TC in the Fresnel zone for a Gaussian beam with original fractional TC, Equation (2.2), calculated using Equation (2.8).

$$E_F(\rho,\theta) = \frac{\sin(\pi\mu)}{\sqrt{2\pi}}\left(\frac{-iz_0}{f}\right)\exp(i\pi\mu)\sqrt{x}\exp(-x)$$

$$\times \sum_{m=-\infty}^{\infty}(-i)^{|m|}\frac{e^{im\theta}}{\mu-m}\left[I_{\frac{|m|-1}{2}}(x)-I_{\frac{|m|+1}{2}}(x)\right],$$

(2.9)

where the argument of Bessel functions is a real value: $x = (z_0/f)^2(\rho /w)^2/2$, and f is the focal length of a lens that generates the far field in the back focal plane. The calculation of TC using Equation (2.9) and Equation (2.1) gives a step function shown in Figure 2.3, which depicts the far-field TC of a Gaussian beam with the original fractional TC, Equation (2.2), calculated using Equations (2.9) and (2.1) for $-5 \le \mu \le 5$ with a step of 0.05. From Figure 2.3, jumps between adjacent integer values of TC occur at even integers, in agreement with [77].

Figures 2.1–2.3 suggest that the source field with original fractional TC (Equation (2.2)) undergoes an interesting evolution. From Figure 2.3 it is seen that in the far field, any original fractional field evolves into one that only has odd TCs. In the far field, the even TC can be generated only if the original field has an even integer TC. Actually, we assume that in the original plane TC is $\mu = 3.3$. Then, in the near field, an OV with $TC = 3$ will be generated (Figure 2.1), farther in the Fresnel zone, a new OV with TC +1 will be born, with the total TC of the beam becoming equal to $TC = 4$ (Figure 2.2). Meanwhile, in the far field another optical vortex with TC −1 will be born, with the total TC of the beam being again $TC = 3$ (Figure 2.3), as in the near field.

The evolution of a field whose original fractional TC is closer to an even integer (e.g. $\mu = 4.3$) follows a different path. In the near field, the beam has $TC = 4$ (Figure 2.1), in the Fresnel zone, $TC = 5$ (Figure 2.2), with TC remaining unchanged in the far field, $TC = 5$ (Figure 2.3).

There are two more evolution scenarios for a field with the original fractional TC, which can be seen in Table 2.1. Table 2.1 gives all four feasible evolution scenarios

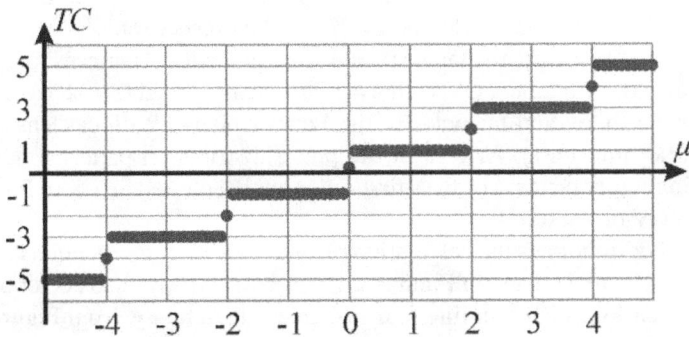

FIGURE 2.3 Fourier-plane TC of a Gaussian beam with the original fractional TC, Equation (2.2), calculated using Equatioins (2.9) and (2.1).

TABLE 2.1

_TC_ evolution scenarios on propagation of a Gaussian beam with the original fractional _TC_, exp($-r^2/w^2 + i\mu\,\varphi$) (_p_ is an arbitrary integer, $0 < \varepsilon < 1/2$).

Original _TC_ ($z = 0$)	_TC_ ($z \ll z_0$)	_TC_ ($z \approx z_0$)	_TC_ ($z \gg z_0$)
$\mu = 2p + \varepsilon$	$2p$	$2p + 1$	$2p + 1$
$\mu = 2p - \varepsilon$	$2p$	$2p$	$2p - 1$
$\mu = (2p + 1) + \varepsilon$	$2p + 1$	$2p + 2$	$2p + 1$
$\mu = (2p + 1) - \varepsilon$	$2p + 1$	$2p + 1$	$2p + 1$

for the initial fractional vortex. If a simple field has an integer _TC_ in the source plane, _TC_ is conserved during propagation. But this is not always the case. For a linear combination of two differently diverging light fields, the original integer _TC_ will change upon propagation [22].

2.1.2 NUMERICAL SIMULATION

The numerical simulation aims to corroborate the values of _TC_ derived from Equations (2.6), (2.8), and (2.9) for different diffraction zones of a Gaussian beam with the original fractional _TC_. The intensity and phase patterns in the near field and Fresnel zone are modeled by the BPM-method (BeamProp software by RSoft) for the following parameters: wavelength $\lambda = 532$ nm, Gaussian beam waist $w_0 = 5\lambda$ ($z_0 \sim 25\pi\lambda$), initial _TC_ $\mu = 3.3$, half-size of the computational domain $R = 50\lambda$, transverse discretization step $\Delta x = \Delta y = \lambda/32$, and longitudinal discretization step $\Delta z = \lambda/16$. Meanwhile, the amplitude and phase patterns in the far field are numerically modeled using a Fourier transform implemented with a spherical lens of focal length $f = 100$ mm for a Gaussian beam with waist radius $w_0 = 0.5$ mm and the same original _TC_ $\mu = 3.3$ in the computational domain with the size of $2R = 1$ mm (the evolution scenario corresponds to line four of Table 2.1). The patterns were calculated for a distance of $z = 3\lambda$ in the near field and $z = 50\lambda$ in the Fresnel zone.

Shown in Figure 2.4 are the intensity (a–c) and phase (d–f) patterns for a Gaussian beam containing a fractional OV with $\mu = 3.3$ in the near field ($z = 3\lambda$) (a, d), Fresnel zone ($z = 50\lambda$) (b, e), and far field (in the Fourier plane of a focal lens with focal length $f = 100$ mm) (c, f). Arrows in the phase distribution patterns mark optical vortices (singularity centers) of the +first order, with a dashed arrow in Figure 2.4(f) marking an OV of −first order.

Figure 2.4 corroborates the calculation results derived from Equations (2.6), (2.8), and (2.9), which are depicted in Figures 2.1–2.3, respectively, also corroborating the evolution scenario depicted in line four of Table 2.1. Actually, from Figure 2.4(d) it is seen that while in the source plane ($z = 0$) _TC_ is $\mu = 3.3$, immediately behind it, at a distance of 3λ, there are only three singularity points and TC is $\mu = 3$. Although some phase distortions are present on the right in Figure 2.4(d), no new singularity

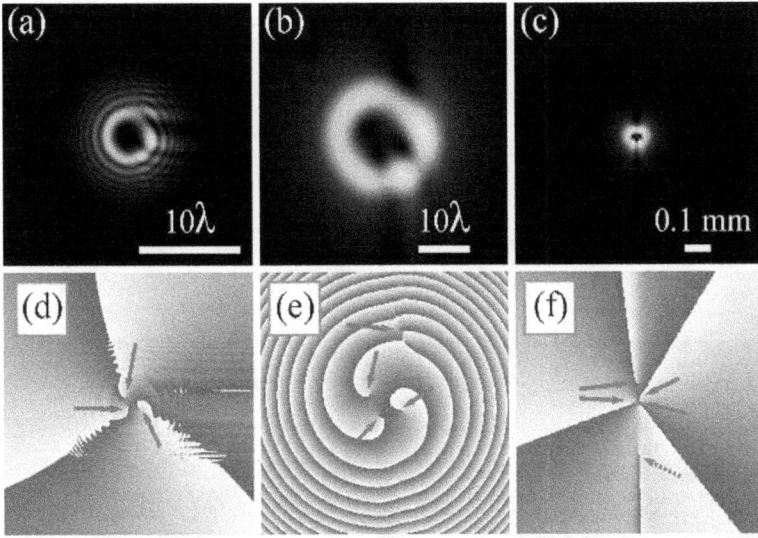

FIGURE 2.4 Intensity (a–c) and phase (d–f) of a Gaussian beam with original fractional TC, $\mu = 3.3$ in the near field (a, d), Fresnel zone (b, e), and far field (c, f). Arrows in the phase distribution patterns mark OVs of the +first order, with a dashed arrow marking an OV of the −first order (f). Dark color denotes a zero phase and white is for a phase of 2π.

points can be seen. In Figure 2.4(e) at a distance of 50λ (while the Rayleigh range is 75λ) a fourth singularity point is seen to form, as is evident from a fringe dislocation for three original singularity points, with TC being $\mu = 4$. Propagating further, the beam is seen to form one more singularity point of the opposite sign (−1) in the focus of a spherical lens (Figure 2.4((f), bottom). This brings the total TC of the beam back to the initial TC of $\mu = 3$.

Below we investigate in more detail how the additional optical vortices are born and propagate. Figure 2.5 shows the TC dependence on the radius R for the initial TC of $\mu = 2.2$ in the near field at distances of $z = 0.01$ μm and $z = 0.1$ μm. We use the following parameters of the field from Equation (2.2) in Figure 2.5 (and below, in Figures 2.6–2.8): field size 8×8 μm (400×400 points), wavelength $\lambda = 0.532$ μm, Gaussian beam waist radius $w = 1.3$ μm, Rayleigh distance $z_R = \pi \omega_0^2 / \lambda = 9.98$ μm. According to Figure 2.5, TC in the near field is equal to 3, starting from the radius $R > 4$ μm. It is also seen that the phase of the field at distances of $z = 0.01$ μm and $z = 0.1$ μm remains the same in the whole field except the central area $R < 4.5$ μm. So, for the initial TC $\mu = 2.2$, TC of the optical vortex in the near field at the distance of just $z = 0.01$ μm is equal to 3. Why this simulation contradicts to the experiments in [75,52,64,78,15]? The answer is that, according to Figure 2.5(c), the intensity of the optical vortex decays almost to zero in the circle with a radius of nearly 2 μm, whereas the radius R of the circle where TC changes from 2 to 3 is about 4 μm. Thus, the additional singularity center is located in the area where it cannot be detected experimentally. Indeed, in Figure 2.5 at $z = 0.1$ μm and $\mu = 2.2$, the intensity on the

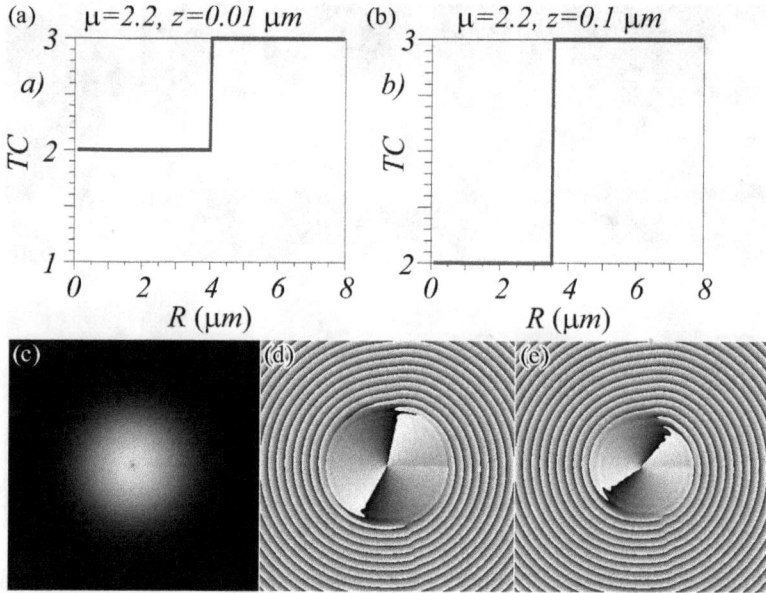

FIGURE 2.5 Dependence of the beam TC on the radius R for the initial beam with $\mu = 2.2$ in the near field at distances $z = 0.01$ μm (a) and $z = 0.1$ μm (b); Amplitude (c) and phase (d) at the distance $z = 0.01$ μm and phase (e) at the distance $z = 0.1$ μm.

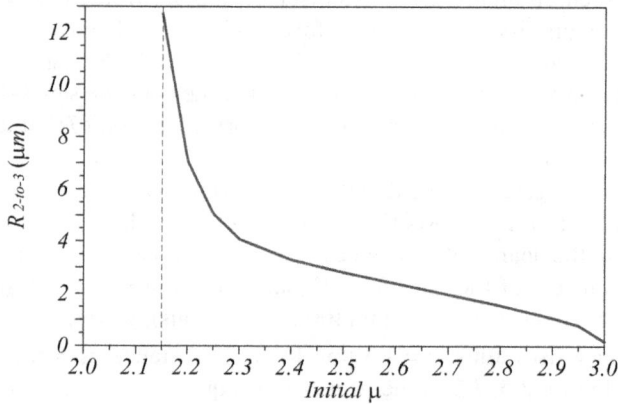

FIGURE 2.6 Radius R where TC jumps from 2 to 3 versus the initial TC μ at $z = 10$ μm.

circle where TC jumps from 2 to 3 (R = 3.5 μm) equals $2.0 \cdot 10^{-6}$ from its maximal value. Such low intensity cannot be measured experimentally.

Figure 2.6 shows that when the fractional part of the initial TC increases, radius of the circle of TC jump from 2 to 3 decreases. Therefore, for an optical vortex in Equation (2.2) with the initial fractional TC in the range $2.12 < \mu < 3$, TC in the Fresnel diffraction zone is equal to 3. The plot in Figure 2.6 shows yet another

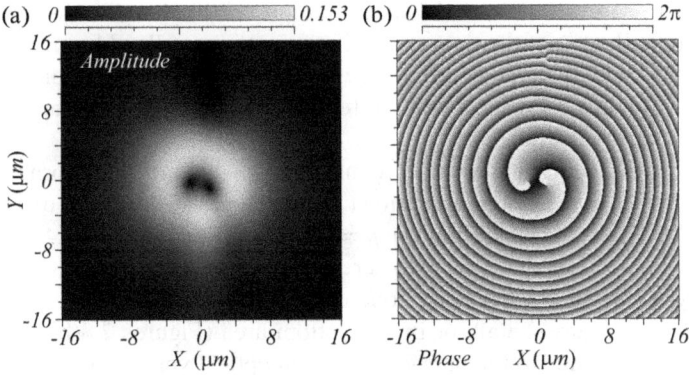

FIGURE 2.7 Amplitude (a) and phase (b) of the field from Equation (2.2) with the initial *TC* $\mu = 2.2$ at the distance of $z = 20$ µm (far field).

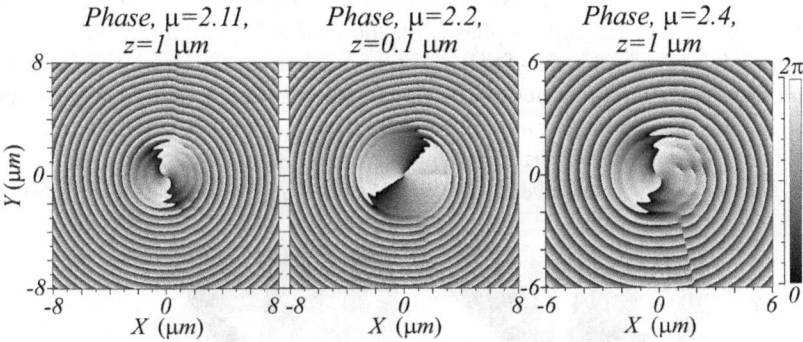

FIGURE 2.8 Phases of the field (2.2) with the different initial TCs in the near field: $\mu = 2.11$, $z = 1$ µm (a), $\mu = 2.2$, $z = 0.1$ µm (b), and $\mu = 2.4$, $z = 1$ µm (c).

interesting feature. If we suppose that R is the distance from the center of the field (i.e. of the Gaussian beam) to the third singularity, then with an increase of the fractional part of the initial *TC* from 0.15 to 0.95 the third singularity is approaching from the periphery (where the intensity is almost zero and thus cannot be detected) to the Gaussian beam center. Starting from *TC* nearly of 2.5, it achieves the distance of circa 3 µm where the intensity can be measured and thus reveal this singularity. Therefore, the plot in Figure 2.6 explains why in the experiments [75,52,64,78,15], *TC* jumps from 2 to 3 at the fractional part of the initial *TC* equal to 0.5. We note that for the initial fractional *TC* in the range from 3.1 to 4 we obtained a plot similar to Figure 2.6. Therefore, we can claim that the plot similar to Figure 2.6 is valid for arbitrary *TC* in the range $m < \mu < (m+1)$ (*m* is an arbitrary integer). In fact, our study made the bridge between the results of the work in [19] (*TC* jumps from *m* to *m*+1 when the fractional part is 0.5) and of the work in [76] (*TC* jumps at any fractional part).

Figure 2.7 shows the amplitude and phase distributions of the initial field in Equation (2.2) at a distance of $z = 20$ µm. The initial field has *TC* $\mu = 2.2$, its size is

32×32 μm (400×400 points). As seen in Figure 2.7(b), there is a fork in the phase distribution (at the top of the figure) at a distance of about $R = 14$ μm. Starting from this fork, TC of the optical vortex changes from 2 to 3. A comparison of amplitude and phase in Figure 2.7 reveals that the fork (third vortex) is in the beam periphery where the intensity is almost zero.

Where do the additional vortices come from? How are they born? Figure 2.8 illustrates phases of the field in Equation (2.2) in the near field for the different initial fractional TCs: $\mu = 2.11$, $z = 1$ μm (a), $\mu = 2.2$, $z = 0.1$ μm (b), $\mu = 2.4$, $z = 1$ μm (c). As seen in Figure 2.8, there are two screw dislocations on the phase distributions in the very center of the beam and one edge dislocation (distorted fringes) in the top phase area (Figure 2.8(a)), or in the bottom area (Figures 2.8(b, c)). This edge dislocation appears due to the interference of an optical vortex with $TC = 2$ and of a boundary wave propagating from the area of the phase break (horizontal dark line in Figure 2.5(a)). On further propagation, this edge dislocation leads to a screw dislocation (additional optical vortex or fork in Figure 2.7(b)).

Yet another question arises as to whether TC in the far field is always odd for any fractional μ. Figure 2.9 depicts phases for the field in Equation (2.2) with the waist radius $w = 0.5$ mm in the rear focal plane of a spherical lens with the focal length $f = 100$ mm for different values of μ : 2.3 (a), 3.3 (b), 4.2 (c), and 5.1 (d). As seen in Figure 2.9 for an even integer $m = 2$ (Figure 2.9(a)), in addition to two initial vortices there is an additional third vortex in the top area, while in the bottom area there is no vortex (i.e. $TC = 3$). At the even value $m = 4$ (Figure 2.9(c)), besides the initial four

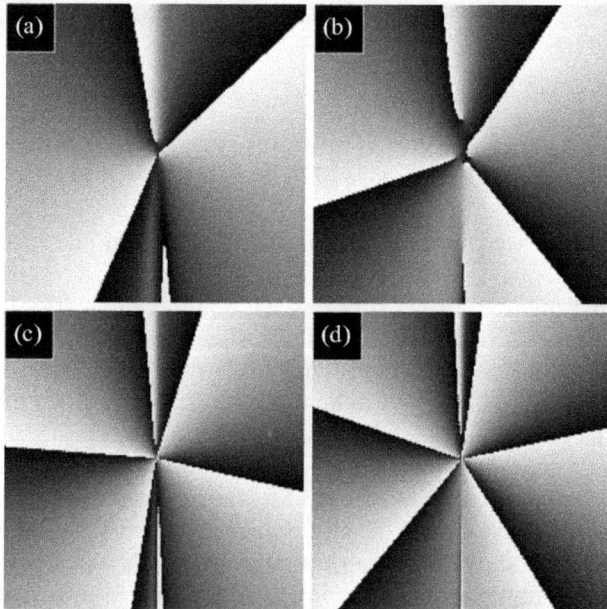

FIGURE 2.9 Phases of the field from Equation (2.2) with the waist radius $w = 0.5$ mm in the focus of the spherical lens ($f = 100$ mm) for different values of μ : 2.3 (a), 3.3(b), 4.2 (c), and 5.1 (d).

vortices, there is an additional fifth vortex in the top area and there is no vortex in the bottom area (i.e. $TC = 5$). In the bottom areas of Figure 2.9(a, c), there is a line of phase jump by 2π. For odd values $m = 3$ (Figure 2.9(b)) and $m = 5$ (Figure 2.9(d)), an additional vortex at the top is compensated by the additional vortex at the bottom. Therefore, TC is equal to 3 in Figure 2.9(b) and 5 in Figure 2.9(d).

As seen in Figure 2.9, additional optical vortices with TC +1 (at the top) and with TC −1 (at the bottom of Figure 2.9(b, d)) are located far from the beam center and these distances are not equal. The vortices with TC −1 are farther than the vortices with TC +1. The distance from the center to the vortex does not affect the TC of the whole beam, but it affects the results of the experiment. Both vortices (+1 and −1) are located at the distance from the beam center where the intensity is almost zero. Therefore, they are hard to register experimentally.

2.1.3 EXPERIMENT

Figure 2.10 shows the experimental setup used in the experiments. A linearly polarized Gaussian laser beam ($\lambda = 532$ nm, w_0 is approximately 5 mm) was expanded and collimated with a combination of a pinhole PH (with an aperture of 40 μm) and a lens $L1$ with a focal length of 350 mm. The mirrors M1 and M2 as well as two beam splitters B1 and B2 were used to implement the Mach–Zehnder interferometric setup. The collimated laser beam was normally incident onto the display of the spatial light modulator SLM (HOLOEYE LC-2012, 1024 × 768 pixels), which was used for the implementation of the phase transmission function of the elements generating Gaussian beam with the original fractional TCs of 2.2, 2.7, 3.3, and 3.7. The 4-f optical system consisted of the lenses $L2$ (focal length $f_2 = 150$ mm) and $L3$ (focal length $f_3 = 150$ mm), and a diaphragm D spatially filtered the laser beam generated by the SLM. The intensity pattern formed in the near-field, Fresnel zone, and focal plane of the lens $L4$ (focal length $f_4 = 350$ mm) was recorded with a video camera Cam camera (3264 × 2448 pixels, 1.67-μm pixel size). A neutral density filter F was used to equalize the intensities of the object and the reference beams.

Figure 2.11 shows the intensity and interferogram patterns obtained in the experiments.

The topological charge of the optical vortices from Figure 2.11 is given in Table 2.2. It is obtained by simply counting the "fork teeth" on the interferograms in Figure 2.11.

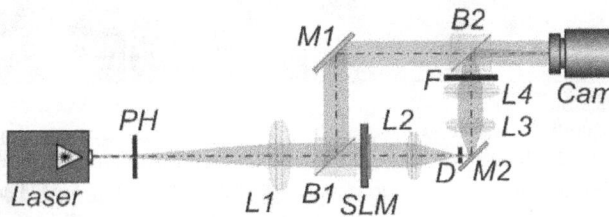

FIGURE 2.10 Experimental setup for investigation of TCs of the generated OVs.

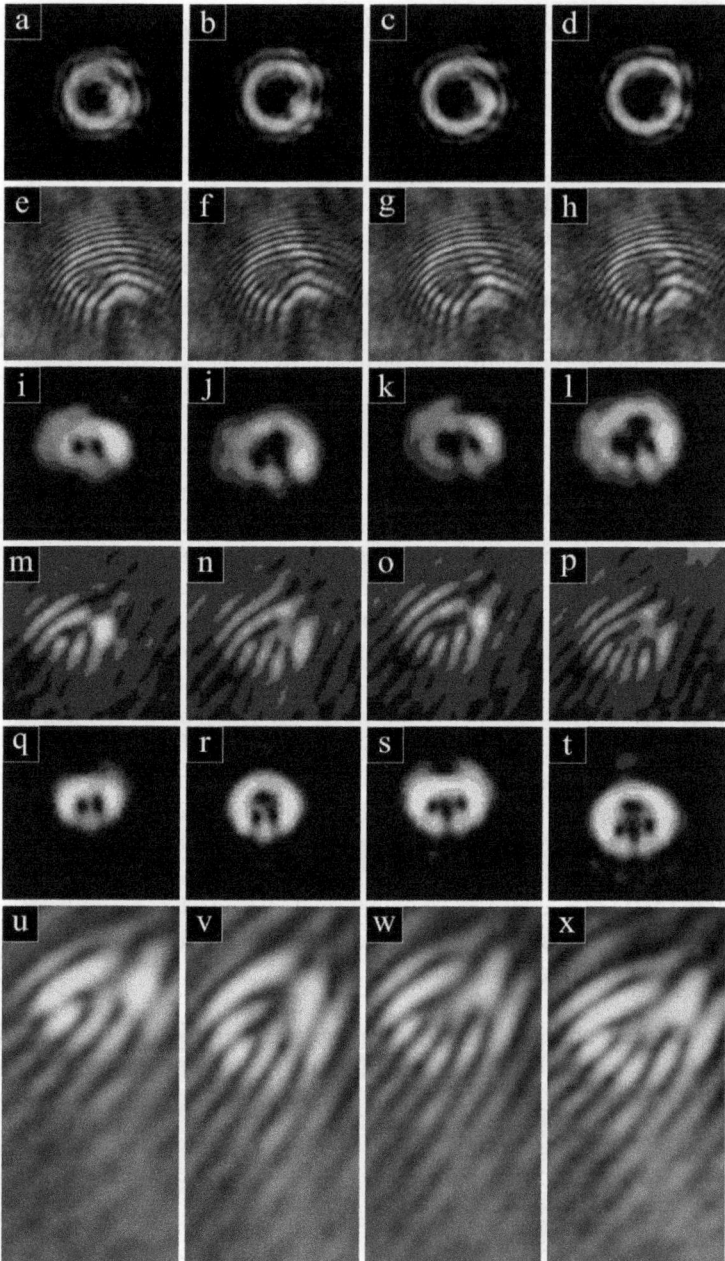

FIGURE 2.11 Intensity (a, b, c, d, i, j, k, l, q, r, s, t) and interferogram (e, f, g, h, m, n, o, p, u, v, w, x) patterns obtained in the case of a Gaussian beam with the original TCs μ : 2.2 (a, e, i, m, q, u), 2.7 (b, f, j, n, r, v), 3.3 (c, g, k, o, s, w), and 3.7 (d, h, l, p, t, x). The image size is 1100×1100 μm (a–h), 200×200 μm (i–t), 150×300 μm (u–x).

TABLE 2.2

Experimental *TC* of different fractional optical vortices in different diffraction zones.

	Topological charge			
Initial field	2.2	2.7	3.3	3.7
Near field	2	3	3	4
Fresnel zone	3	3	4	4
Lens focus	3	3	3	3

The third row of Table 2.2 contains *TC* values obtained from the interferograms (Figure 2.11(e–h)) of the optical vortices in the near field, whose intensity is shown in Figure 2.11(a–d). It is seen in the third row of Table 2.2 that in the near field (at a distance 40 mm from the focus of the lens *L4* in Figure 2.10) the experiment matches the theory (Figure 2.1) and the simulation (Figure 2.4(d)), but contradicts with the simulation from Figure 2.5. We gave an explanation for this contradiction above: the additional vortex in Figure 2.5 for the initial *TC* in Table 2.2 is generated far from the optical axis and thus cannot be registered experimentally. The fourth row of Table 2.2 contains TCs in the Fresnel diffraction zone (at a distance of 4 mm from the focus of the lens *L4* in Figure 2.10). These values were measured from the interferograms (Figure 2.11(m–p)) for the optical vortices, whose intensity is shown in Figure 2.11(i–l). For the Fresnel zone, *TC* from Table 2.2 coincides with the theory in Equation (2.8) and with the simulation in Figure 2.4(e) and Figure 2.6. The last (fifth) row of Table 2.2 contains TCs in the far field (in the focus of the spherical lens *L4* in Figure 2.10). These values were found from the interferograms (Figure 2.11(u–x)) of the optical vortices with intensity shown in Figure 2.11(q–t). The experimental results shown in the fifth row of Table 2.2 agree with the theory in Equation (2.9) and with the simulation in Figure 2.4(f) and Figure 2.9. Indeed, for an even integer part of the *TC* value $m = 2$, as seen in Figure 2.11(u, v), the forks on the interferogram contain three additional teeth. For the odd integer part of the *TC* value $m = 3$, interferograms analysis in Figure 2.11(w, x) is more complicated. In the beam center, there are four additional fork teeth clearly seen on the interferogram. However, in the bottom part of the interferogram, the characteristic defect of fringes can be seen (although with a low contrast), which reveals the additional vortex with *TC* –1. Therefore, for fractional vortices with the odd value $m = 3$, *TC* in the far field equals 3. The weak contrast of fringes in the bottom parts of the interferograms is due to the low intensity in the beam periphery. Thus, in the far field, the experiment also confirms the theory in Equation (2.9) and the simulation in Figures 2.4(f) and 2.9.

2.1.4 STABILITY OF THE TOPOLOGICAL CHARGE TO PHASE NOISE

We also made the experimental study of the stability of an OV's *TC* against phase distortions. Specifically, we experimentally study the conservation of the integer *TC* following random phase distortions of a vortex laser beam. The experimental setup is similar to that in [82], but the phase distortions are generated on a spatial light modulator (SLM) instead of the ground glass plate, thus allowing us to change the distortions magnitude and to define such threshold distortions for which TC is still conserved. We also adopt a more accurate method of TC measurement, using for this purpose a cylindrical lens [15]. A Gaussian beam with waist radius $w = 1.1$ mm is incident on a spatial light modulator, where a vortex phase $m\varphi$ with $m = 5$ is recorded. Each pixel of the SLM is being distorted by adding $2\pi \alpha$ to its phase $m\varphi$, where α is a random number from the interval $[0,1]$. Figure 2.12 depicts distorted phases of the original light field recorded on SLM (left column), intensity patterns (600×600 μm) measured in the focal plane of a spherical lens with focus $f = 150$ mm (central column), and intensity patterns (1900×1900 μm) measured at a double focal length from a cylindrical lens with focus $f = 100$ mm (right column) for different phase distortions α. In our experiment, we use the SLM (HOLOEYE LC-2012, 1024×768 pixels), with a pixel size of 8 μm. Random phase jumps on the SLM are seen as dark (phase is 0) and light (phase is 2π) dots in Figure 2.12(d, g, j, m, p). The phase on the SLM is changed in each pixel by a proportional change of the refractive index of the liquid crystal. Therefore, the power spectrum of the random phase distortions is not that of white-noise. Instead, the correlation function of the distortions has the width defined by the SLM pixel size.

Figure 2.12 suggests that at $\alpha = 0.6$ (i.e. the phase is distorted by a random value from the interval $(0, 1.2\pi)$), six peaks are still clearly seen on the diagonal at an angle of -45 degrees (either 5 dark fringes or 5 intensity nulls), meaning that the optical vortices have a TC of 5. However, at $\alpha \geq 0.8$ TC is nowhere to be seen (Figure 2.12). Therefore, TC of an optical vortex remains equal to the integer 5 until random phase distortion of the initial vortex gets as large as about half-wavelength ($\lambda/2$).

Summing up, it has been theoretically shown that a Gaussian beam with an embedded OV with original fractional TC does not conserve the original TC upon propagation [84,85]. TC of the Gaussian beam with the original fractional TC can take different values in different diffraction zones, such as near field ($z \ll z_0$), Fresnel zone ($z \approx z_0$), and far field ($z \gg z_0$). This conclusion has been corroborated via numerical simulation using the BeamProp software for the near field and Fresnel zone and a Fourier transform for the far field. Four evolution scenarios for the fractional Gaussian beam have been described. In one scenario, if the source beam has TC 3.3, it has been shown to have TC 3 in the near field, TC 4 in the Fresnel zone, and again TC 3 in the Fraunhofer zone (in the focus of a spherical lens). Thus, in this evolution scenario, an OV with TC +1 is produced in the Fresnel zone, before being compensated for in the far field by a newly born OV with TC −1. The experiment on determining TC by using the interferograms matches the theory and the simulation. We have also shown experimentally that given weak random phase distortions (a phase shift smaller than π) TC is conserved. Based on these properties, it becomes possible to identify OVs in wireless communications by measuring TC, alongside OAM measurements.

Additional OVs in the Fresnel and Fraunhofer diffraction zones (see Table 2.1) are explainable since the initial fractional-TC light field contains the whole angular

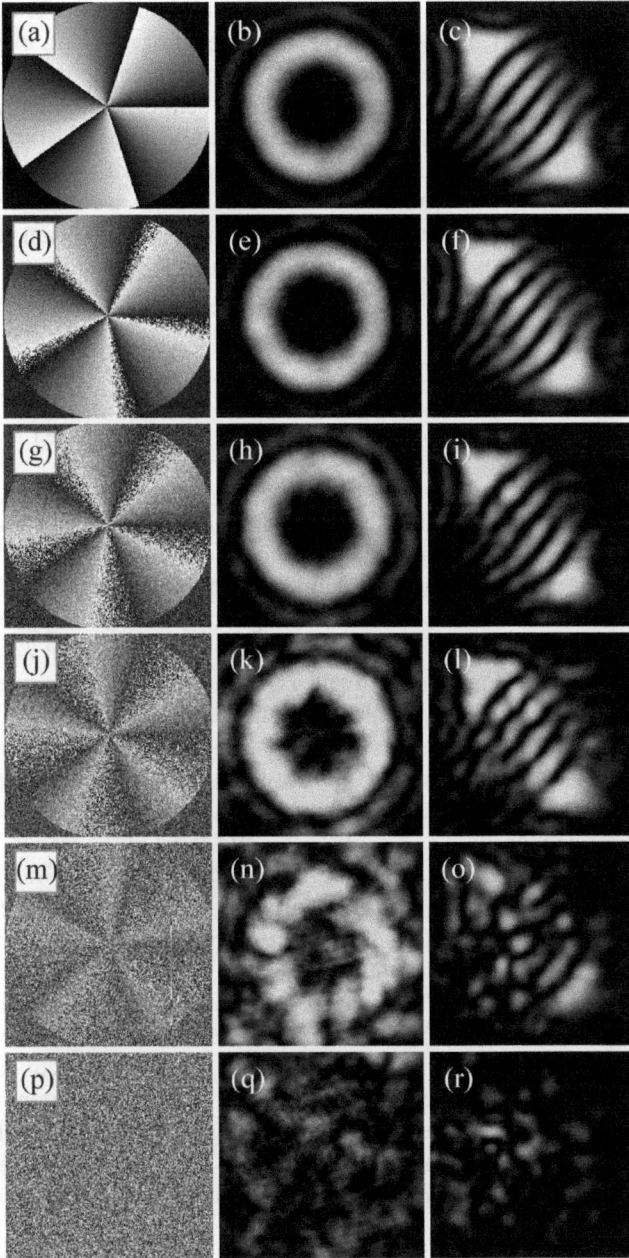

FIGURE 2.12 Distorted phase patterns (a, d, g, i, m, p), intensity patterns (600 × 600 μm) in the focal plane of a spherical lens with focus $f = 150$ mm (b, e, h , k, n, q), and intensity patterns (1900 × 1900 μm) at a double focal length of a cylindrical lens with focus $f = 100$ mm (c, f, i, l, o, r) for different degree of distortion: $\alpha = 0$ (a, b, c), $\alpha = 0.2$ (d, e, f), $\alpha = 0.4$ (g, h, i), $\alpha = 0.6$ (j, k, l), $\alpha = 0.8$ (m, n, o), and $\alpha = 1.0$ (p, q, r).

harmonics spectrum and thus carries all these additional OVs. Two well-known examples confirm this. If the Hermite-Gaussian beam of order (0, n) passes in the initial plane through a cylindrical lens rotated by 45 degrees to the Cartesian axes, its initial TC is zero. However, at the double focal length, the beam transforms to a vortex LG beam with the order (0, n) [83], i.e. an OV with TC of n was born. As another example, if in the initial plane there is a combined beam composed of a Gaussian beam with the waist radius w_1 and of a LG vortex with the order (0, n) and waist radius w_2, and if $w_1(0) < w_2(0)$, then TC of such combined beam equals n up to a distance z_1 [22]. At $z > z_1$, though, $w_1(z) > w_2(z)$ and TC becomes zero, i.e. n OVs were born with TC of −1 and compensated n OVs with TC of +1, present in the initial plane.

2.2   NONPARAXIAL MODELING OF THE EVOLUTION OF AN OPTICAL VORTEX WITH AN INITIAL FRACTIONAL TC

Recent years have seen an active interest of researchers in the study of optical vortices. Their characteristic features include the orbital angular momentum (OAM) [12,13,34,35,36], OAM-spectrum [86], and topological charge [11,55]. The topological charge (TC) of an optical vortex is defined as an integer number of phase jumps by 2π that occur on a full circle around the singularity (uncertainty) center. If multiple optical vortices are embedded into a Gaussian beam, the total TC of the beam will obviously equal the sum of constituent TCs as it is possible to make a full circle around each of the embedded vortices and count the number of phase jumps by 2π. But here a nontrivial question arises: when determining the total TC of the beam, is it necessary to account for the TC of constituent vortices possibly embedded on the beam periphery with near-zero intensity? On the one hand, such peripheral singularities cannot be measured as e.g. characteristic forks in the interference pattern due to near-zero intensity of the beam. On the other hand, when defining the TC using Berry's formula [19], the radius of a circle of TC measurements needs to tend to infinity. Thus, the total TC of the beam is contributed to by all constituent singularities affecting the phase of the entire vortex beam. The TC can be measured by counting the number of forks in an interferogram [87] or using aperture diaphragms [64,78] and cylindrical lenses [15]. However, the said techniques are only suitable for measuring phase singularities (or isolated intensity nulls). In his seminal work [19], Michael Berry for the first time analyzed the free-space propagation of an optical vortex with the initial fractional TC, μ, deriving a formula for calculating TC of the fractional optical vortices in the near field. The TC was found to be equal to an integer nearest to the fractional μ, which means that TC changes in steps depending on the value of μ: if $\mu = m+\alpha$, where $\alpha < 0.5$, then $TC = m$, whereas if $0.5 < \alpha < 1$, then $TC = m+1$ ($m>0$). This relationship between the TC of the propagating fractional vortex and the initial value of $TC = \mu$ was repeatedly corroborated in the experiments [75,52]. However, later publications have cast some doubt on the stepwise dependence with the TC jumping at $\alpha = 0.5$ [76,77].

In this section, using the numerical simulation based on the Rayleigh-Sommerfeld integral [88] and Berry's formulae for calculating the TC [19], we reveal that the integer

TC of the propagating optical vortex with the initial fractional *TC* μ experiences a stepwise change at the fractional part value of $\alpha = 0.12$ rather than at $\alpha = 0.5$.

It should be noted that as a topological object, optical vortices occur in many divisions of physics. For instance, for the first time, fractional optical vortices were described in a pure superfluid [89] before being experimentally detected in an inhomogeneous superfluid [90]. A hidden phase complementing the fractional *TC* of the vortex to an integer number was found to exist in [90]. The phase has become known as the Pancharatnam–Berry phase and occurs in vector quantum fields when measured around a singularity. A fractional *TC* ceases to exist in vector states of the field. Recently, A. Volyar and colleagues have shown [53] that when propagating in an inhomogeneous photonic crystal, the fractional *TC* of an optical vortex becomes integer in each component of the vector field thanks to a phenomenon of nonadiabatic polarization following. Propagation of fractional optical vortices in an "exciton polariton condensate" has also been studied [91] and half-integer quantum vortices propagating in such media have been shown to be stable [92].

In this section, we numerically simulate the evolution of an initial fractional-charge optical vortex upon its free-space propagation based on the nonparaxial scalar Rayleigh–Sommerfeld integral.

2.2.1 TOPOLOGICAL CHARGE OF AN INITIAL FRACTIONAL-CHARGE VORTEX IN THE NEAR FIELD

This work aimed to numerically simulate the evolution of an initial fractional-charge Gaussian vortex:

$$E_x(r,\varphi) = \exp\left(-r^2/w^2 + i\mu\varphi\right), \tag{2.10}$$

where μ is *TC* (a real number) and w is the radius of the Gaussian beam waist. The *TC* was calculated using Berry's formula [19]:

$$TC = \frac{1}{2\pi}\lim_{r\to\infty}\int_0^{2\pi} d\varphi \frac{\partial}{\partial\varphi}\arg E(r,\varphi) = \frac{1}{2\pi}\lim_{r\to\infty}\mathrm{Im}\int_0^{2\pi} d\varphi \frac{\partial E(r,\varphi)/\partial\varphi}{E(r,\varphi)}. \tag{2.11}$$

We note that from the definition of *TC* in Equation (2.11), the radial variable of the light field amplitude is seen to tend to infinity. Thus, this definition suggests that when calculating the *TC* of a light field whose amplitude is an analytical integer function that tends to (but never equals) zero at infinity, singularities located at any point of the two-dimensional space need to be accounted for, including peripheral points of near-zero intensity.

In this section, the amplitude and phase of the light field of interest were calculated based on the Rayleigh–Sommerfeld integral [88]. The numerical simulation was conducted for the following parameters: a 8 × 8-μm initial field, a region of 400 × 400 pixels (with a pixel size of about λ /30), a wavelength of $\lambda = 0.532$ μm, the radius of the Gaussian beam waist, $w = 1.3$ μm, and the initial *TC*, $\mu = 2.2$. The amplitude and phase of the initial field are depicted in Figure 2.13. We note that in the near field, at a smaller-than-wavelength distance, the scalar Rayleigh–Sommerfeld integral certainly gives a coarse approximate solution since it disregards the vector character of the field and the decaying waves. However, the purpose of its use was to find the location of additional centers of singularities in the near field region.

From Figure 2.14, the *TC* in the original plane is seen to equal *TC* = 2.2. Shown in Figure 2.15 are the amplitude and phase of field form Equation (2.10) in the near field at a distance of $z = 0.03$ μm, which were calculated using the Rayleigh–Sommerfeld formulas.

Figure 2.16 shows the *TC* versus the radius *R* of a circle along which the *TC* was calculated using Equation (2.11) for the phase in Figure 2.15(b).

Figure 2.16 suggests that in the near field, the *TC* equals $\mu = 2$, which is in agreement with the results reported in a number of experimental studies [64,78,15,75,52], where the *TC* of fractional optical vortices was measured using apertures [78] or a cylindrical lens [15].

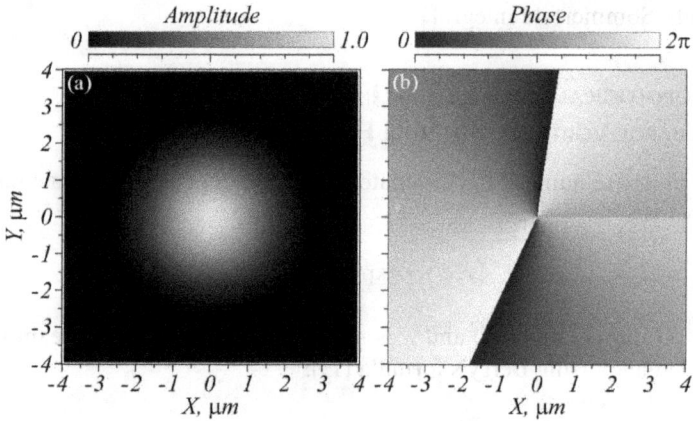

FIGURE 2.13   Original field: amplitude (a) and phase (b).

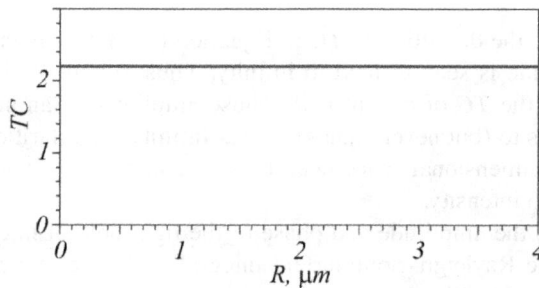

FIGURE 2.14   The value of *TC* versus the radius *R* of a circle on which the *TC* of the field from Figure 2.13 was calculated using Equation (2.11).

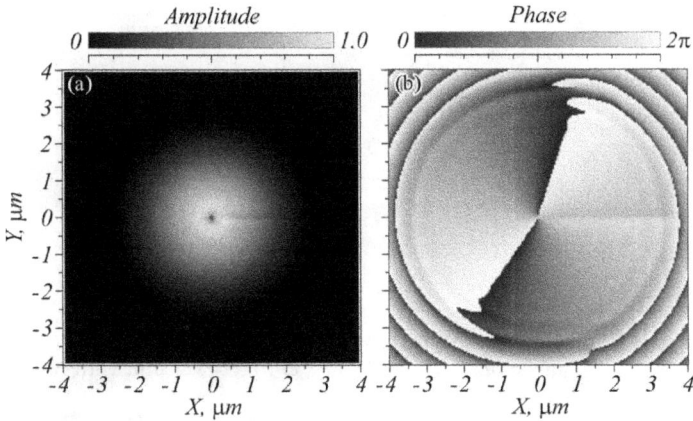

FIGURE 2.15   Field in Equation (2.10) in the near field at a distance of $z = 0.03$ μm. Field size: 8×8 μm.

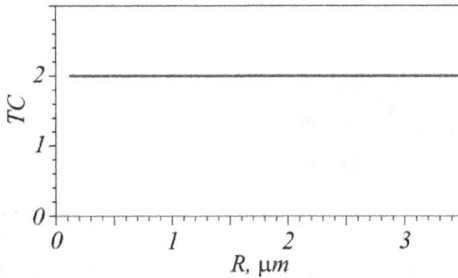

FIGURE 2.16   *TC* of the field from Figure 2.15 versus the radius ($z = 0.03$ um)

But is it possible to check if this is really the case? Below, we analyze what will happen if we increase the radius R at which the *TC* is calculated. Figure 2.17 depicts the *TC* versus the radius given the initial *TC* $\mu = 2.2$ at distances of $z = 0.01$ um and $z = 0.1$ μm.

From Figure 2.17 the *TC* of the beam in the near field is seen to become equal to 3 starting from $R > 4$ μm. The phase of the field at distances of $z = 0.01$μm and $z = 0.1$ μm is seen to remain unchanged across the entire field except for the central part at $R{<}4.5$ μm. Hence, assuming the initial *TC* $\mu = 2.2$, the *TC* of the optical vortex becomes equal to 3 at a distance of just $z = 0.01$ μm.

Why do the above-described numerical simulation results contradict the experiments in [78,15]? The fact is that from Figure 2.13, the intensity of the optical vortex is seen to drop to a near-zero value in a circle of radius ~2 μm, whereas at a distance of ~4 μm, the *TC* value changes from $\mu = 2$ to $\mu = 3$. Thus, we infer that an additional singularity center is located in a region where it cannot be experimentally detected. Actually, in igure 2.15 ($z = 0.03$ μm, $\mu = 2.2$) the intensity on the radius $R = 4$ μm, where *TC* changes from $\mu = 2$ to $\mu = 3$, is $1{,}6{\cdot}10^{-7}$-times that of the maximum

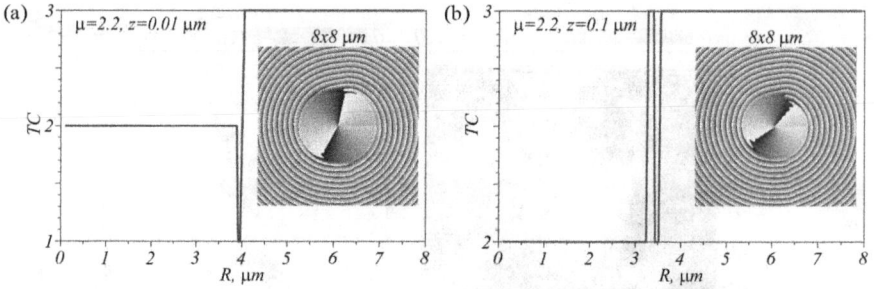

FIGURE 2.17 *TC* of an optical vortex vs the radius *R* for the initial fractional-*TC* $\mu = 2.2$ in the near field at a distance of (a) $z = 0.01$ μm and (b) $z = 0.1$ μm, with respective phases shown in insets.

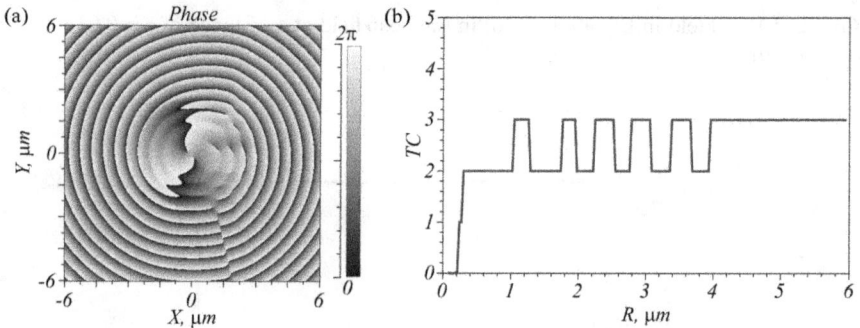

FIGURE 2.18 (a) Phase pattern of field (1) at $z = 1$ um and (b) *TC* versus the radius *R* of a circle of *TC* calculation. Initial *TC*: $\mu = 2.4$.

intensity in Figure 2.15. Meanwhile in Figure 2.17 ($z = 0.1$ μm, $\mu = 2.2$), the intensity on the radius $R = 3.5$ μm, where the *TC* changes from 2 to 3, is $2,0 \cdot 10^{-6}$-times that of the maximum value, which cannot be measured experimentally.

Obviously, an integer initial *TC*, $\mu = 2$ of an optical vortex propagating in free space will remain unchanged. In this work, we seek to find a threshold value of the fractional part of *TC* $\mu > 2$ in the near zone at which *TC* becomes equal to 3. The numerical simulation has shown that in the near field at $z = 0.01$ μm, *TC* of the field equals 2 for $2 < \mu < 2.12$. Thus, we can conclude that in the near field ($z < \lambda$), for the original fractional-charge Gaussian beam in Equation (2.10), in the interval $2 < \mu < 2.12$ *TC* remains equal to 2, becoming equal to 3 in the interval $2.12 < \mu < 3$.

2.2.2 TOPOLOGICAL CHARGE OF AN ORIGINAL FRACTIONAL-CHARGE VORTEX IN THE FRESNEL ZONE

Figure 2.18 depicts the phase of the vortex field in Equation (2.10) at a distance of $z = 1$ μm, given the initial *TC* $\mu = 2.4$. From Figure 2.18(b), in the Fresnel zone such an optical vortex is seen to have *TC* $\mu = 3$.

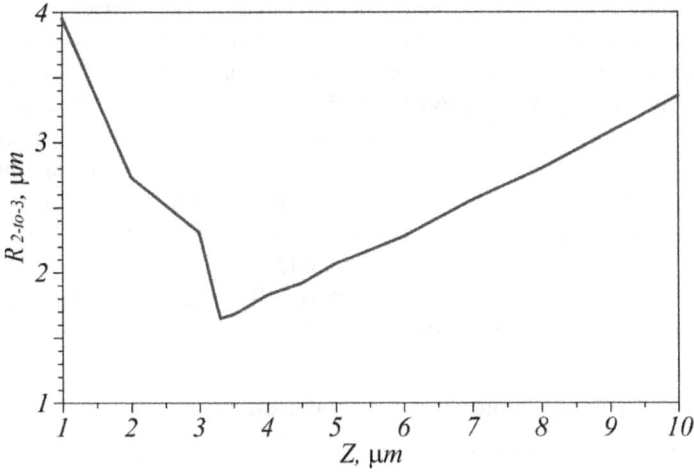

FIGURE 2.19   Radius of *TC* change from 2 to 3 versus the distance z, assuming the initial *TC* $\mu = 2.4$.

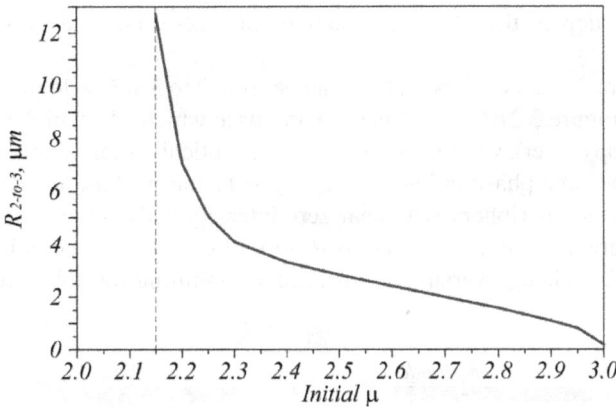

FIGURE 2.20   The radius R of 2-to-3 *TC* change versus the initial *TC* μ at $z = 10$ μm.

Figure 2.19 shows at which radius the *TC* value changes from 2 to 3 depending on the distance z in the Fresnel zone.

Figure 2.20 shows that with an increasing fractional part of the initial *TC*, the radius R of 2-to-3 *TC* change decreases.

Hence, we can infer that for the optical vortex in Equation (2.10) whose initial fractional *TC* lies in the interval $2.12 < \mu < 3$, its *TC* in the Fresnel zone equals $\mu = 3$. The plot in Figure 2.20 reveals one more interesting peculiarity. May R be a distance from the field (Gaussian beam) center to the third singularity. Then, with the fractional part of the initial *TC* increasing from 0.15 to 0.95, the third singularity will be moving from the "near-zero intensity" periphery (where the singularity is undetectable) towards the Gaussian beam center. Then, starting with the initial fractional *TC* of about 2.5, the third singularity is found ~3 μm away from the center, where it can be detected and measured. Thus, the plot in Figure 2.20 explains why in the experiments reported in [64,78,15,75,52] the 2-to-3 *TC* change was detected at the *TC* fractional part of 0.5. In the conclusion of this section, a similar plot (Figure 2.20) was numerically simulated for an initial fractional *TC* ranging from 3.1 to 4. Therefore, we may expand our finding and state that a similar plot (Figure 2.20) will be observed at the initial *TC* found in the interval $m < \mu < (m+1)$, where m is an arbitrary integer. In fact, our study has reconciled the results found in [19], in which the *TC* was found to change at a fractional part of 0.5, and those found in [16], which claimed that *TC* changed at any fractional part.

2.2.3 TOPOLOGICAL CHARGE OF AN INITIAL FRACTIONAL-CHARGE VORTEX IN THE FAR FIELD

In this subsection, we analyze *TC* of the vortex in the far field. Figure 2.21 depicts a field amplitude and phase at a distance of $z = 20$ μm, given the initial field in Equation (2.10) of Figure 2.13. The *TC* of the initial field is $\mu = 2.2$ and the field size is 32 × 32 μm (400 × 400 pixels).

Figure 2.22 depicts the *TC* of the field in Figure 2.21 versus the radius R of a circle of *TC* calculation.

From Figure 2.22, the *TC* is seen to change from 2 to 3 at $R > 14$ μm, which is also evident from Figure 2.21(b) – as this is the radius at which a fork in the phase pattern is observed (top center), wherefrom the *TC* of the optical vortex changes. An analysis of the amplitude and phase in Figure 2.21 suggests that the fork (third singularity) is found on the beam periphery with near-zero intensity and cannot be experimentally detected. Figure 2.23 depicts the radius R (fork-to-center distance in Figure 2.21(b)) of the 2-to-3 *TC* change versus the distance z, assuming the following simulation

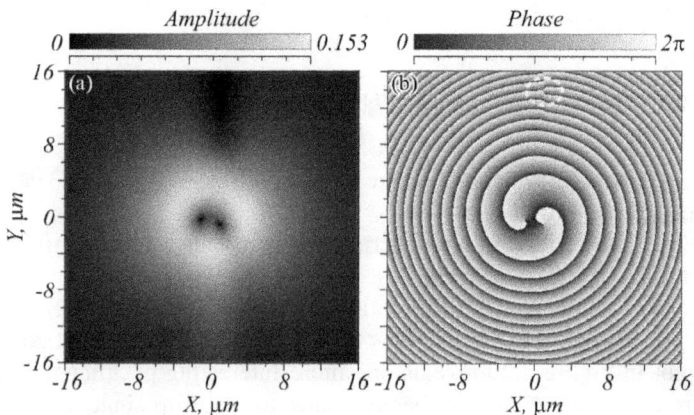

FIGURE 2.21   (a) Amplitude and (b) phase of the field at $z = 20$ μm.

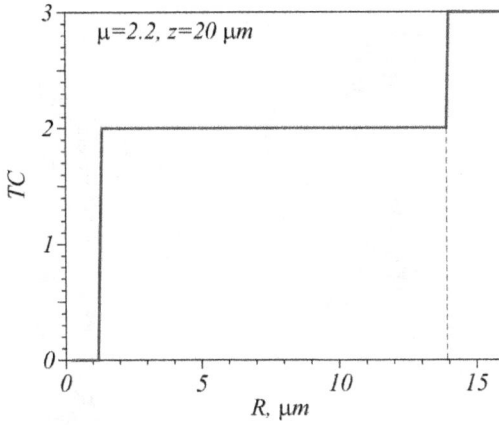

FIGURE 2.22   *TC* versus the radius *R* for the field in Figure 2.21.

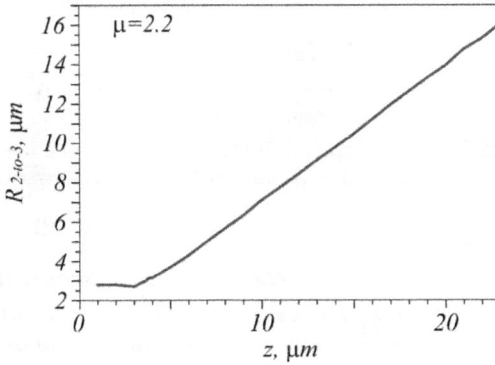

FIGURE 2.23   The radius *R* of the 2-to-3 *TC* change versus the distance z to the initial plane for the fractional *TC* $\mu = 2.2$.

parameters: a 8 × 8-μm field, 400 × 400 pixels for $\lambda = 0.532$ μm, $w = 1.3$ μm, and μ = 2.2. The Rayleigh range is $z_R = \dfrac{\pi w_0^2}{\lambda} = 9{,}98$ μm. From Figure 2.23, with increasing distance z, the fork is seen to move away from the center, resulting in the increased radius R of the 2-to-3 *TC* change.

From Figure 2.23, for the initial $\mu = 2.2$, the radius of *TC* change is seen to linearly depend on the distance z:$R_{2-to-3} \approx 0{,}71+0{,}665z$ at $z > 3$ μm. That is, at a distance of $z = 100$ μm, it is logical to expect a "fork" to be found at a radius of 67.2 μm from the center.

Let's check it out. In Figure 2.24 shows the calculated phase of the field in Equation (2.10) at a distance of $z = 100$ μm in coordinates $-16 < x < 16$ μm, $51.2 < y < 83.2$ μm, the center of the image is at the point with coordinates (0; 67.2 μm).

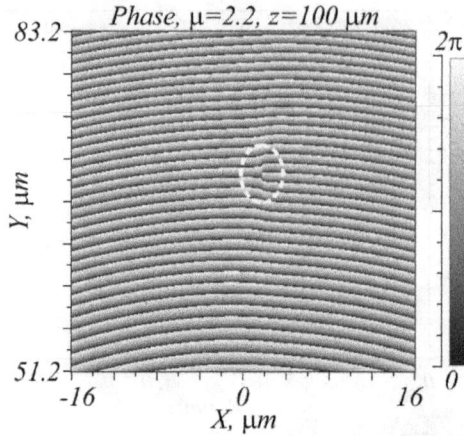

FIGURE 2.24 Phase of field in Equation (2.10) for the initial TC $\mu = 2.2$ in the near field at $z = 100$ μm in the interval $-16 < x < 16$ μm, $51.2 < y < 83.2$ μm, 400×400 pixels.

The fork (singularity center) is seen to be located near the center of the pattern in Figure 2.24, having the coordinates (1.6 μm; 69.84 μm), which corroborates the assumption that the "fork" shifts linearly with distance z.

Thus, if the initial TC of the optical vortex in Equation (2.10) lies in the interval $2.12 < \mu < 3$, its TC will be $\mu = 3$ in any diffraction zone upon free-space beam propagation. Fractional-charge vortices with the initial TC in the interval $2 < \mu < 2.12$ propagate by a different scenario, with the TC in all diffraction zones remaining equal to $\mu = 2$. Figure 2.25 depicts phases of the optical vortex with the initial TC $\mu = 2.11$ at different distances z: (a) 1 μm, (b) 3 μm, and (c) 20 μm. For all the phases, TC equals $\mu = 2$. Here, no additional singularity center (fork) occurs, or it may be formed too far from the optical axis to be detected in the course of numerical simulation. From Figure 2.25(d) it is seen that at the initial TC $\mu = 2.12$, an edge dislocation in the form of a vertical line of zero intensity is formed. In the course of propagation, the edge dislocation is not transformed into a screw dislocation (isolated intensity null) or, alternatively, the screw dislocation may be formed too far away from the center to be detected during simulation. We note that in this work, the numerical simulation included peripheral regions of the optical field where the intensity was 10^{-14} times the beam maximum intensity.

Interestingly, the evolution scenarios for the optical vortices with even and odd TC are different. We conducted the numerical modeling for a wavelength of $\lambda = 532$ nm and a waist radius of $w_0 = 3$ μm, with the initial fractional TC of the beam in Equation (2.10) assumed to change in the interval $3 < \mu < 4$. The initial Gaussian beam was multiplied by a spherical function of radius $f = 10$ μm. The real focal length turned out to be $z = 9.236$ μm (at this distance the on-axis intensity is maximum at $\mu = 0$). Thus, all values at the lens focus were calculated for the said distance z. The simulation results have shown that in the interval $3.0 < \mu < 3.11$ neither upper nor lower additional vortices are formed yet. Accordingly, TC of the vortex equals m

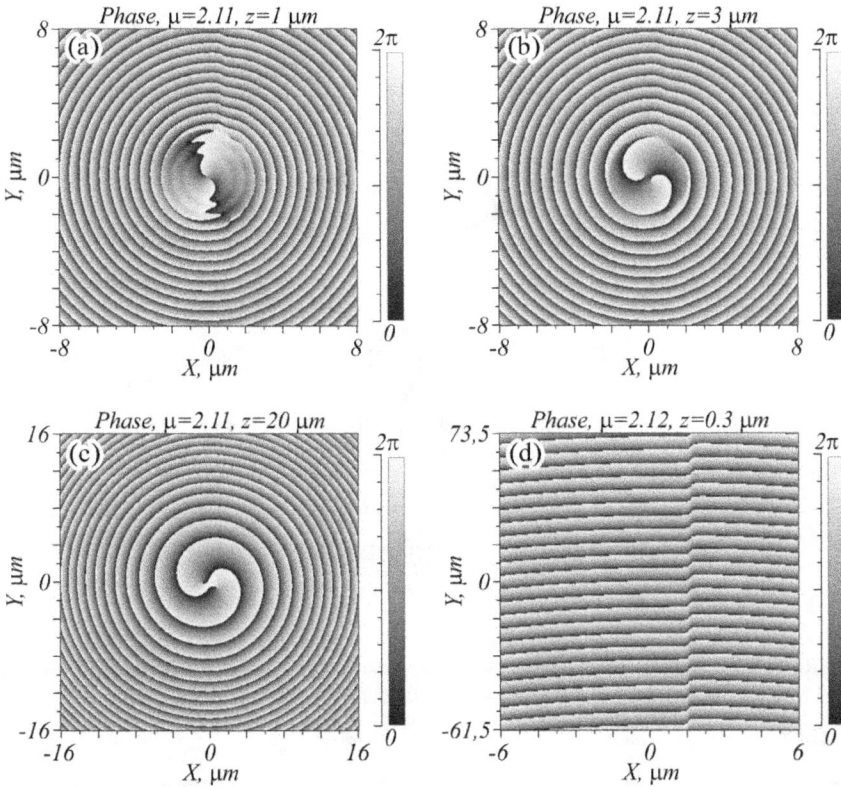

FIGURE 2.25 Phase of the field at distances (a, b, c) $z = 1$, 3, and 20 μm for the field with the initial fractional TC $\mu = 2.11$. The images are of sizes 16×16 μm, 16×16 μm, and 32×32 μm, respectively. (d) phase at a distance of $z = 0.3$ μm on the beam periphery at $\mu = 2.12$ (frame size $-6 < x < 6$ μm; $60 < y < 75$ μm).

$= 3$. Shown in Figure 2.26 are 16×16-μm phase patterns (insets) and the value of TC versus the radius R of a circle of TC measurement (a, b, c).

From Figure 2.26 it is seen that the evolution scenario of a fractional-charge optical vortex for an odd initial TC ($m = 3$) is different from the earlier discussed even initial TC. Actually, if the fractional part of TC is less than 0.5, the total TC is $m = 4$ (which is in agreement with [76]), see Figure 2.26(a). If the fractional part equals 0.5, the total TC in the far field is $m = 3$ (in agreement with [77]), see Figure 2.26(b). Finally, if the fractional part is larger than 0.5, the TC again equals $m = 4$ (in agreement with [19,76]).

Using the numerical simulation, we revealed that if an optical vortex has a non-integer initial TC, $\mu = m+\alpha$, $\alpha \ll 1$, an additional optical vortex of charge +1 ($m > 0$) is generated in all diffraction zones on the beam periphery with near-zero intensity. Thus, given the initial fractional TC, the total TC of the propagating beam will be equal to $m+1$. The additional singularity cannot be detected experimentally. With

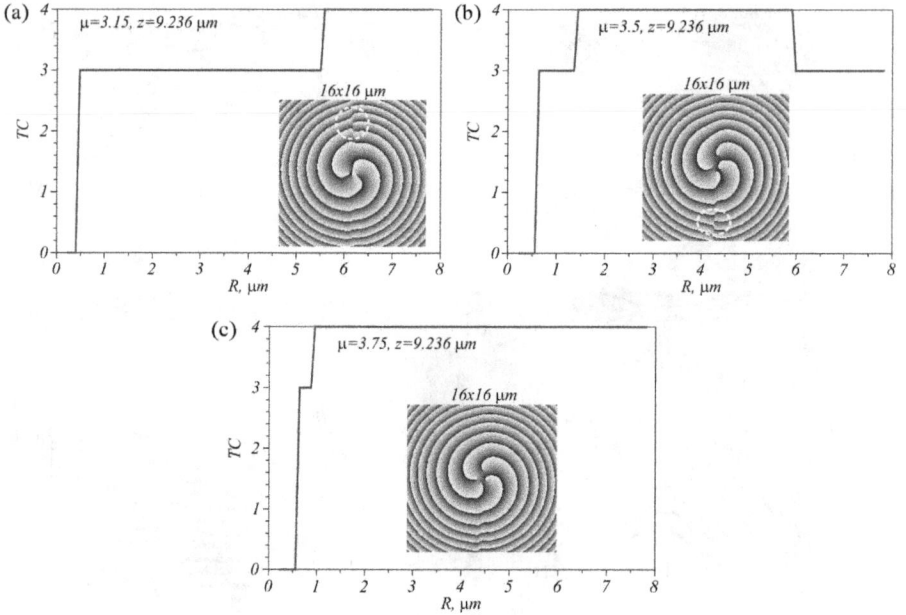

FIGURE 2.26 (a, b, c) Phase patterns of size 16×16 μm and plots of the *TC* versus the radius *R* of *TC* measurements for values of μ : (a) 3.15, (b) 3.5, and (c) 3.75.

increasing *z*, this additional vortex moves further away from the optical axis. With increasing fractional part α , the additional singularity occurs nearer to the optical axis, at $\alpha > 0.5$ getting close enough to be detected. This accounts for why in the well-known experiments [64,78,15,75,52] the *m*-to-*m*+1 change of the *TC* was found to occur at $\alpha = 0.5$. Our guess is that the said *m*-to-*m*+1 change of *TC* is likely to occur at any $\alpha > 0$, but the numerical simulation gives the change threshold at $\alpha = 0.12$. The results reported here are in good agreement with those of [76], but being more comprehensive than [76], our numerical simulation additionally included the determination of the threshold of *TC* change ($\alpha = 0.12$), an analysis of the beam propagation in the near field at $z \ll \lambda$, finding a formula to link the distance *R* of the beam center to the newly emerging singularity (screw dislocation) with the fractional part of the initial *TC* μ (Figure 2.20). In fact, our simulation has made it possible to reconcile the research findings reported in [64,78,15,75,52] and those of [76], which were previously contradictory. At the same time, our numerical modeling produced new research findings never reported before, which include the description of the *TC* behavior at the focus of a spherical lens in the case of an initial fractional *TC* that has an odd integer part, $m = 2k+1$. In particular, we have numerically shown that at $\alpha < 0.1$, the total *TC* equals *m*; at $0.1 \, ^{\zeta} \, \alpha \, ^{\zeta} \, 0.5$, *TC* equals *m*+1; at $\alpha = 0.5$, *TC* again equals *m*; and, finally, at $\alpha > 0.5$, *TC* again equals *m*+1. Such an α -dependence of the *TC* in the far field for an odd fractional part of the initial *TC* is in disagreement with all the results discussed in the above-mentioned papers.

2.3 ORBITAL ANGULAR MOMENTUM AND TOPOLOGICAL CHARGE OF A MULTI-VORTEX GAUSSIAN BEAM

Free-space communications have presently been the focus of active research [93]. One way to increase the data transmission capacity is through light beam multiplexing through the use of orbital angular momentum (OAM). This has prompted numerous publications dealing with the propagation of various optical vortices through a turbulent atmosphere. The use of vortex Bessel beams [94] is convenient because they retain their shape upon propagation and self-reconstruct after passing through an obstacle. However, for such beams to be generated and utilized for free-space optical communications high-aperture axicons are required [95]. When speaking of Laguerre–Gaussian (LG) [4,96,97,98], Bessel–Gaussian (BG) beams [40,99,100,101] or simple optical vortices, which are generated through diffracting a Gaussian beam by a spiral phase plate (SPP) [102], it is noteworthy that with increasing vortex order (topological charge), the intensity ring radius also increases, necessitating the use of larger reception devices. Because of this, alternative light fields, including fields devoid of radial symmetry, are also of practical interest as candidates for data transmission. In asymmetric light fields, OAM is no more equal to the topological charge. Nonetheless, with OAM being also suited for beam identification, techniques for OAM measurement are discussed in a number of publications [12,13].

There are papers showing that interference of the vortex and vortex-free Bessel beams leads to the split of the optical vortex into unitary-charge vortices located on a circle [103]. Elliptic perturbation of a vortex beam splits the singularity center into several optical vortices with TC of 1 located on a straight line [29,104]. In addition to the vortex strength (or topological charge) related to the phase singularities, other quantum numbers are known (the Poincaré index) related to the saddle points [105,106,107], though we don't study them here.

In this section, we study a Gaussian beam composed of several single-phase singularities uniformly arranged on a circle. We derive exact relationships to describe some integral characteristics of the vortex Gaussian beam, namely, power, OAM, and topological charge (TC). In addition to TC, several other quantities are shown (both theoretically and numerically) to be propagation-invariant. The intensity pattern of such a beam is a large intensity spot containing a set of shadow spots, so that the beam can be identified based on the number of spots. We numerically model the propagation of such beams through a turbulent medium (a screen with a random phase).

2.3.1 COMPLEX AMPLITUDE OF A MULTI-VORTEX GAUSSIAN BEAM

A relationship to describe a multi-vortex Gaussian beam has been derived [49]. If m phase singularities of a multi-vortex beam are located at points (r_p, φ_p) ($p = 0, \ldots, m - 1$) (in polar coordinates), its complex amplitude is given by:

$$E(r,\varphi,z) = \frac{1}{\sigma}\left(\frac{\sqrt{2}}{w_0}\right)^m \exp\left(-\frac{r^2}{\sigma w_0^2}\right)\prod_{p=0}^{m-1}\left(\frac{re^{i\varphi}}{\sigma} - r_p e^{i\varphi_p}\right), \qquad (2.12)$$

where (r, φ, z) are the cylindrical coordinates, w_0 is the waist radius of the Gaussian beam, $\sigma = 1 + iz/z_0$, and $z_0 = kw_0^2/2$ is the Rayleigh range (k is the wavenumber). From Equation (2.12) the beam is seen to be structurally stable and retain its transverse phase distribution upon propagation, only changing in scale and rotating about the optical axis. In [108], dynamics were studied of an off-axis optical vortex with TC of 1, embedded into an LG beam with the radial index of 1 and the azimuthal index of 0. It was shown that TC of such beam equals 1 and is conserved on propagation.

Below, we analyze a particular case when phase singularities are uniformly arranged on a circle of radius a, with their coordinates given by:

$$x = a\cos\varphi_p, \quad y = a\sin\varphi_p, \qquad (2.13)$$

where $\varphi_p = 2\pi p/m$, $p = 0, \ldots, m-1$. Writing down Equation (2.12) in an explicit form for several small values of m, we assume that for an arbitrary number of vortices m the complex amplitude is given by:

$$E(r,\varphi,z) = \frac{1}{\sigma}\left(\frac{\sqrt{2}}{w_0}\right)^m \exp\left(-\frac{r^2}{\sigma w_0^2}\right)\left(\frac{r^m e^{im\varphi}}{\sigma^m} - a^m\right). \qquad (2.14)$$

To verify that Equation (2.14) is valid at any m, we need to prove that: first, in the source plane the intensity is zero at points in Equation (2.13) and, second, beam in Equation (2.14) satisfies a paraxial propagation equation.

The first requirement is proven trivially, because the polar coordinates of points in Equation (2.13) are $r = a$ and $\varphi_p = 2\pi p/m$. After substituting these magnitudes in Equation (2.14), the expression in the round brackets becomes zero at $z = 0$ (i.e. $\sigma = 1$). The validity of the second assumption follows from the fact that beam in Equation (2.14) is a superposition of a Laguerre–Gaussian beam with zero radial index and a Gaussian beam, both of which present solutions to the paraxial propagation equation. Hence, following interference with the Gaussian beam, TC carried by the Laguerre–Gaussian beam can be identified by simply counting the number of nulls on the intensity distribution. Such combined beams, which are a coaxial superposition of Laguerre–Gaussian beams, and in particular, Laguerre–Gaussian and Gaussian beams, were considered in [109,110,111,112].

2.3.2 ORBITAL ANGULAR MOMENTUM AND THE TOPOLOGICAL CHARGE OF THE MULTI-VORTEX GAUSSIAN BEAM

Let us find power W and OAM J_z of beam in Equation (2.14), which can be derived using standard formulae:

$$W = \int_0^\infty \int_0^{2\pi} |E|^2\, rdrd\varphi, \qquad (2.15)$$

$$J_z = -i\int_0^\infty \int_0^{2\pi} E^* \frac{\partial E}{\partial \varphi}\, rdrd\varphi, \qquad (2.16)$$

where the complex amplitude distribution $E(r, \varphi, z)$ can be taken at any transverse plane. Substituting complex amplitude from Equation (2.14) in the source plane ($z = 0$) into Equation (2.15) yields:

$$W = \frac{\pi w_0^2}{2}\left[m! + \left(\frac{2a^2}{w_0^2}\right)^m \right].$$ (2.17)

In a similar way, by substituting complex amplitude from Equation (2.14) in the source plane ($z = 0$) into Equation (2.16) for OAM of the beam, we obtain:

$$J_z = 2\pi m \frac{2^m}{w_0^{2m}} \int_0^\infty \exp\left(-\frac{2r^2}{w_0^2}\right) r^{2m} r \, dr.$$ (2.18)

Integral in Equation (2.18) can be expressed via the Γ-function (6.1.1 in [113]), or a factorial (since m is integer), so that OAM is:

$$J_z = 2\pi m \frac{2^m}{w_0^{2m}} \frac{m!}{2\left(2/w_0^2\right)^{m+1}} = m \frac{\pi w_0^2}{2} m!.$$ (2.19)

Expression (2.19) was obtained earlier in [22]. But note that if we do not consider the OAM normalized to the beam power, then it is impossible to see how the OAM depends on the distance between the singularities. Really, dividing Equation (2.18) by Equation (2.17) yields the value of OAM normalized to beam power:

$$\frac{J_z}{W} = m \frac{m!}{m! + \left(2a^2/w_0^2\right)^m}.$$ (2.20)

There is no Equation (2.20) for the normalized OAM in [22]. Equation (2.20) is important in that it shows that the multi-vortex beam OAM in Equation (2.14) decreases with an increase in the radius of the circle on which the optical vortices are embedded. Also, Equations (2.17) and (2.20) show that for the combined beam in Equation (2.14) to have singularity centers at a very large distance from the center of the beam (a >> w0), the energy of the Gaussian beam (the second term in Equation (2.17)) must also be very big. From Equation (2.20), normalized OAM is seen not to exceed the number of OVs in the beam, decreasing to zero with increasing distance of the vortices from the optical axis. In [19], a relationship for TC of a vortex light field (vortex strength) was given:

$$TC = \lim_{r \to \infty} \frac{1}{2\pi} \int_0^{2\pi} \frac{\partial}{\partial \varphi} \left[\arg E\left(r, \varphi, z\right)\right] d\varphi.$$ (2.21)

It can be rearranged to a simpler form:

$$TC = \frac{1}{2\pi} \lim_{r\to\infty} \mathrm{Im}\left\{ \int_0^{2\pi} \frac{1}{E} \frac{\partial E}{\partial \varphi} d\varphi \right\}. \tag{2.22}$$

Substituting complex amplitude from Equation (2.14) in Equation (2.22) yields:

$$TC = \frac{1}{2\pi} \lim_{r\to\infty} \mathrm{Im}\left\{ \int_0^{2\pi} \frac{im\sigma^{-m} r^m e^{im\varphi}}{\sigma^{-m} r^m e^{im\varphi} - a^m} d\varphi \right\}. \tag{2.23}$$

Since at $r \to \infty$ the denominator's term a^m can be neglected, we infer that TC of beam in Equation (2.14) is independent of both the distance z passed and the radius a of the circle of OVs, being equal to the number of constituent OVs in the beam:

$$TC = m. \tag{2.24}$$

Equation (2.24) suggests that unlike OAM, TC remains unchanged with the distance a, being equal to the number of vortices. Using Equation (2.22), TC can be derived not only for the beam from Equation (2.14), but for a more general beam with multiple singularities as well. Indeed, we can obtain TC of a beam with several degenerate singularities (isolated intensity nulls) located unevenly in the cross-section of an arbitrary rotationally symmetric beam. Such a multi-vortex beam has the complex amplitude of the following form:

$$E_m(r,\varphi,z=0) = A(r) \prod_{p=1}^m \left(re^{i\varphi} - r_p e^{i\varphi_p} \right)^{m_p}. \tag{2.25}$$

Substitution of Equation (2.25) into Equation (2.22) yields the topological charge of the light field from Equation (2.25):

$$TC = \frac{1}{2\pi} \lim_{r\to\infty} \mathrm{Im}\left\{ \int_0^{2\pi} ire^{i\varphi} \sum_{p=1}^m \frac{m_p}{re^{i\varphi} - r_p e^{i\varphi_p}} d\varphi \right\} = \sum_{p=1}^m m_p. \tag{2.26}$$

It is seen in Equation (2.26) that TC of the multi-vortex beam is equal to the sum of TCs of all the constituent optical vortices (with their degeneracy, or multiplicity, taken into account). If there are m positive-charge vortices and n negative-charge vortices embedded into the beam in the initial plane, then instead of Equation (2.25) we have:

$$E_{m,n}(r,\varphi,z=0) = A(r) \prod_{p=1}^m \left(re^{i\varphi} - r_p e^{i\varphi_p} \right)^{m_p} \prod_{q=1}^n \left(re^{-i\varphi} - r_q e^{-i\varphi_q} \right)^{n_q}. \tag{2.27}$$

Similarly to Equation (2.26), topological charge of the field in Equation (2.27) equals the sum of TCs of all positive-charge vortices minus the sum of TCs of all negative-charge vortices:

$$TC = \sum_{p=1}^{m} m_p - \sum_{q=1}^{n} n_q . \qquad (2.28)$$

We note that, in contrast to the field in Equation (2.25), the field in Equation (2.27) is not structurally stable and its transverse shape changes on space propagation.

2.3.3 Asymptotic Phase Invariants of the Multi-Vortex Gaussian Beam

It is worth noting also that not only the whole integral in Equation (2.23) is independent of z, but the integrand also does not depend on z, i.e. for the multi-vortex beam in Equation (2.14) Im$\{E^{-1}\partial E/\partial \varphi \}$ tends to m at $r \to \infty$. So, in addition to TC, other propagation-invariant quantities may also be constructed for the beam in Equation (2.14), like, e.g.:

$$\mu_g = \frac{1}{2\pi} \lim_{r \to \infty} \left\{ \int_0^{2\pi} g(\varphi) \frac{\partial}{\partial \varphi} \left[\arg E(r,\varphi,z) \right] d\varphi \right\}, \qquad (2.29)$$

with $g(\varphi)$ being an arbitrary function. Below we test two such invariants and study whether or not the following values:

$$\mu_1 = \frac{1}{2\pi} \lim_{r \to \infty} \left\{ \int_{-\pi/4}^{\pi/4} \frac{\partial}{\partial \varphi} \left[\arg E(r,\varphi,z) \right] d\varphi \right\}, \qquad (2.30)$$

$$\mu_2 = \frac{1}{2\pi} \lim_{r \to \infty} \left\{ \int_0^{2\pi} \exp\left(-\frac{\varphi^2}{\varphi_0^2} \right) \frac{\partial}{\partial \varphi} \left[\arg E(r,\varphi,z) \right] d\varphi \right\}, \qquad (2.31)$$

conserve on propagation. In Equation (2.30), the angle integration covers only a quarter of the circle. But it can be shown that if we integrate over any part of the circle, then the resulting value will also be invariant.

2.3.4 Multi-Singularity Spiral Phase Plate

The beam in Equation (2.12) is not easy to generate practically. The easiest way is to use, for this purpose, a phase-only optical element illuminated by a Gaussian beam. Optical multi-vortices can be generated by using a spiral phase plate (SPP) with the transmittance function:

$$E(r,\varphi) = \text{circl}\left(\frac{r}{R}\right)\exp\left(i\Psi(r,\varphi)\right),$$ (2.32)

$$\Psi(r,\varphi) = \sum_{p=1}^{M} n_p \arctan\left(\frac{r\sin\varphi - r_p\sin\varphi_p}{r\cos\varphi - r_p\cos\varphi_p}\right),$$ (2.33)

where (r, φ, z) are the cylindrical coordinates, n_p is TC of each single singularity, $\text{circl}(r/R) = \{1, r<R; 0, r>R\}$ is the function of a circular aperture with the radius R, which bounds the SPP. If the SPP in Equation (2.32) is illuminated by a light beam with the complex amplitude $A(r)$, then immediately beyond the SPP, an optical vortex is generated with its TC obtained by Equation (2.26):

$$TC = \sum_{p=1}^{M} n_p, r_p < R, p = 1, 2, \ldots M.$$ (2.34)

As seen from Equation (2.34), TC of the field beyond the SPP from Equation (2.34) equals the sum of all TCs. If the singularity center of the SPP is on the diaphragm edge $(r_p = R)$, $n_p/2$ should be used in Equation (2.34) instead of n_p. Note that the beam formed by SPP from Equation (2.32) differs from the beam in Equations (2.12) or (2.14), although they are similar [114]. Substituting Equations (2.32) and (2.33) into Equations (2.15) and (2.16), and using the amplitude of the incident beam $A(r)$, we get the OAM of the light field beyond the SPP from Equation (2.33):

$$\frac{J_z}{W} = \sum_{p=1}^{M} n_p \int_{r_p}^{R} |A(r)|^2 r dr \left(\int_{0}^{R} |A(r)|^2 r dr\right)^{-1}.$$ (2.35)

According to Equation (2.35), if the singularity center is on the diaphragm edge $(r_p = R)$, this optical vortex does not contribute to the total OAM of the beam.

2.3.5 Stability of Shape, Orbital Angular Momentum, Topological Charge, and Asymptotic Phase Invariants of the Multi-Vortex Gaussian Beam to Random Phase Distortions

In optical communication systems, conventional radially symmetric OVs are utilized as data carriers, with OV carrying different TCs generating different radius intensity rings. But after passing through a random medium the intensity rings get distorted, rendering inaccurate the results of TC measurement based on the ring radius. Because of this, OVs devoid of radial symmetry may also be of practical interest for free-space communications. For instance, beams in the form of a superposition of two Gaussian beams that acquire different TCs after passing two different spiral phase plates (SPP) have been reported [115]. Instead of a ring, the intensity pattern of such beams is composed of bright intensity spots whose number equals the modulus of the difference of the constituent TCs. In this section, we analyze beams of Equation (2.14) that also produce m intensity spots in the transverse plane, except that the spots are not bright but dark – with intensity nulls found at their centers.

Figure 2.27 depicts the intensity and phase patterns of the beams in Equation (2.14) in the source plane and following free-space propagation through a medium

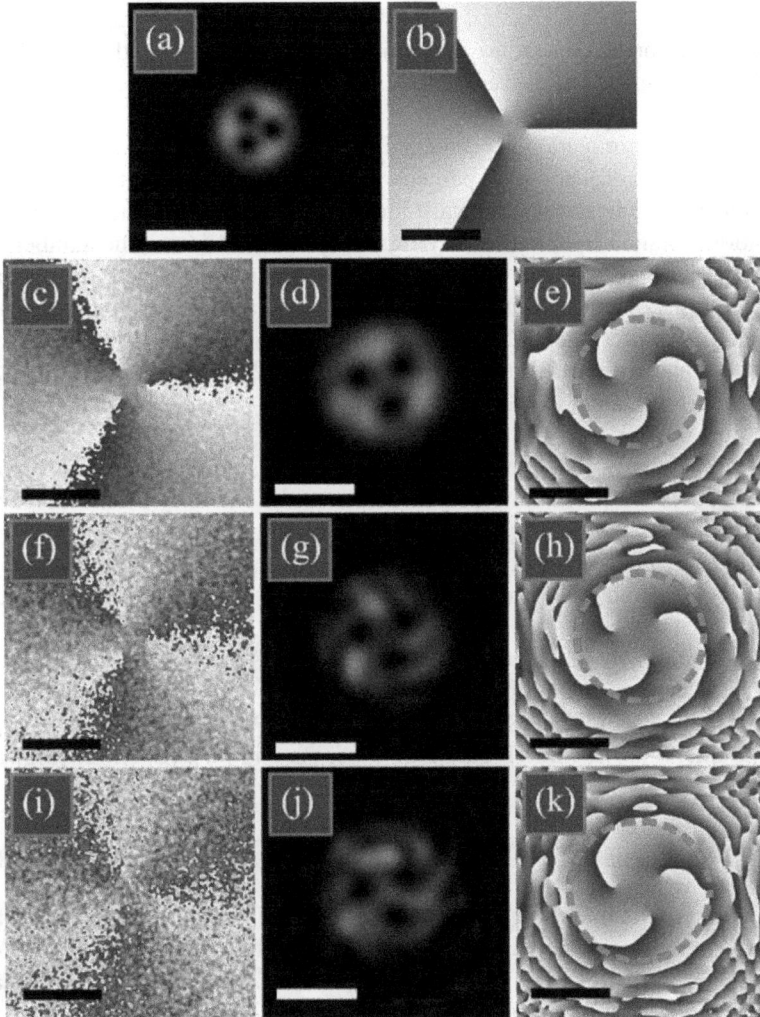

FIGURE 2.27 Patterns of (a) intensity and (b) phase for beam (2.14) at $m = 3$ in the source plane, (c, f, i) phase patterns after passing through a random phase diffuser, and patterns of (d, g, j) intensity and (e, h, k) phase upon free-space propagation. Random phase variations are on the interval $[-\pi/2, \pi/2]$ (c–e), $[-\pi, \pi]$ (f–h), and $[-3\pi/2, 3\pi/2]$ (i–k). Dashed circle (e, h, k) shows the area where TC and OAM were computed. The scale bar in each figure means 5 mm.

with random distortions (turbulent atmosphere). In the simulation, the random medium is replaced with a diffuser with a random phase $\psi(x, y)$, which is placed in the source plane. The correlation function of the diffuser can be approximated by a Gaussian exponential function:

$$\langle\exp\left(i\psi(\mathbf{r}) - i\psi(\mathbf{r}')\right)\rangle = \exp\left(-\left|\mathbf{r} - \mathbf{r}'\right|^2 / \delta^2\right), \tag{2.36}$$

where δ is the correlation radius of a uniform phase diffuser and the angle brackets in (2.36) designate the averaging over an ensemble of statistically similar phase diffusers.

The numerical simulation is conducted for a wavelength of $\lambda = 1.55$ μm (the summer atmosphere passes about 70 percent of light at this wavelength and it is widely used in telecommunications), the waist radius of the Gaussian beam is $w = 1.5$ mm, propagation distance $z = 5$ m (the Rayleigh range is $z_0 = 4.56$ m), the number of constituent OVs with unity TC $m = 3$, distances of the vortex centers to the optical axis $a = 0{,}7w_0$ (at this distance, all three intensity nulls do not merge and are clearly seen in the intensity pattern), and the diffuser correlation radius $\delta = 100$ μm. Across the diffuser, the phase varied in the interval $[-\pi/2, \pi/2]$ (Figure 2.27(c–e)), $[-\pi, \pi]$ (Figure 2.27 (f–h)), and $[-3\pi/2, 3\pi/2]$ (Figure 2.27 (i–k)). The computation domain in all figures is $-R \leq x, y \leq R$, where $R = 8$ mm.

Figure 2.27 shows that even despite essential phase distortion (Figure 2.27 (i)), three dark spots are present around the center of the intensity pattern (Figure 2.27 (j)) and three phase singularity points in phase patterns (Figure 2.27 (k)).

Figure 2.28 shows intensity patterns for beam (2.14) for a different number of constituent OVs and different phase variations on the diffuser. The simulation parameters are the same as in Figure 2.27, except for the number of vortices: $m = 2$ (Figure 2.28 (a, b)), $m = 3$ (Figure 2.28 (c, d)), $m = 4$ (Figure 2.28 (e, f)). The distance of the vortex centers to the optical axis is $a = 0.6w_0$ (Figure 2.28 (a, b)), $a = 0.7w_0$ (Figure 2.28 (c, d)), and $a = 0.8w_0$ (Figure 2.28 (e, f)). Phase variations on the diffuser are in the intervals $[-\pi, \pi]$ (Figure 2.28 (a, c, e)), and $[-3\pi/2, 3\pi/2]$ (Figure 2.28 (b, d, e)).

From Figure 2.28 it is seen that while the intensity nulls are clearly seen if phase variations $\Delta\psi$ on a random diffuser are on the interval $[-\pi, \pi]$, they become poorly discernible on the interval $[-3\pi/2, 3\pi/2]$ at $m \geq 4$.

In addition to the intensity patterns, OAM, TC, and phase invariants μ_1 and μ_2 of the beam were also calculated. OAM was calculated by Equations (2.15) and (2.16), whereas TC, μ_1, and μ_2 were obtained by Equation (2.22) along a circle containing 80% of the beam energy (dashed circles in Figure 2.27). Table 2.3 gives these values upon free-space propagation after passing a random phase diffuser.

As can be seen from Table 2.3, if the phase variations on the diffuser are ranging from $-\pi$ to π, the deviation of numerically simulated OAM from the theoretically derived OAM is no more than 10%. The topological charge, rounded to an integer value, was always coinciding with its theoretical value for phase distortions $|\Delta\psi| \leq \pi$. The same holds for the propagation-invariants μ_1 and μ_2. Both of them

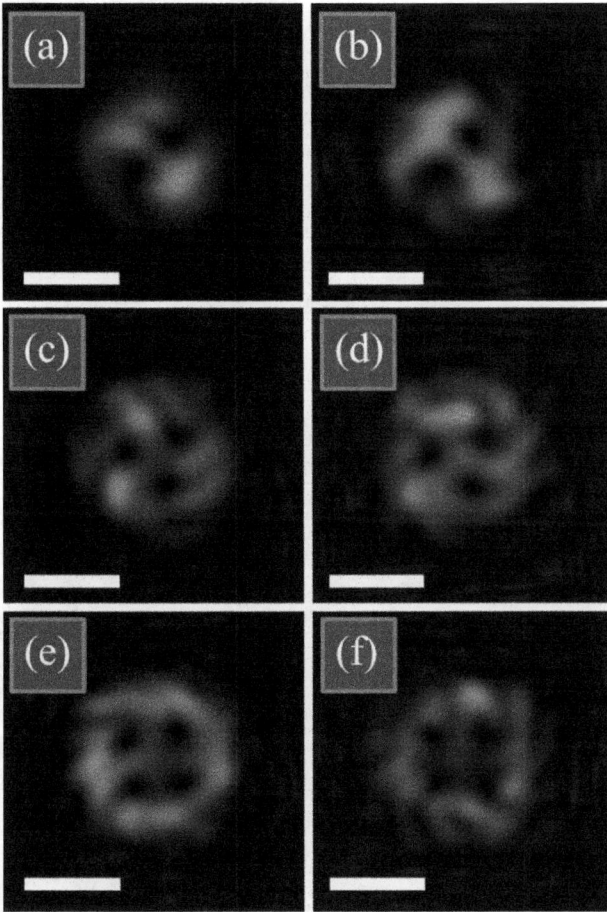

FIGURE 2.28 Intensity patterns for beam in Equation (2.14) at distance $z = 2$ m for a different number of constituent OVs in the beam and differently varying random phase over the diffuser: $m = 2$ (a, b), m = 3 (c,d), and m = 4 (e, f); $|\Delta \psi| \le \pi$ (a, c, e) and $|\Delta \psi| \le 3\pi/2$ (b, d, f). The scale bar in each figure means 5 mm.

approximately retain their initial value on propagation and at $|\Delta \psi| \le \pi$, the error was 9% only in one case (at $m = 2$, $\mu_2 = 0.6310$ at $z = 5$ m, while $\mu_2 = 0.5785$ at $z = 0$), whereas in other cases it rarely achieved 6%.

2.3.6 NUMERICAL SIMULATION OF ASYMPTOTIC PHASE INVARIANTS

In this subsection, we study numerically that the invariants in Equations (2.29)–(2.31) indeed conserve on propagation. Using Equation (2.14), we calculated the complex amplitude distributions for three propagation distances: near field, Fresnel diffraction zone, and the far field. The following parameters were used: wavelength λ

TABLE 2.3

Normalized OAM, *TC*, and phase invariants μ_1 and μ_2 of beam in Equation (2.14) upon free-space propagation after passing a random phase diffuser at different number of constituent beams (*m*) and different phase variations $\Delta \psi$ on the diffuser (with r.m.s. error given in brackets), calculated theoretically using Equations (2.20) and (2.24) and numerically using Equations (2.16) and (2.22).

	m = 2	m = 3	m = 4	m = 5
OAM (theory)	1.59	2.59	3.60	4.57
OAM (num.), $\lvert\Delta\psi\rvert \leq \pi$	1.45	2.37	3.47	4.31
	(9%)	(8%)	(4%)	(6%)
TC (theory)	2.0000	3.0000	4.0000	5.0000
TC (num.), $\lvert\Delta\psi\rvert \leq \pi$	2.0012	2.9991	4.0002	4.9976
Invariant μ_1 at $z = 0$	0.5000	0.7500	1.0000	1.2500
μ_1 at $z = 5$ m, $\lvert\Delta\psi\rvert \leq \pi/2$	0.4810	0.7667	0.9921	1.2307
Invariant μ_2 at $z = 0$	0.5785	0.8862	1.1817	1.4770
μ_2 at $z = 5$ m, $\lvert\Delta\psi\rvert \leq \pi/2$	0.5878	0.8885	1.1885	1.4819

= 1.55 μm, Gaussian beam waist radius $w_0 = 1.5$ mm, number of vortices $m = 5$, distance from the vortices to the optical axis in the initial plane $a = 0.9w_0$, propagation distances $z = 0.5$ m $<< z_0$ ($z_0 = 4.560$ m), $z = 5$ m $\approx z_0$, $z = 20$ m $>> z_0$. Calculation area was $\lvert x\rvert$, $\lvert y\rvert \leq R$ with $R = 10$ mm ($z = 0.5$ m), $R = 15$ mm ($z = 5$ m), $R = 30$ mm ($z = 20$ m). Figure 2.29 shows the intensity and phase distributions of such beam at all three planes.

We tested the following asymptotic phase invariant:

$$\mu_3 = \frac{1}{2\pi} \lim_{r \to \infty} \left\{ \int_{-\pi/6}^{\pi/6} \frac{\partial}{\partial \varphi} \left[\arg E\left(r, \varphi, z\right) \right] d\varphi \right\}, \tag{2.37}$$

i.e. the same integral as the topological charge, but evaluated over the angle $\pi/3$ (dashed circles in Figure 2.29).

Numerically, the invariant μ_3 was evaluated along a circle with the radius $R_1 = 0.9R$. The following values of μ_3 were obtained: $\mu_3 = 0.83349$ ($z = 0.5$ m), $\mu_3 = 0.83303$ ($z = 5$ m), $\mu_3 = 0.83201$ ($z = 20$ m).

Therefore, simulation confirms that the asymptotic phase invariant μ_3 is conserved on propagation. In conclusion, we have analyzed the integral characteristics of a Gaussian beam composed of several vortices located on a circle [112]. Relationships to describe power, the OAM, and the *TC* of the multi-vortex beam have been deduced. The normalized OAM of the beam has been shown to be less

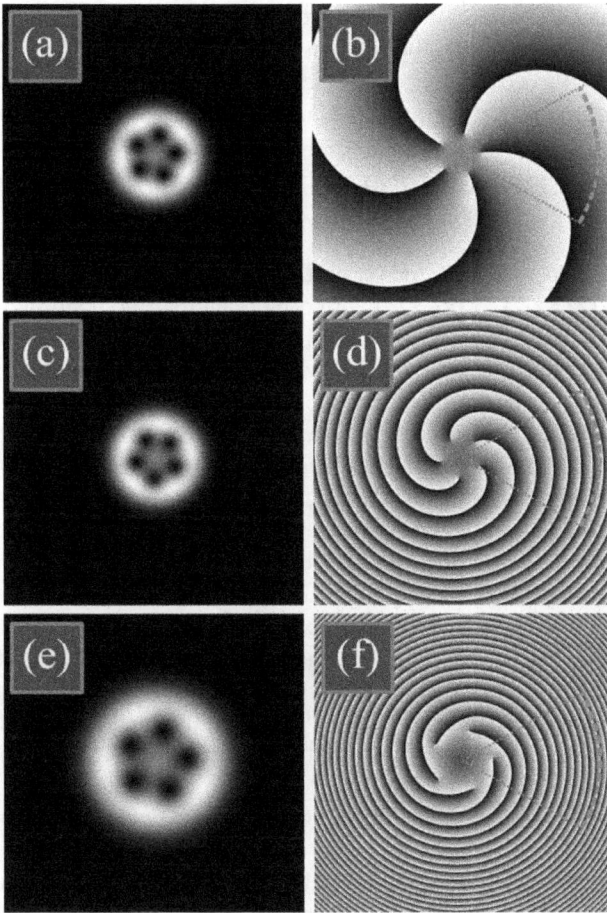

FIGURE 2.29 Intensity (a, c, e) and phase (b, d, f) distributions of the beam (2.14) at the distances $z = 0.5$ m (a, b), $z = 5$ m (c, d), $z = 20$ m (e, f). Dotted sector denotes the angles over which the asymptotic phase invariant from Equation (2.37) is evaluated.

than the number of constituent vortices, with its value decreasing with increasing distance from the optical axis to the vortex centers. By contrast, TC has been shown to be distance-independent and equal to the number of vortices. In addition to TC, other propagation-invariant quantities have been constructed theoretically and tested numerically [116]. Using a Fresnel transform and a random phase screen (diffuser) described by the correlation function in the form of a Gaussian exponential function, the propagation of the multi-vortex Gaussian beam in a random medium has been numerically simulated. Given moderate phase variations on the diffuser (from $-\pi$ to π), local intensity minima (optical vortex centers) can be seen in the intensity patterns. The r.m.s. deviation of the numerically simulated OAM was not larger than 10% for $m < 5$. TC is unchanged if $|\Delta \psi| \leq \pi$. Thus, alongside the number of dark spots, OAM and *TC* can also be utilized for beam identification. In optical data transmission, the multi-vortex beams can offer an alternative to conventional radially symmetric OVs because of the simplicity of an OV identification by counting the number of local intensity maxima in a Gaussian beam.

2.4   INFLUENCE OF OPTICAL "DIPOLES" ON THE TOPOLOGICAL CHARGE OF A FIELD WITH A FRACTIONAL INITIAL CHARGE

Many current works deal with the generation and detection of vortex laser beams [1 17,118,119,120,121,122,123,124,125] with orbital angular momentum. Further, many reviews [126,127,128,129,130,131,132] and monographs [56,133,134] reflect the progress in this field of knowledge and the widespread use of singular optics. Most of them are experimental, and different methods to obtain laser beams with orbital angular momentum and topological charge *TC* are demonstrated in these works. Although vortex beams are more than 40 years old, there are still many unresolved theoretical problems associated with them. For example, optical vortices with an initial fractional *TC* were investigated in the pioneering work of Berry [19]. Using the asymptotics of the complex amplitude of such a field, Berry showed that in the near field (and in the image plane), several additional screw dislocations were formed with a center not on the optical axis, in addition to the main screw dislocations (phase singularities) of the OV. The centers of these dislocations lay above and below the horizontal axis (if the edge dislocation was on the horizontal axis), and the *TC* alternated between +1 and −1. Together, these additional screw locations constituted the "Hilbert Hotel" [135]. With an increase in the fractional part of the initial *TC*, the nearest screw dislocation from the Hilbert Hotel approached the main axial screw dislocation, and the remaining additional vortices were displaced to the beam periphery.

The following question arose: what is the *TC* of the beam that has so many screw dislocations? In [19], this question was answered thus: If the fractional part of the initial *TC* is less than 0.5, then the *TC* of the beam is equal to the nearest integer; if the fractional part is greater than 0.5, then the *TC* of the beam is increased by 1. Such dynamics of the *TC* of a beam with an initial fractional charge have been experimentally confirmed in a number of works [64,78,15,75,52]. However, some works have also experimentally shown that in the Fresnel diffraction zone and in the far field, the *TC* of optical vortices with an initial fractional charge behaves differently [76,77]. In our earlier studies [55,85], we made attempts to theoretically explain how the *TC* is formed in the diffraction zone for such fractional OVs. However, it was not completely clear where the additional screw dislocations in the beam came from, which changed its *TC* at different values of the fractional part of the charge.

In this section, we continued investigating the formation of an entire *TC* of an optical vortex with an initial fractional *TC* in the far field. To find additional OVs located far at the beam periphery, where the light intensity was close to zero, a simulation was carried out using the nonparaxial Rayleigh–Sommerfeld integral [88]. Nonparaxial calculation of the beam with a waist radius of 3 μm (wavelength 0.633 μm) in a field with a radius greater than 138 μm helped us determine the dynamics of peripheral phase singularities in the beam with a change in the fractional part of the initial *TC*. Calculation of the *TC* of the beam in the far field was performed along a circle with a large (but finite) radius. When the fractional *TC* was in the range 2–3, the whole *TC* of the beam in the simulation changed abruptly from 2 to 3 and from 3 to 2 five times. When the fractional *TC* was in the range 3–4, the whole *TC* of the beam changed abruptly from 3 to 4 and from 4 to 3 three times. The causes of this behavior of the *TC* are shown in this section.

2.4.1 FORMULATION OF THE PROBLEM

We considered an initial light field with linear polarization along the *x*-axis, in which the complex amplitude was described by a Gaussian exponent and the optical vortex with a fractional *TC*, μ [19]:

$$E_x(r,\varphi) = \exp\left(-r^2/w^2 + i\mu\varphi\right). \tag{2.38}$$

The nonparaxial propagation of the light field in Equation (2.38) was simulated using the scalar Rayleigh–Sommerfeld integral [88],

$$E(u,v,z) = -\frac{z}{2\pi} \iint_\Sigma E(x,y)\frac{e^{ik\ell}}{\ell^2}\left(ik - \frac{1}{\ell}\right)dxdy, \tag{2.39}$$

where $E(x,y)$ is an initial field in Equation (2.38), $\ell = \sqrt{(u-x)^2 + (v-y)^2 + z^2}$ that is at a distance from a point in the initial plane to a point of view, and Σ is a function describing an aperture limiting the initial field (if there is an aperture).

The *TC* was calculated using the Berry formula [19]:

$$TC = \frac{1}{2\pi}\lim_{r\to\infty}\int_0^{2\pi} d\varphi \frac{\partial}{\partial\varphi}\arg E(r,\varphi)$$

$$= \frac{1}{2\pi}\lim_{r\to\infty}\mathrm{Im}\int_0^{2\pi} d\varphi \frac{\partial E(r,\varphi)/\partial\varphi}{E(r,\varphi)}. \tag{2.40}$$

Decomposing the field (2.38) in a series of angular harmonics with an integer *TC*, we obtained the following:

$$E(r,\varphi,z=0) = \exp\left[-i\mu\varphi - \left(\frac{r}{w}\right)^2\right]$$

$$= \frac{e^{i\pi\mu}\sin\pi\mu}{\pi}\sum_{n=-\infty}^{\infty}\frac{e^{in\varphi - r^2/w^2}}{\mu - n}. \tag{2.41}$$

In a paraxial case, the field in the Fresnel diffraction zone for any z is given by ($B = z$, $A = D = 1$):

$$E(\rho,\theta,z) = \frac{1}{\sqrt{2\pi}} \left(\frac{-iz_0}{q_1 z} \right) \exp\left(\frac{ik\rho^2}{2z} + i\pi\mu \right)$$

$$\times \sin(\pi\mu)\sqrt{\xi}\exp(-\xi) \qquad\qquad (2.42)$$

$$\times \sum_{m=-\infty}^{\infty} (-i)^m (\operatorname{sgn} m)^{|m|} \frac{e^{im\theta}}{\mu - m} \left[I_{\frac{|m|-1}{2}}(\xi) - I_{\frac{|m|+1}{2}}(\xi) \right],$$

where $I_n(x)$ is a modified Bessel function, and:

$$\xi = \left(\frac{z_0}{z} \right)^2 \left(\frac{\rho}{w} \right)^2 \left(\frac{1}{2q_1} \right),\ q_1 = 1 - i\frac{z_0}{z}. \qquad\qquad (2.43)$$

In the far field, q_1 in Equation (2.43) is equal to 1, and instead of Equation (2.42), we obtained an expression at the focal distance $z = f$:

$$E_F(\rho,\theta) = \frac{\sin(\pi\mu)}{\sqrt{2\pi}} \left(\frac{-iz_0}{f} \right) \exp(i\pi\mu)$$

$$\times \sqrt{x}\exp(-x) \qquad\qquad (2.44)$$

$$\times \sum_{m=-\infty}^{\infty} (-i)^{|m|} \frac{e^{im\theta}}{\mu - m} \left[I_{\frac{|m|-1}{2}}(x) - I_{\frac{|m|+1}{2}}(x) \right],$$

where $x = (z_0/f)^2(\rho/w)^2/2$. Series (2.44) shows that the light field with an optical vortex with an initial fractional TC is a superposition of an infinite number of optical vortices with both positive and negative integer charges. Further, the denominator in (2.42) shows that the largest contribution is made by the vortex with the number m closest to the fractional μ. It follows from Equation (2.44) that for $\mu = n + \alpha$, $\alpha \ll 1$, the TC of the vortex in Equation (2.44) will be equal to n, and with $\mu = n + 1 - \alpha$, $\alpha \ll 1$, the TC of the beam in Equation (2.44) will be equal to $n + 1$. If $\mu = n + 1/2$ is a half-integer, then two vortices with the numbers n and $n + 1$ will make almost the same contribution. Therefore, if the initial TC of the vortex in Equation (2.38) is in the range $n \leq \mu \leq (n+1)$, then the TC of the optical vortex in the Fresnel diffraction zone and in the far zone will jump abruptly between n and $n + 1$.

To test this hypothesis, we carried out a simulation and calculated the TC of the beam using the Equation (2.40). In Equation (2.40), the TC is calculated along a circle of infinite radius, but in the simulation, we calculated the TC on circles of different radii, the values of which were indicated each time. The following were the simulation parameters: the size of the initial field = 8 × 8 µm, 400 × 400 points, wavelength, λ = 0.532 µm, and waist radius of the Gaussian beam, w = 3 µm. The initial fractional *TC* of the beam in Equation (2.38) varied in the range 3 < μ < 4. The initial Gaussian beam was multiplied by a spherical lens function with a focal distance f = 10 µm. The real focal length, z = 9.236 µm (at this distance, the maximum intensity was observed on the optical axis at μ = 0). Thus, at the focus of the lens, all values were calculated at this z. Since the *TC* of the beam was calculated using Equation (2.39) on circles of large radii (up to 140 µm), which were greater than the distance from the waist to the output plane (10 µm), a nonparaxial calculation of the complex amplitude of the light field using the Rayleigh–Sommerfeld integral in Equation (2.39) was used to obtain adequate *TC* values.

2.4.2   TOPOLOGICAL CHARGE OF THE BEAM IN THE FAR FIELD, INITIAL TC 3 < µ < 4

Figure 2.30 shows the amplitude (a) and phase (b) of the light field in Equation (2.38) in the near field (μ = 2.2). The change in the *TC* of a beam with an initial fractional *TC* was shown by the simulation to occur in different ways if the initial *TC* changed from odd to even and from even to odd. Therefore, we separately considered two cases of changing the *TC* μ in the ranges 3–4 and 2–3. Further, all calculations were carried out at a distance z = 9.236 µm, where the focus was formed at μ = 0 (far field). Figure 2.31 shows the dependence of the *TC* on µ with a step of 0.01 in the range 3 < µ < 4, calculated for comparison on the circles of two radii: R = 8 µm (solid line) and R = 138 µm (dashed line).

Figure 2.31 shows that for different values of μ , the *TC* of the optical vortex took either of two integer values, 3 and 4. The transition from 3 to 4 and from 4 to 3 for circles of different radii occurred at different values of μ . Table 2.4 shows the exact values of µ at which the *TC* jumped from 3 to 4 and from 4 to 3.

A detailed analysis of the phase patterns obtained for different values of µ helps explain the graph in Figure 2.32. To understand which vortices exist in the field at

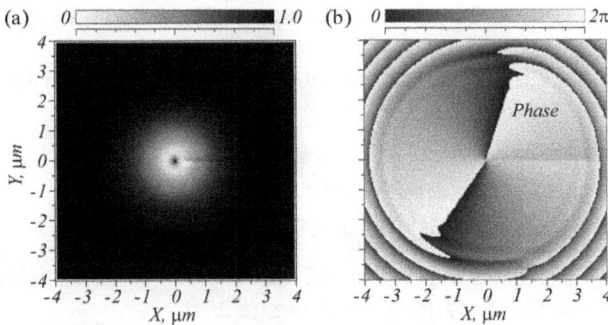

FIGURE 2.30   The amplitude and phase of the beam in Equation (2.38), μ = 2.2 in the near field at a distance z = 0.03 µm, at field size 8 × 8 µm

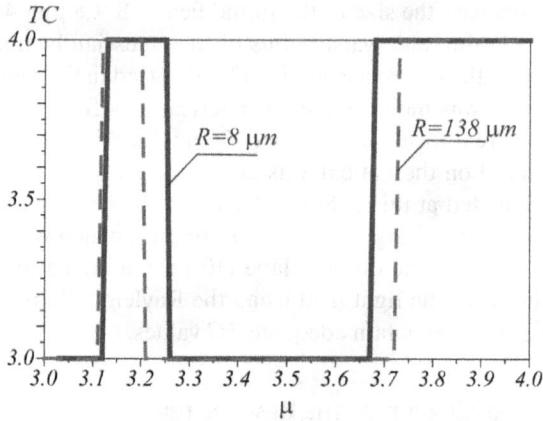

FIGURE 2.31 Dependence of the *TC* on the value of the fractional charge μ of the initial field in the range $3 < \mu < 4$ when calculated along the circles of two radii: $R = 8$ μm (solid line) and $R = 138$ μm (dashed line)

TABLE 2.4

Values of μ at which there is a transition of the *TC* from 3 to 4 and from 4 to 3.

R = 8 μm		R = 138 μm	
μ	*TC*	μ	*TC*
3.0–3.12	3	3.0–3.11	3
3.14–3.25	4	3.12–3.2	4
3.26–3.67	3	3.21–3.72	3
3.68–4	4	3.73–4	4

large radii, the fields of 3000×3000 points (276×276 μm) were calculated. Table 2.5, compiled using them, describes all the transitions of the *TC* from 3 to 4 and from 4 to 3 in the range of the initial *TC* $3 < \mu < 4$.

The transitions of the *TC* from 3 to 4 and from 4 to 3 are described in Table 2.5 and illustrated in Figure 2.32.

2.4.3   Topological Charge of the Beam in the Far Field when TC Is $2 < \mu < 3$

Shown in Figure 2.33 is the dependence of the *TC* of the beam μ , calculated along a circle of radius $R = 138$ μm, analogous to Figure 2.30.

A comparison of Figure 2.31 and Figure 2.33 shows that in both cases, near the initial and final edges of the ranges of μ values, the *TC* was equal to the value of the integer *TC* at the ends of the segments [2,3] and [3,4]. The difference was that for the

TABLE 2.5

Dependence of the *TC* on μ at *R* = 138 μm

μ	TC	Comment
3.0–3.11	3	An optical vortex in the center on the optical axis with $TC = 3$ was shown in the phase picture. There were two edge dislocations of different signs on the vertical axis at $y > 0$ and at $y < 0$. Therefore, the TC of the entire optical vortex was 3.
3.11–3.12	3→4	The upper edge dislocation ($y > 0$) generated an optical "dipole": two adjacent optical vortices with charges +1 and –1 above the center (centered at the point $y = 20$ μm, $x = 0$). The TC of the beam did not change and remained equal to 3. When μ increased, the upper vortex in the dipole with a TC of –1 moved beyond the boundary of the considered field ($y > R = 138$ μm), and the lower vortex in the dipole with TC +1 descended to the center. Further, for $\mu = 3.12$, only the lower vortex with TC +1 situated in $y = 9.02$ μm remained. The TC of the beam became equal to 4.
3.12–3.18	4	The optical vortex with TC +1 remaining in a circle of radius $R = 138$ μm still moved to the center of the beam, and the TC of the entire beam remained equal to 4.
3.18–3.21	4→3	Now, the lower edge dislocation ($y < 0$) generated an optical "dipole": two close vortices with TCs of –1 and +1 (centered at the point $y = -26.1$ μm, $x = 0$). When μ increased, these two vortices diverged in different directions: the lower one with TC +1 moved to the boundary of the region ($y = -138$ μm), and the upper one with TC –1 moved toward the center of the beam. At $\mu = 3.21$, a vortex with $TC + 1$ left the field boundary, and only a vortex with TC –1 remained at $y = -11$ μm. The TC of the entire beam again became equal to 3.
3.21–3.72	3	When μ increased, the vortex above the center ($y > 0$) with TC = +1 continued to gradually approach the center. The vortex from below ($y < 0$) with $TC = -1$ first approached the center and then, at $\mu > 3.5$, began to move away from the center.
3.73–3.74	3→4	Two vortices with TCs of –1 and +1 which diverged at the μ value of 3.5 began to approach each other at $\mu > 3.73$. A vortex with $TC = -1$ went along the $y < 0$ coordinate from the center, and a vortex with $TC = +1$ from the beam periphery shifted to the center. These vortices connected and mutually "annihilated" approximately at a distance of $y = -25$ μm. There remained only an additional vortex with $TC = +1$ from above ($y > 0$), which had practically already approached the center of the beam. Therefore, the TC of the entire beam became equal to 4.
3.74–4.00	4	An additional vortex with $TC = +1$ on top of the beam ($y > 0$) merged with the original vortices in the center of the beam with $TC = 3$. Further, the TC of the entire beam remained equal to 4.

interval $2 < \mu < 3$, the TC changed from 2 to 3 thrice, whereas for the interval $3 < \mu < 4$, the TC changed from 3 to 4 twice. As can be assumed from Figure 2.33, three optical "dipoles" were involved in the formation of the whole TC, not two, which were generated by edge dislocations. The evolution of optical vortices that were part of the "dipoles" with a change in the initial TC of the field in Equation (2.38) in the range $2 < \mu < 3$ is shown in Figure 2.34.

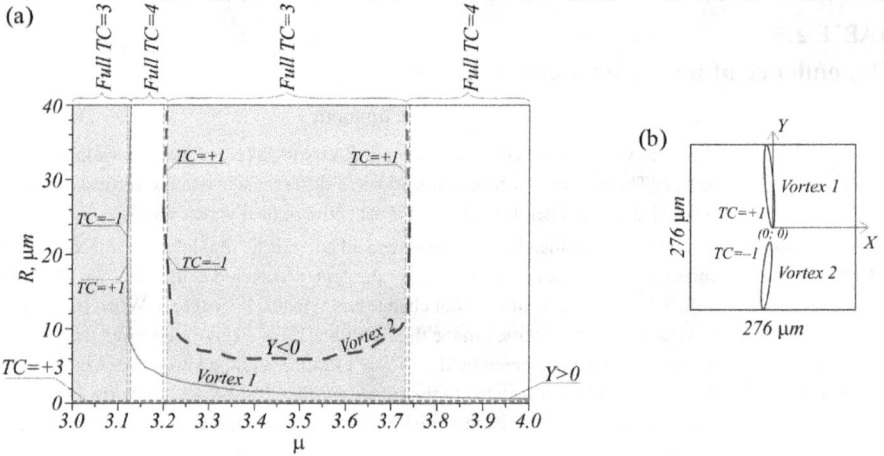

FIGURE 2.32 Dependence of the TC of the beam on the TC of individual vortices and the radii of their location depending on μ (a). In the center, in the entire range of μ values, there was already a vortex with $TC = +3$, and optical "dipoles" were added to it (pairs of vortices with $TC = +1$ and -1 appearing "at the bend" of two edge dislocations above ($y > 0$) and below ($y < 0$) the center at radii $R = 20$–26 μm. After one vortex from each "dipole" "left" (or "came") beyond the observation radius ($R = 138$ μm), the TC of the entire beam changed. The schematic location of these vortices on the field is shown in (b).

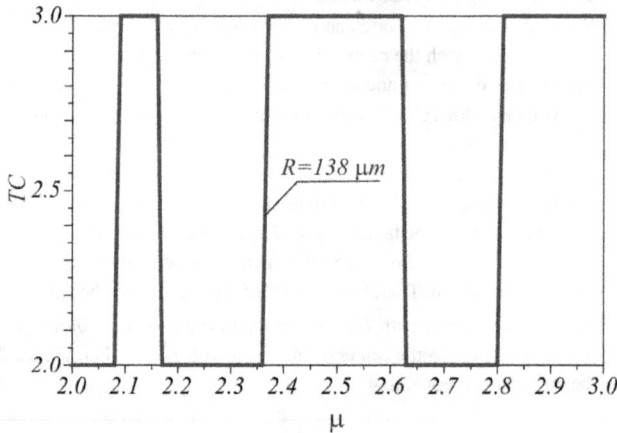

FIGURE 2.33 Dependence of the TC of the beam on μ for the range $2 < \mu < 3$, calculated along a circle of radius $R = 138$ μm.

If we determined the TC of the beam along a circle of radius $R < 20$ μm, we would see only one additional vortex with $TC = +1$, which would change the TC of the beam from 2 to 3 for $\mu > 2.1$. If we measured the TC along a circle of radius $R < 60$ μm, then we would see only two additional vortices with $TC = +1$ above ($y > 0$) and with $TC = -1$ below ($y < 0$). As shown in Figure 2.34, the TC of the beam was measured

(a)

(b)

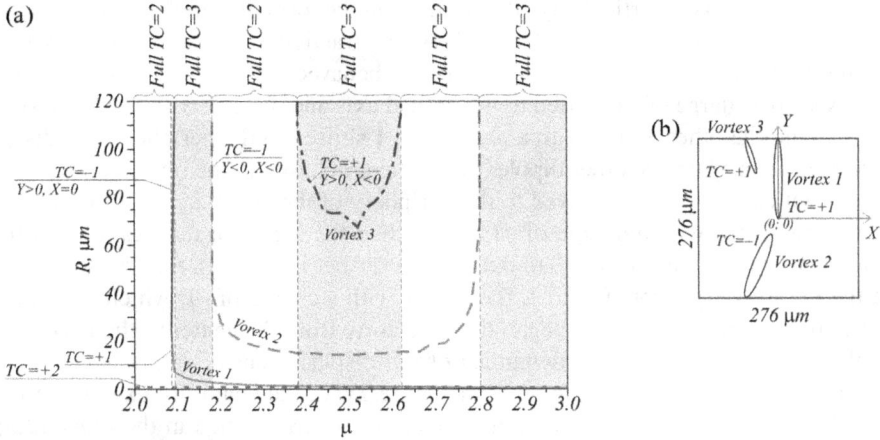

FIGURE 2.34 Distribution of three vortices with TCs of +1 and –1 appearing in the field at $2 < \mu < 3$ (a) and their schematic arrangement in the field (b)

along a circle of radius $R = 138$ μm, and therefore three vortices were visible, generated by three optical "dipoles". These three vortices (solid, dashed, and dash-dotted curves in Figure 2.34) led to jumps of the TC of the beam in the range $2 < \mu < 3$.

In the range $2 < \mu < 2.1$ inside a circle of radius $R = 138$ μm, only a base vortex was seen in the center with $TC = 2$. In the range $2.1 < \mu < 2.2$, an additional vortex with $TC = +1$ appeared in the field, and the TC of the entire beam became equal to 3. In the range $2.2 < \mu < 2.37$, two additional vortices with TCs of +1 and –1 were present in the beam, which mutually compensated each other and led to the entire TC becoming equal to 2. Next, in the range $2.37 < \mu < 2.63$, the field contained the main vortex with $TC = 2$ and three additional ones, two of which mutually compensated for each other, and the final TC of the beam was 3. In the range $2.63 < \mu < 2.8$, the third (distant) additional vortex with $TC = +1$ left the calculation area (a circle of radius $R = 138$ μm). The two remaining additional vortices mutually compensated their contribution, and the TC of the beam became equal to 2. Finally, in the range $2.8 < \mu < 3$, the lower ($y < 0$) additional vortex with $TC = -1$ left the calculation area. Only the main vortex with $TC = 2$ and the additional vortex almost merging with it ($TC = 1$) remained, and thus the entire TC was equal to 3. The third dipole, which was the most distant from the center of the beam, as well as the other two dipoles, was formed from an edge dislocation. Most likely, if the radius of the circle R were increased, it would have been possible to detect other optical "dipoles" generated by either the upper ($y > 0$) or the lower ($y < 0$) edge dislocations.

In this section, dynamics of the motion of additional vortices depending on a fractional TC in the initial field were determined in two ranges, [2,3] and [3,4]. The simulations of light propagation were carried out using the nonparaxial Rayleigh–Sommerfeld integral, and the TC was calculated along a large-radius circle passing along the periphery in the beam section. The intensity of the light field at the beam periphery was almost equal to zero. We found that additional optical vortices were formed from the two vertical edge dislocations in the form of "dipoles" consisting of a pair of vortices with charges +1 and –1. As the fractional part of the *TC* increased from 0 to 1, the optical vortices of the dipoles behaved differently. In one dipole, a vortex with a charge of +1 shifted to the optical axis and merged with the main axial vortex, and the other vortex with a charge of –1 shifted to the periphery and disappeared at infinity. In the other dipole, the vortices behaved in the opposite way: The vortex with a charge of +1 moved to the periphery of the beam and went to infinity, while the vortex with a charge of –1 approached the center of the beam when the fractional part increased from 0 to 0.5. With a further increase in the fractional part of the initial charge from 0.5 to 1, the vortex with a charge of –1, which was shifting to the center of the beam, began to move away from the center to the periphery of the beam. The discovered dynamics of the displacement of additional vortices in the beam determined the jumps of the entire *TC* of the beam, for example, from 2 to 3 and from 3 to 2, when the initial fractional charge varied in the range [2,3]. Moreover, in the range [2,3], there were five such jumps, and in the range [3,4], there were three jumps. This is attributed to the fact that during the transition from an even to an odd integer charge in the beam field, three dipoles were formed, not two, and three additional vortices were present in the beam cross-section. Thus, the nonparaxial calculation of the phase distribution in a beam with an initial fractional *TC* helped us understand in detail the changes (jumps) in the whole charge in the far field of the beam. In theory [19] and experiments [64,78,15,75,52], only the "dipole" closest to the beam center was considered. In [52,76], two "dipoles" were implicitly considered, but the motion of optical vortices entering these dipoles was not studied. Further, the influence of the third "dipole" on the formation of the beam charge was examined for the first time in this study.

3 Topological Charge Superposition of only Two Laguerre–Gaussian or Bessel–Gaussian Beams with Different Parameters

3.1 OPTICAL PHASE SINGULARITIES "GOING TO" INFINITY WITH A HIGHER-THAN-LIGHT SPEED

While the orbital angular momentum of paraxial vortex beams has been known to be preserved upon free-space propagation [33], the same cannot be said so far of the topological charge (TC) of optical vortices (OV). All radially symmetric OVs (e.g. Laguerre–Gaussian and Bessel–Gaussian beams), as well as some asymmetric OVs, preserve their TC upon free-space propagation [73]. No general proof of TC conservation has so far been offered; however, some publications discussed examples of OVs whose TC was not preserved upon propagation. M. Soskin et al. were the first to demonstrate this effect in 1997 [22]. Their study looked into a simple superposition of a Gaussian beam and a Laguerre–Gaussian (LG) mode $(0,n)$, with the Gaussian beams' waists having different radii. With the constituent components of the combined beam diverging differently, the TC was shown to change upon free-space propagation. If in the original plane, the Gaussian beam waist was larger than the LG mode waist, at first, the TC of the combined beam was zero. Upon propagation, the beams radii were getting closer, with the LG mode radius becoming larger than that of the Gaussian beam after passing through the same radius plane. From that plane onwards, the TC of the combined beam was shown to be equal to n. Upon free-space propagation, a fractional Gaussian beam was theoretically [19] and experimentally [52,76] shown to have in the near-field an integer TC equal to the one closest to the fractional number. On further propagation, the TC of such beams was shown to experience other, previously unknown, changes. For instance, it was numerically and experimentally demonstrated [77] that in the Fresnel zone the TC of an original fractional Gaussian OV becomes equal to the closest larger integer. A similar study was conducted in [85], where the TC was measured in the Fourier-plane (in the focus of a spherical lens), with the reported results being different from those of [77]. The study has shown that the *TC* does not always conserve during propagation. Thus, we can infer that if the fractional part of the original *TC* is smaller than 1/2, the *TC* of the propagating beam equals the closest smaller integer. Meanwhile, the larger-than-1/2 original fractional part leads to a propagating OV with *TC* equal to the closest larger integer. It would be interesting to find out what integer value the *TC* will take for a propagating OV whose original fractional part exactly equals 1/2.

So, OVs with the half-integer original *TC* deserves special attention. The superposition $r\exp(-r^2)\big(\exp(-i\varphi)-1\big)$ of two beams (comprising a screw and an edge dislocation) with $TC = -1/2$ has been discussed [136]. But considering that the second component in the superposition was not a mode and no far-field intensity null was observed, the resulting *TC* was found to be $TC = -1$. Half-integer original OVs were also studied in [53] and have been given the name Gamma–Gaussian beams, but explicit analytical formulae were derived only for the original plane.

For certain OVs, *TC* may be indeterminate. For instance, in the on-axis superposition of two diffraction-free same-weight Bessel beams, *TC* is indeterminate because Bessel functions have intensity nulls at different radii. Thus, being equal to the number of one constituent Bessel function at some radii, the superposition's *TC* will be equal to the number of the other constituent Bessel function at other radii.

As a follow-up to the pioneering research by M. Soskin et al. [22], in this section, we show that in the superposition of two different-waist LG beams, *TC* does not conserve during propagation because phase singularities partly either "go to" or "come from" infinity. Interestingly, the phase singularities "go to and come from" infinity with a higher-than-light speed. Strictly speaking, "going to and coming from" infinity does not mean that the singularities literally disappear or exist at infinity. Thus, while in practice the *TC* of interest does not conserve (because with the peripheral intensity being near-zero, a peripheral intensity null is not possible to measure), it does conserve theoretically, just partly "hiding at infinity". A similar effect was described and termed the "hidden phase" in [53].

3.1.1 THEORETICAL BACKGROUND

The topological charge can be calculated using a formula proposed by Berry [19]:

$$TC = \frac{\lim}{r \to \infty} \frac{1}{2\pi} \int_0^{2\pi} d\varphi \frac{\partial}{\partial \varphi} \arg E(r,\varphi)$$

$$= \frac{1}{2\pi} \frac{\lim}{r \to \infty} \operatorname{Im} \int_0^{2\pi} d\varphi \frac{\partial E(r,\varphi)/\partial \varphi}{E(r,\varphi)}. \tag{3.1}$$

Below, we discuss an example of an OV (superposition of two vortices) whose *TC* does not conserve. In [22], using a combined beam as an example, the *TC* of the combination was shown not to conserve upon propagation. However, no straightforward calculation of the *TC* using Equation (3.1) was done in [22]. Below, we generalize the results reported in [22] and calculate the *TC* of a sum of two different-waist LG modes through the straightforward use of Equation (3.1). Assume that the amplitude of a combined field composed of two different-waist LG modes $(0, n)$ and $(0, m)$ is given by:

$$
E(r,\varphi,z) = a(z) \left(\frac{r}{w_1(z)} \right)^{|n|} \exp\left(-r^2 / w_1^2(z) + ikr^2 / 2R_1(z) + in\varphi \right)
$$

$$
+ b(z) \left(\frac{r}{w_2(z)} \right)^{|m|} \exp\left(-r^2 / w_2^2(z) + ikr^2 / 2R_2(z) + im\varphi \right),
$$

(3.2)

where $w_1(z)$ and $w_2(z)$ are waist radii of two Gaussian beams, $R_1(z)$ and $R_2(z)$ are curvature radii, a and b are z-dependent constants, and n and m are integer TCs of the OVs. Substituting (3.2) into (3.1) yields:

$$
TC = \frac{1}{2\pi} \lim_{r\to\infty} \mathrm{Im} \int_0^{2\pi} d\varphi \, \frac{inar^n w_2^m(z)e^{-r^2/w_1^2(z)+iA+in\varphi} + imbr^m w_1^n(z)e^{-r^2/w_2^2(z)+iB+im\varphi}}{ar^n w_2^m(z)e^{-r^2/w_1^2(z)+in\varphi} + br^m w_1^n(z)e^{-r^2/w_2^2(z)+im\varphi}}
$$

$$
= \frac{1}{2\pi} \lim_{r\to\infty} \mathrm{Re} \int_0^{2\pi} d\varphi \, \frac{nar^{n-m} w_2^m(z)e^{-r^2\left(1/w_1^2(z)-1/w_2^2(z)\right)+i(A-B)+i(n-m)\varphi} + mbw_1^n(z)}{ar^{n-m} w_2^m(z)e^{-r^2\left(1/w_1^2(z)-1/w_2^2(z)\right)+i(A-B)+i(n-m)\varphi} + bw_1^n(z)}
$$

$$
= \begin{cases} m, & w_1(z) \le w_2(z), \\ n, & w_1(z) > w_2(z). \end{cases}
$$

(3.3)

To make Equation (3.3) less cumbersome, we introduced designations: $A = \dfrac{kr^2}{2R_1(z)}, B = \dfrac{kr^2}{2R_2(z)}$. The integral in Equation (3.3) was calculated as follows. Putting $w_{10} > w_{20}$, where $w_{10} = w_1(0)$, $w_{20} = w_2(0)$, we obtain that $w_1(z) > w_2(z)$ at $z_1 < kw_{10}w_{20}/2$ and the Gaussian exponent in the integrand in Equation (3.3) is positive. Thus, at $r \to \infty$, the exponents in the numerator and denominator in the integrand in Equation (3.3) tend to infinity faster than the power. Hence, the second terms in the numerator and denominator can be neglected and the remaining first terms are mutually canceled so that the only retained term is n. Thus, $TC = n$. However, at $z > kw_1w_2/2$, the radius of the first LG mode becomes smaller than that of the second LG mode: $w_1(z) < w_2(z)$, and the Gaussian exponent in the integrand in Equation (3.3) becomes negative. Hence, we see that at $r \to \infty$, the first terms in the numerator and denominator in the integrand in Equation (3.3) tend to zero. Then, the second terms can be mutually canceled and the only term retained is m. Thus, $TC = m$. And, vice versa, if $w_{10} < w_{20}$, then at $z_1 < kw_{10}w_{20}/2$, $TC = m$ and at $z_1 > kw_{10}w_{20}/2$, $TC = n$. Thus, we can infer that upon free-space propagation of beam in Equation (3.2), $|n-m|$ OVs with $TC = -1$ are born (or annihilated) at some distance.

3.1.2 Movement of Phase Singularities in the Propagating Beam

Equating amplitudes of the two constituent LG beams in the combination (3.2) yields:

$$a(z)\left(\frac{r}{w_1(z)}\right)^{|n|} e^{-r^2/w_2^2(z)} = b(z)\left(\frac{r}{w_2(z)}\right)^{|m|} e^{-r^2/w_2^2(z)}. \tag{3.4}$$

Grouping the exponential functions on the right-hand side and the power functions on the left-hand side of Equation (3.4) (putting $n > m > 0$), we obtain:

$$\frac{a(z)}{b(z)}\left(\frac{w_2^m(z)}{w_1^n(z)}\right) r^{n-m} = \exp\left[-r^2\left(\frac{1}{w_2^2(z)} - \frac{1}{w_1^2(z)}\right)\right]. \tag{3.5}$$

At any z, the left-hand side of Equation (3.5) is seen to be a power function of argument r, which grows from zero to infinity. On the right-hand side of Equation (3.5) is a Gaussian exponential function of r, which, either grows from unity to infinity (at $w_2(z) > w_1(z)$) or drops from unity to zero (at $w_2(z) < w_1(z)$). Increasing with the distance z differently:

$$w_1(z) = w_{10}\sqrt{1 + \frac{z^2}{z_1^2}}, \quad w_2(z) = w_{20}\sqrt{1 + \frac{z^2}{z_2^2}}, \tag{3.6}$$

where $z_1 = \dfrac{kw_{10}^2}{2}, z_2 = \dfrac{kw_{20}^2}{2}$, at some distance z, the waist radii in Equation (3.6) become equal to each other:

$$w_1(z_0) = w_2(z_0), \quad z_0 = \frac{kw_{10}w_{20}}{2}. \tag{3.7}$$

Then, Equations (3.7) and (3.5) suggest that the evolution of intensity nulls, also termed vortex dislocations or phase singularity centers, of the propagating field (3.2) will be as follows.

Because the two LG beams in Equation (3.2) have isolated on-axis intensity nulls of multiplicity n and m, the near-axis TC of the combined OV (3.2) will be equal to the TC of the larger-amplitude constituent OV [52]. We note that the lower-order OV ($n > m$) has the larger near-axis amplitude:

$$a(z)\left(\frac{r}{w_1(z)}\right)^n < b(z)\left(\frac{r}{w_2(z)}\right)^m, \quad r \ll 1. \tag{3.8}$$

Hence, at any distance z and any values of the constants a, b, w_{10}, and w_{20}, there will occur a near-axis OV with $TC = m$. Other singularity centers will form on radii of circles on which equality from Equation (3.5) holds. Putting in the initial plane $n > m$, $w_2 > w_1$, the exponential function in Equation (3.5) will grow to infinity, twice intersecting a curve r^{n-m}. On the first intersection with the curve, $n-m$ intensity nulls with $TC +1$ will form on a circle of radius R_1, whereas the second intersection of the exponential function (3.5) with the degree function on a circle of radius $R_2 > R_1$ will also produce n-m intensity nulls, each having $TC -1$. Thus, in the initial plane and the near field, there will occur an on-axis OV with $TC = m$, with the singularities on the circles with radii R_2 and R_1 mutually canceling each other.

But most interesting effects occur on further propagation of the combined beam in Equation (3.2), because the smaller waist $w_1(z)$ grows faster than the larger waist $w_2(z)$, catching up with the latter. As the beam propagates, the exponential function in the right-hand side of Equation (3.5) will tend to infinity at a slowing rate and the radius R_2 will grow. As a result, n-m intensity nulls with $TC -1$, which have formed on the circle of radius R_2, will move from the optical axis towards the beam periphery. As the beam approaches the plane $z_1 = (kw_{10}w_{20})/2$, where the radii of the constituent LG beams in Equation (3.2) become the same, the exponential function in Equation (3.2) gets equal to unity, meaning that the radius R_2 becomes infinitely large. This may be interpreted as $(n$-$m)$ intensity nulls with $TC = -1$ "going to" infinity. While the laser beam is traveling from zero ($z = 0$) to $z_1 = (kw_{10}w_{20})/2$, $(n$-$m)$ intensity nulls with $TC = -1$ move from a finite distance on the circle of radius R_2 to infinity. Hence, the intensity nulls may be thought of as traveling with a larger-than-light speed, which does not contradict the relativity theory as this effect has to do with the movement of intensity nulls due to interference of two waves. Similar effects are known in optics. By way of illustration, a diffraction-free Bessel beam propagates along the optical axis with a phase velocity that is $1/\cos\alpha$ times higher than the light of speed in a vacuum, where α is half the vertex angle of a conic wave that produces the Bessel beam. In [137], it was shown that a "light flash", generated when a partially coherent light reflects off a grating and when there is a nonzero angle between the constant-phase planes and the coherence layers, can propagate with a faster-than-light speed velocity. In [138], an electromagnetic field of four interfering plane waves was studied and it was demonstrated that, depending on their polarization, the energy flow velocity can achieve the speed of light. Thus, at $z > z_1$, the beam in Equation (3.2) will have $TC = m + (n$-$m) = n$, because $(n$-$m)$ OVs with $TC = -1$ "have departed to" infinity. At $n > m$, $w_2 < w_1$, a similar evolution scenario for the singularity points in the beam from Equation (3.2) will be unfolding in the opposite order. In this case, in the initial plane and the near field, the exponential function in Equation (3.5) will decrease from 1 to 0, only once intersecting on a circle of radius R_1 the power function in Equation (3.5), which grows from zero to infinity. On the radius R_1, $(n$-$m)$ OVs, each with $TC = +1$, will be formed. Therefore, at $z < z_1$, beam in Equation (3.2) will have $TC = m + (n$-$m) = n$. However, with the beam propagating beyond the distance z_1, i.e. at $z > z_1$, the second waist $w_2(z)$ will "catch up with" and then "overtake" the first waist $w_1(z)$, so that the exponential function in Equation (3.5) will again grow from 1 to infinity, twice intersecting the power function in Equation (3.5). In other words, $(n$-$m)$ OVs, each having $TC = -1$, will be generated on a circle radius R_2, with the combined beam having $TC = m$ at $z > z_1$. Hence, before the plane z_1, beam in Equation (3.2) has $TC = n$, but as soon as it passes the plane, n-m OVs with $TC = -1$ immediately "come from" infinity with a larger-than-light speed velocity, with the combined beam becoming defined by $TC = n - (n - m) = m$.

In [55], analyzing the case of same-waist constituent beams at $n > m$, $w_2 = w_1$, it was shown that in the superposition of an arbitrary number of same-waist LG beams, TC would equal the largest number of the LG mode. Hence, in our case, superposition in Equation (3.2) has $TC = n$, which will conserve upon propagation at $w_2 = w_1$.

Summing up, we can infer that at $w_2 \neq w_1$, beam (3.2) serves as a proving example of the non-conservation of TC of the propagating OVs. However, it is worth noting that although the OVs partly "go to" infinity this does not mean that they fully disappear. If the OVs at infinity are taken into account, the total TC of the superposition will be conserved.

3.1.3 NUMERICAL MODELING

The initial field was derived from the formula:

$$E = E_1 \left(\frac{\sqrt{2}r}{w_1} \right)^n \exp\left(in\varphi - \frac{r^2}{w_1^2} \right) + E_2 \left(\frac{\sqrt{2}r}{w_2} \right)^m \exp\left(im\varphi - \frac{r^2}{w_2^2} \right). \qquad (3.9)$$

We analyzed a situation when $n > m$ and $w_1 < w_2$. The simulation was conducted for $E_1 = E_2 = 1$, $n = 5$, $w_1 = 3$ μm, $m = 2$, $w_2 = 4$ μm, and $\lambda = 532$ nm, the entire 20×20 μm field was confined by a 10-μm pupil and calculated for 400×400 pixels. The amplitude and phase of the initial field are shown in Figure 3.1, from which it is seen that $TC = 2$, because at the opposite ends of three segments shown in Figure 3.1(b) opposite-sign singularities are centered. Thus, three OVs centered at the closed ends of three segments on a circle of radius R_1 have $TC = +1$. The other three OVs are centered at the remote ends of the segments in Figure 3.1(b) on a circle of R_2 have $TC = -1$. The downward wavy curve in Figure 3.1 passing through the origin marks phase jumps by 2π of an OV with $TC = +2$.

From Equation (3.7) and given the specified simulation parameters, the waists of the constituent LG beams in Equation (3.9) are seen to become mutually equal at

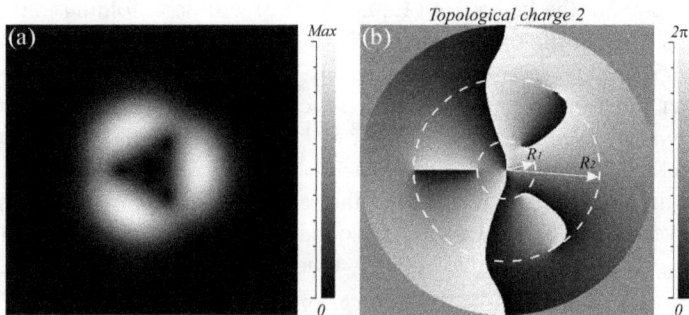

FIGURE 3.1   (a) Amplitude and (b) phase of the initial field in Equation (3.9). Regions with zero phase are marked black and those with phase 2π – light cyan. The images are of size 20 × 20 μm.

the distance $z_0 = \dfrac{\pi}{\lambda} w_{10} w_{20} = 70,86$ µm. *TC* was found to change from the value of 2 to 5 on the radius $R_1 = 1.85$ µm, and from 5 to 2 on the radius $R_2 = 10.5$ µm. At the distance z, the field produced by each LG beam was derived from the relation:

$$E(\rho,\theta,z) = \frac{w(0)}{w(z)} \left[\frac{\sqrt{2}\rho}{w(z)} \right]^n \exp\left[-\frac{\rho^2}{w^2(z)} + \frac{ik\rho^2}{2R(z)} + in\theta - i(n+1)\varsigma(z) \right], (3.10)$$

where:

$$w(z) = w_0 \sqrt{1 + \left(\frac{z}{z_R} \right)^2}, R(z) = z \left[1 + \left(\frac{z_R}{z} \right)^2 \right],$$

$$\varsigma(z) = \arctan\left(\frac{z}{z_R} \right).$$

(3.11)

Intensity nulls of two interfering beams (3.10) (with topological charges n and m, wavefront curvature radii $R_1(z)$ and $R_2(z)$, and with Gouy phases $\varsigma_1(z)$ and $\varsigma_2(z)$) appear at the following polar angles:

$$\varphi(z) = \frac{1}{n-m} \left[\pi + 2\pi p + \frac{kr^2}{2} \left(\frac{1}{R_2} - \frac{1}{R_1} \right) - (m+1)\varsigma_2 + (n+1)\varsigma_1 \right] \quad (3.12)$$

with $p = 0, 1,$ and 2. With increasing distance z, the radii R_1 and R_2 at which *TC* changes its value also grow. Shown in Figure 3.2 are the amplitude and the phase in the cross-section of the beam in Equation (3.9) at a distance of $z = 20$ µm. In comparison with Figure 3.1, the amplitude and the phase of field in Equation (3.9) in Figure 3.2 are seen to rotate anticlockwise at a certain angle. And similar to

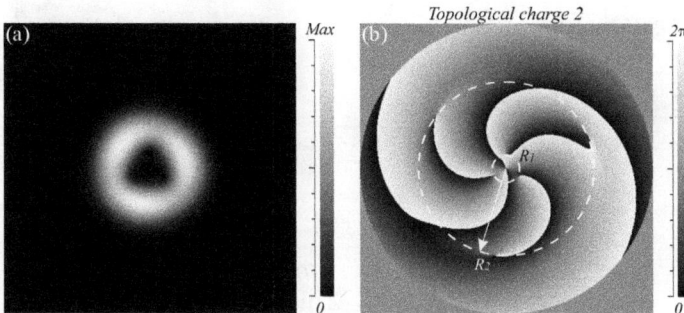

FIGURE 3.2 (a) Amplitude and (b) phase of field (3.9) at the distance $z = 20$ µm. Regions with zero phase are marked black and regions with 2π phase are marked light cyan. The images are of size 40×40 µm.

Figure 3.1, three OVs are located on a circle of radius R_1, each having $TC = +1$ (with the OVs centered at the closed ends of three segments in Figure 3.2(b)), with three other OVs, each having $TC = -1$ and found on a circle of radius R_2 (at the remote ends of the three segments) in Figure 3.2(b). The OVs with $TC = -1$ are seen to be shifted further from the optical axis. For instance, at $z=20$ μm in Figure 3.2, the radii of circles on which the three OVs with $TC = +1$ and the three OVs with $TC = -1$ are located, respectively are, $R_1=2.05$ μm and $R_2=12.04$ μm.

Shown in Figure 3.3 are plots for the radii R_1 and R_2 versus the distance z.

From Figure 3.3, the radius R_2 of a circle on which three OVs with $TC = -1$ are located, is seen to grow increasingly faster with distance, tending to infinity at $z \approx 71$ μm. If the plot for R_2 in Figure 3.3 is approximated as:

$$R_2 = \frac{1}{c_1 + c_2 z} + b, \qquad (3.13)$$

Then, at $c_1 = 0.152$ μm^{-1}, $c_2 = -0.0021$ μm^{-2}, $b = 3.95$ μm, the plot of function in Equation (3.13) will agree well with that shown by the solid line in Figure 3.3. Both functions are presented in Figure 3.4.

Equation (3.13) suggests that given properly fitted coefficients, the radius R_2 will tend to infinity at $z \simeq 73.2$ μm.

Now we show that dislocations move faster than light, in more detail. We suppose that at a distance z_1 or z_2 from the waist plane, the dislocation is located respectively on a circle with the radius R_{21} or R_{22}. From Equation (3.13), $R_{21} = 42.41$ μm at $z_1 = 60$ μm and $R_{22} = 68.47$ μm at $z_2 = 65$ μm. From the center of the waist, the light travels with light velocity to the center of the vortex at a distance z_1, i.e. the distance

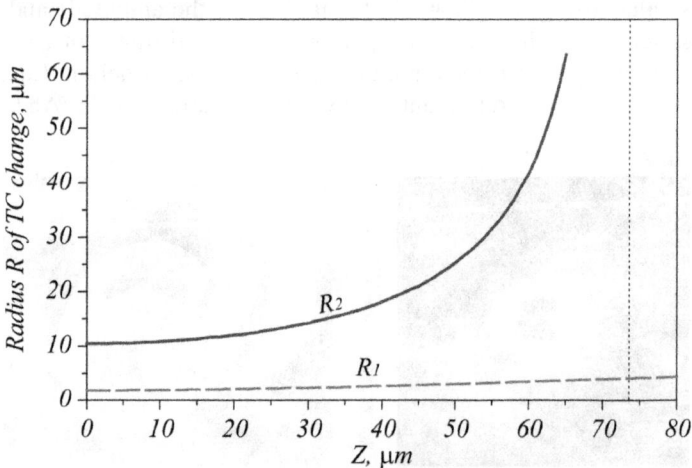

FIGURE 3.3 Radii (μm) at which the value of TC changes from 2 to 5 (R_1, dashed line) and from 5 to 2 (R_2, solid line) versus the distance z from the waist plane (μm).

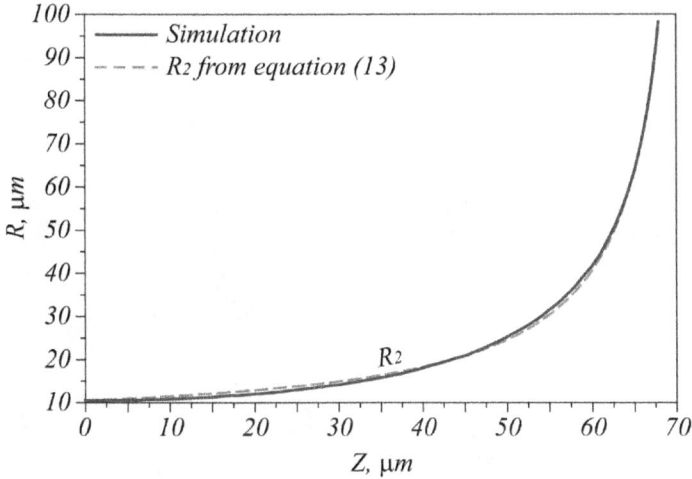

FIGURE 3.4 Comparative plots for the radius R_2 of a circle of location of three OVs with $TC = -1$, derived from a Fresnel transform and formula in Equation (3.13) with coefficients: $c_1 = 0.152 \ \mu m^{-1}$, $c_2 = -0.0021 \ \mu m^{-2}$, $b = 3.95 \ \mu m$.

$L_1 = (R_{21}^2 + z_1^2)^{1/2} = 73.47 \ \mu m$, and to the center of the vortex at a distance z_2, i.e. the distance $L_2 = (R_{22}^2 + z_2^2)^{1/2} = 94.41 \ \mu m$. Thus, by equal time intervals, the light travels the distance $L_2 - L_1 = 20.94 \ \mu m$, while the optical vortex moves by the distance $[(R_{22} - R_{21})^2 + (z_2 - z_1)^2]^{1/2} = 26.53 \ \mu m$. Therefore, at $60 \ \mu m < z < 65 \ \mu m$, the optical vortex moves faster than light $26.53/20.94 \approx 1.27$ times.

Figures 3.3 and 3.4 contain plots showing how the radius R_2 of the circle with the phase singularities of the order $TC = -1$ grows with the distance from the waist plane. However, the phase singularities, located on a circle with the radius R_2, have different polar angles in Equation (3.12), which depend on the propagation distance z. Thus, screw dislocations with the $TC = -1$ move to the beam's periphery along a spiral path with a velocity exceeding that that can be obtained from Figures 3.3 and 3.4. Figure 3.5 depicts the phase distributions at different propagation distances z as well as a segment of the propagation trajectory (curved line) of one of the three phase dislocations with $TC = -1$ lying on the circle of the radius R_2. Figure 3.5 reveals that the screw dislocation moves from the optical axis at a growing speed. For instance, on a segment $30 \ \mu m \leq z \leq 40 \ \mu m$, the phase singularity moves along the polar angle by nearly 45 degrees, whereas at $40 \ \mu m \leq z \leq 50 \ \mu m$, it moves by nearly 315 degrees.

Below we give several examples of light fields propagating with a phase velocity exceeding the speed of light. However, the light energy, that carries a signal, still propagates with the light speed. Figure 3.6 demonstrates three examples including one from the current work (Figure 3.6(c)). Figure 3.6(a) shows the generation of a diffraction-free Bessel beam with a refractive axicon. The refractive axicon transforms

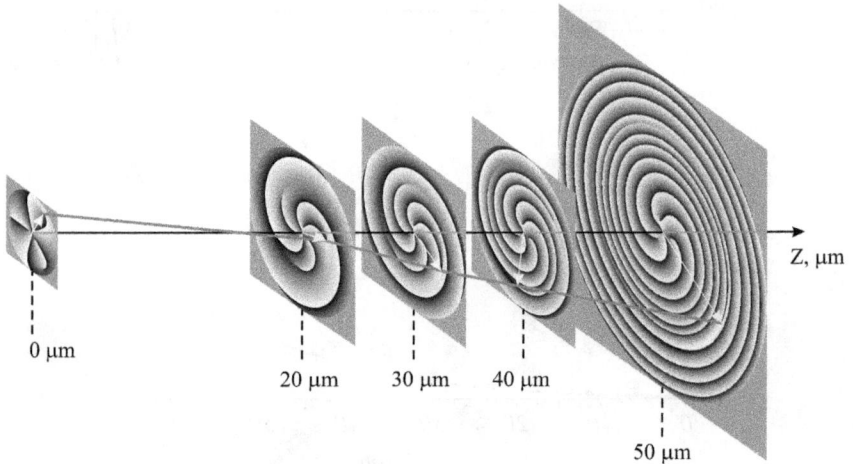

FIGURE 3.5 Transverse phase distributions at different propagation distances z (0, 20, 30, 40, and 50 μm) and 3D segment of the trajectory (curveline), along which propagates one of the three screw dislocations with the $TC = -1$, lying on the circle with the radius R_2 (Figure 3.4).

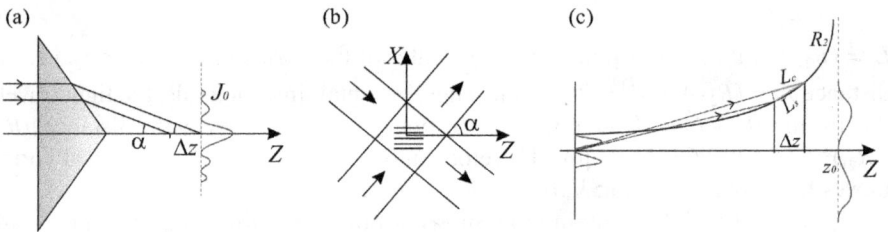

FIGURE 3.6 Examples of light fields propagating with a higher-than-light phase velocity: (a) diffraction free Bessel beam, (b) two interfering plane waves, and (c) schematic generation of the phase dislocations on a circle with the radius R_2 in Figures 3.3 and 3.4.

the incident plane wave with the wavenumber k into a converging conical wave with an angle of 2α at its vertex. Along the optical axis, the Bessel beam is generated with the following complex amplitude:

$$E(r,z,t) = E_0 J_0(k_r r) \exp(ik_z z - iwt),\qquad(3.14)$$

where $k^2 = k_r^2 + k_z^2$, $k_z = k\cos\alpha$, $k_r = k\sin\alpha$, w is the cyclic frequency of the light wave, t is the time, $J_0(x)$ is the 0th-order Bessel function. According to Equation (3.14), the phase velocity of the Bessel beam is given by:

$$v = \frac{z}{t} = \frac{w}{k_z} = \frac{c}{\cos\alpha} \geq c.\qquad(3.15)$$

Figure 3.6(b) illustrates two monochromatic plane waves tilted symmetrically to the horizontal axis by an angle α. In an area where these two waves intersect, an interference pattern occurs with the following complex amplitude:

$$E(r,z,t) = E_0 \cos(k_x x) \exp(ik_z z - iwt),$$ (3.16)

It is seen from Equation (3.16) that the interference pattern propagates along the axis z with the phase velocity exceeding the speed of light and equal to Equation (3.15).

Figure 3.6(c) schematically depicts the generation of phase dislocations on a circle with the radius R_2 (Figures 3.3 and 3.4). When the beam propagates between two transverse planes, divided by a distance Δz, the singularities (Figure 3.6(c)) move by a distance L_s, while, by the same time interval, the light from the waist center goes straight by a distance equal to L_c. According to Figure 3.6(c), $L_c = L_s \cos\alpha$ with α being an angle between the segment L_s and the straight line with the segment L_c. Therefore, the phase singularity propagates with a velocity $v = c/(\cos\alpha)$, similarly to Equation (3.15). In all three cases, the light does not propagate from a single point on the optical axis (Figure 3.6a) or on the curve R_2 (Figure 3.6c) to another point. The light incomes to the start and to the end point of the considered segments (Δz in Figure 3.6(a) and L_s in Figure 3.6(c)) from the side by an angle α to these segments.

Next, let us analyze the reverse situation of the first waist being larger than the second one: $n > m$, $w_1 > w_2$. The simulation parameters remained the same as in Figure 3.1, except for $w_1 = 4$ μm and $w_2 = 3$ μm. The amplitude and the phase of field in Eqution (3.9) in the initial plane (at $z = 0$) are depicted in Figure 3.7.

From Figure 3.7, field in Eqution (3.9) is seen to have $TC = 5$, with a near-vertical wavy line of phase jump by 2π passing through the origin in Figure 3.7 belonging to an OV with $TC = +2$ and three rays outgoing from a near-origin circle of radius R_1 belonging to three OVs, each having $TC = +1$. As field in Eqution (3.9) propagates, its total TC remains equal to 5, but the radius R_1 of a circle where three OVs with $TC = +1$ are formed grows. However, after the beam has traveled the distance in Eqution (3.7), $z_0 = \dfrac{\pi}{\lambda} w_{10} w_{20} = 70,86 \mu m$, at which the waists of the two LG beams

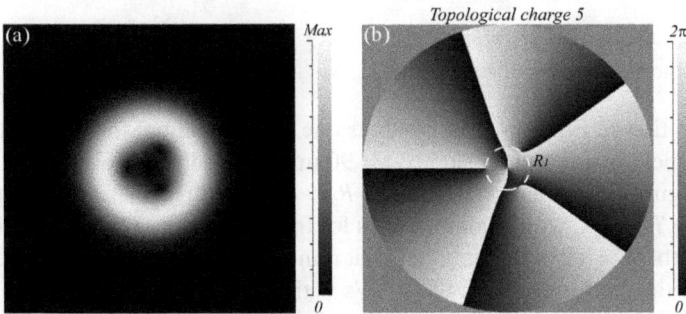

FIGURE 3.7 (a) Amplitude and (b) phase of the initial field. Regions of zero phase are marked black and those of phase 2π –white. The images are of size 20×20 μm.

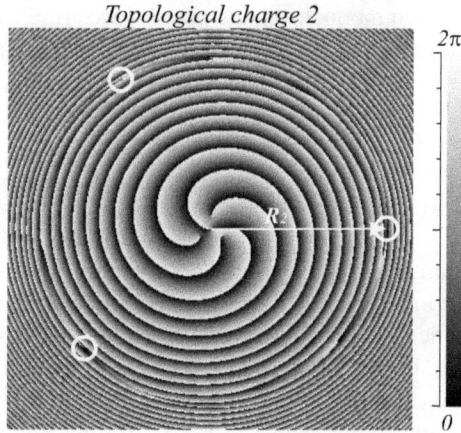

FIGURE 3.8 Pattern of the field phase at the distance $z = 90$ μm, the field size is 100×100 μm.

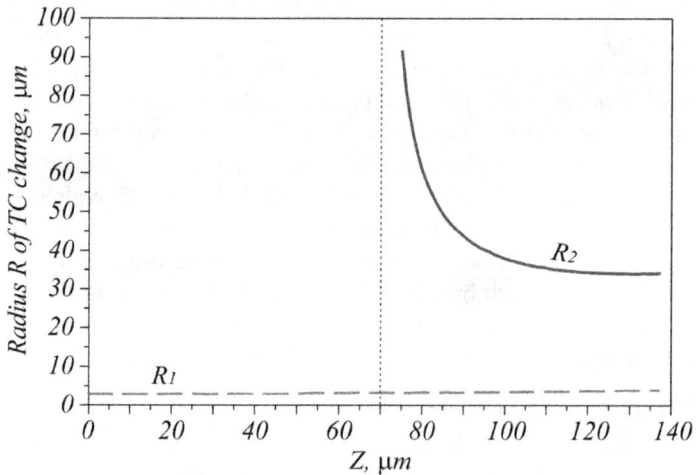

FIGURE 3.9 Radii (μm) of TC change from 2 to 5 (R_1, dashed line) and from 5 to 2 (R_2, solid line) versus the distance z to the waist (μm).

in superposition in Eqution (3.9) get the same, TC gets equal to $TC = 2$. Figure 3.8 depicts the field phase at the distance $z = 90$ μm, derived fromEqution (3.10).

From Figure 3.8, on a circle of radius $R_2 \approx 45$ μm, three OVs are seen to appear, each having $TC = -1$, which compensates for three OVs with $TC = +1$. So that an on-axis OV with $TC = +2$ is the only one that remains uncompensated. Figure 3.9 plots the radii of the circles at which three OVs with $TC = +1$ (R_1, dashed line) and three OVs with $TC = -1$ (R_2, solid line) are formed versus the distance to the waist.

As before, the solid line in Figure 3.9 can be approximated by Equation (3.13). Putting in Equation (3.13) $c_1 = -0.258$ μm^{-1}, $c_2 = 0.00366$ μm^{-2}, $b = 30$ μm, we find

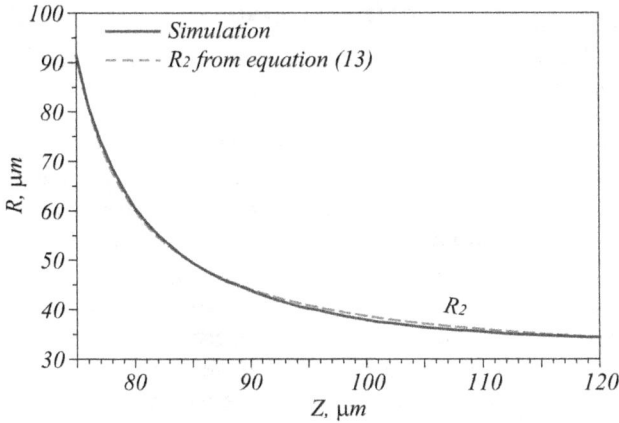

FIGURE 3.10 Comparative plots for the radius of *TC* change, where three OVs, each with $TC = -1$, are formed, derived from Equations (3.10) and (3.13) with coefficients $c_1 = -0.258$ μm^{-1}, $c_2 = 0.00366$ μm^{-2}, b = 30 μm.

that the plot for R_2 will agree well with the solid curve in Figure 3.9. A comparison of the two plots is given in Figure 3.10.

From Figure 3.10, the radius R_2 is seen to tend to infinity at $z \approx 69.8$ μm, with this distance being approximately 1.5% different from that derived from Equation (3.13).

Summing up, we have shown [23] in this section that in an on-axis superposition of two different-waist LG beams with the numbers (0, *n*) and (0, *m*), *TC* remains equal to *m* until the beam reaches a plane where the waists are equalized, given that the LG mode (0, *m*) has a larger waist, after which the superposition acquires the total $TC = n$. This is because in the initial plane, the superposition contains an on-axis OV with $TC = m$, with extra (*n-m*) Ovs with $TC = +1$ and (*n-m*) OVs with $TC = -1$ centered at different-radius circles. On approaching an equal-waists plane, the OVs with $TC = -1$ "go to" infinity with a larger-than-light speed, with the superposition acquiring $TC = n$. Note that the relationship between the radii of OV location and the distance *z* can be approximated by a function $R \sim 1/z$ up to coefficients. If the opposite is the case, i.e. the LG beam (0, *m*) has a smaller waist, from the initial plane up to an equal-waists plane, the superposition has $TC = n$, changing to $TC = m$ afterwards. In this case, on the contrary, *n-m* OVs with $TC = -1$ may be said to "come from" infinity with a larger-than-light speed. We note that the phase singularity is generated due to interference of two vortex diverging light waves and under its propagation velocity we mean the phase velocity, which can exceed the speed of light (Figure 3.5). We also note that a dielectric particle with a diameter approximately equal to the diameter of the beam in Eqution (3.2) should rotate due to the orbital angular momentum, which is partially transferred to the particle. And, since the orbital angular momentum of the beam in Eqution (3.2) is conserved on propagation, then, despite the fact that the *TC* can change, the particle rotation velocity is the same in the arbitrary cross-section of the beam.

3.2 CONSERVATION OF THE HALF-INTEGER TOPOLOGICAL CHARGE ON PROPAGATION OF A SUPERPOSITION OF TWO BESSEL-GAUSSIAN BEAMS

Laser beams are known to be defined by two key characteristics: orbital angular momentum (OAM) and topological charge (TC). These are two different quantities characterizing the light field: TC depends only on the phase of the light field [19], whereas OAM depends on both the phase and the amplitude [20]. The light field can have nonzero TC and zero OAM at the same time [139]. And vice versa, optical fields are known with the nonzero OAM, but with zero TC [17]. For some light fields, TC is undefined. For example, for a coaxial superposition of two diffraction-free Bessel beams, the zeros of two different-order Bessel functions are at different radii. Therefore, the TC of such a superposition is different, when calculated at different distances from the optical axis.

Note that while publications dealing with an analysis of OAM carried by vortex beams are abundant (e.g. [35,140,36,120,141,142,143,144,145]), papers concerned with the study of TC of the OVs are not too many [146,147,11]. However, giving closer consideration to TC may be important because, say, upon propagation in a turbulent medium, amplitude and phase distortions of an OV lead to continuous variations of OAM (which normally decreases). The magnitude of OAM variations is proportional to the magnitude of the distortions while leading to discrete variations in TC since it is supposed to remain integer. This means that for TC to change by unit, the distortion value needs to be essential, making it more stable against OV distortions in comparison with OAM.

Whereas the OAM of any paraxial light field is conserved on propagation, TC can be also conserved (for instance, TC of the Laguerre–Gaussian or of the Bessel–Gaussian beams) or not conserved [22,76,77,74]. For example, as Berry showed [19] if an optical vortex (OV) has an initial fractional TC, then after free space propagation, TC becomes an integer. If the fractional part of the TC is less than 1/2, TC on propagation becomes equal to the nearest lower integer. Otherwise, if the fractional part of the TC exceeds 1/2, TC becomes equal to the nearest higher integer [19]. On further propagation, this integer TC can change by the unit in the Fresnel diffraction zone and in the far-field [76,77].

The case of a half-integer TC, when the fractional part is exactly 1/2, stands alone, because in this case its OAM equals half-integer TC [19,52] and remains unchanged upon propagation. Up to now, it was not known whether the half-integer TC can be conserved on propagation. We may mention a work [136], where the following superposition of two beams was studied with its TC being equal to $-1/2$: $r\exp(-r^2)\left(\exp(-i\varphi)-1\right)$ (mix of screw and edge dislocations). However, since the second term does not describe a mode and there is no intensity null on the optical axis in the far field, the TC of the superposition in the far field is equal to -1.

What is important, both the main characteristics of the vortex laser beams, OAM and TC, are nonlinear, i.e. OAM or TC of a superposition of two beams is not a sum of their OAMs or TCs. The OAM is nonlinear since it is proportional to the dot product of the beam complex amplitude with its derivative by the angular polar coordinate.

For example, the superposition of two Hermite–Gaussian beams with zero OAM can have a nonzero OAM [148]. The TC is nonlinear since it is contributed by the phase of the light beam along some contours. So, the TC of a superposition is typically not a sum of TCs of the constituent beams [55]. So, deriving the OAM or TC of some superposition is not a trivial task.

It is worth noting that there are works studying superpositions of the Laguerre–Gaussian beams [149,150,151] and of the diffraction-free Bessel beams [152,153,154]. However, TC of the whole superposition was not obtained in these works.

The closest to our work is the paper by S. Orlov and A. Stabinis [155], which also studies a coaxial superposition of two different Bessel–Gaussian (BG) beams. Similar to [22], the work [155] demonstrates that if the waist radii of the Gaussian beams are different, then TC of the superposition is not conserved on propagation. However, TC of the superposition was not obtained exactly in [155] by using the Berry's formula [19] and it was not shown that if two BG beams are different only by their TCs, then TC of the superposition can be half-integer and this half-integer TC is conserved on propagation.

In this section, we show for the first time that a superposition of two BG beams with different TCs m and n can have a half-integer TC $(n + m)/2$, which is conserved on propagation. Such a superposition has an m-fold degenerate intensity null on the optical axis, while $n - m$ optical vortices with TC +1 are located on the infinite-radius circle, and therefore their TC equals $(n - m)/2$. We also propose a novel approach to controlling TC and OAM of the superposition of laser vortex beams using a linear combination of two coaxial BG beams characterized by different amplitudes, TCs, and scaling factors (radial projections of the wave-vector) as an example. It is shown that upon the beam propagation, TC is conserved and equal to TC of the constituent beam with the larger scaling factor. In a similar way, given the same scaling factors of the constituent BG beams, the TC of the superposition is shown to be equal to the TC of the beam with higher amplitude (power). We also derive explicit expressions for OAM carried by the superposition of two BG beams and an expression to describe in which manner the superposition rotation angle varies with distance.

3.2.1 TOPOLOGICAL CHARGE OF A COAXIAL SUPERPOSITION OF BESSEL–GAUSSIAN BEAMS

Let us analyze an on-axis superposition of BG beams in any transverse plane at distance z [40]:

$$E(r,\varphi,z) = \sum_{p=1}^{N} \frac{E_p}{q(z)} J_{n_p}\left(\frac{\alpha_p r}{q(z)}\right) \exp\left(ikz - \frac{i\alpha_p^2 z}{2kq(z)} - \frac{r^2}{w^2 q(z)} + i n_p \varphi\right), \quad (3.17)$$

where (r, φ, z) are the cylindrical coordinates, E_p are the (positive real) amplitude (weight) coefficients, $J_{n_p}(x)$ is the n_pth-order Bessel function, $\alpha_p = k \sin \theta_p$ is the scaling factor of the Bessel function, k is the wavenumber, θ_p is the half-angle of a conical wave that forms the Bessel beam, w is the Gaussian beam waist radius, n_p is the (integer) topological charge of an OV, and $q(z) = 1 + iz/z_0$ and $z_0 = kw^2/2$ is the Rayleigh range. The light field in Equation (3.17) obeys the paraxial Helmholtz equation, which in Cartesian coordinates (x, y, z) reads as:

$$\left(2ik\frac{\partial}{\partial z} + \frac{\partial^2}{\partial x^2} + \frac{\partial^2}{\partial y^2} \right) E(x, y, z) = 0. \tag{3.18}$$

For simplicity, below we assume the superposition of Equation (3.1) at any distance z to be composed of just two BG beams ($N = 2$). In practice, the superposition of two Bessel beams (Equation (3.17)) can be generated by a spatial light modulator (SLM) with its phase being equal to the sum of phases of two helical axicons [154] ($\alpha_1 = \alpha$, $\alpha_2 = \beta$, $n_1 = n$, $n_2 = m$):

$$\Psi(r, \varphi) = \frac{\pi}{2}\left\{ 1 - \text{sgn}\left[\cos\left(\frac{(\alpha - \beta)r + (n - m)\varphi}{2} \right) \right] \right\} + \left(\frac{(\alpha + \beta)r + (n + m)\varphi}{2} \right),$$

$$\tag{3.19}$$

where sgn is the sign of a number. *TC* can be derived using a familiar Berry formula [19]:

$$TC = \frac{1}{2\pi} \lim_{r \to \infty} \text{Im} \int_0^{2\pi} d\varphi \frac{\partial E(r, \varphi)/\partial \varphi}{E(r, \varphi)}, \tag{3.20}$$

where Im is the imaginary part of the number. This expression for the *TC* can be obtained from the following:

$$TC = \frac{1}{2\pi} \oint_\gamma \frac{\partial}{\partial \varphi} \arg E(r, \varphi) d\varphi \tag{3.21}$$

with γ being a contour in the transverse plane passing far enough (theoretically, in infinity) and encompassing all the optical vortices of the light field. Equation (3.21) is similar to the closed-path Berry phase [156], obtained for an adiabatic evolution of a quantum system. We use it in the form of Equation (3.20) since it is convenient to replace the contour γ with an infinite-radius circle to obtain *TC* analytically.

We note that *TC* in Equations (3.20) and (3.21) account for all the points of phase singularities of the light field. For the vectorial light fields, polarization singularities also exist, which are characterized by the Poincaré–Hopf index. In this section; however, we consider only a paraxial scalar field from Equation (3.18) and derive its *TC* in the form given by Equation (3.20). Under scalar field we mean one transverse component of a linearly polarized light beam.

At a far distance from the optical axis ($r \to \infty$), the argument of the Bessel function in Equation (3.17) is large and we can use the following approximation [81] (index p is omitted for brevity):

$$J_n\left(\frac{\alpha r}{q(z)}\right) \approx \sqrt{\frac{2q(z)}{\pi\alpha r}}\cos\left(\frac{\alpha r}{q(z)} - \frac{\pi n}{2} - \frac{\pi}{4}\right).$$ (3.22)

The cosine is a sum of two exponents. However, at $z > 0$, the argument of the Bessel function has a nonzero imaginary part and, for large r, one exponent is greatly overwhelming another one. Thus, substituting Equation (3.17) into Equation (3.20) at $N = 2$ and considering asymptotical properties of the complex-argument Bessel function at $r \to \infty$:

$$J_n\left(\frac{\alpha r}{q(z)}\right) \approx \sqrt{\frac{q(z)}{2\pi\alpha r}}\exp\left(\frac{\alpha rz}{z_0|q(z)|^2} + \frac{i\alpha r}{|q(z)|^2} - \frac{in\pi}{2} - \frac{i\pi}{4}\right),$$ (3.23)

we obtain ($\alpha_n = \alpha$, $\alpha_m = \beta$):

$$TC = \frac{1}{2\pi}\lim_{r\to\infty}\mathrm{Re}\int_0^{2\pi}d\varphi\left[\frac{nE_n(-i)^n}{\sqrt{\alpha}}\exp\left(in\varphi - \frac{i\alpha^2 z}{2kq(z)}\right)\right.$$

$$+\frac{mE_m(-i)^m}{\sqrt{\beta}}\exp\left(im\varphi - \frac{i\beta^2 z}{2kq(z)} + \frac{\beta-\alpha}{|q(z)|^2}\left(i+\frac{z}{z_0}\right)r\right)\right]$$

$$\times\left[\frac{E_n(-i)^n}{\sqrt{\alpha}}\exp\left(in\varphi - \frac{i\alpha^2 z}{2kq(z)}\right) + \frac{E_m(-i)^m}{\sqrt{\beta}}\exp\left(im\varphi - \frac{i\beta^2 z}{2kq(z)}\right)\right.$$

$$\left.+\frac{\beta-\alpha}{|q(z)|^2}\left(i+\frac{z}{z_0}\right)r\right]^{-1}.$$ (3.24)

The exponent function in Equation (3.23) grows faster at $r \to \infty$, when the scaling factor α is larger. Therefore, at $\alpha > \beta$, only the first terms are seen to be retained in the numerator and denominator of Equation (3.24) at $r \to \infty$, which, after the cancellation of like terms, yields $TC = n$. And vice versa, with just second terms retained in the numerator and denominator of Equation (3.24) at $\alpha < \beta$ and $r \to \infty$, after the cancellation of like terms, we obtain $TC = m$. Hence, the TC of the superposition of two BG beams will be defined by the "competition" between the Bessel function scaling factors α and β, irrespective of the amplitudes of the constituent beams (E_n, E_m).

In a particular case of $\alpha = \beta$, Equation (3.24) is replaced with:

$$TC = \frac{1}{2\pi}\mathrm{Re}\int_0^{2\pi}d\varphi\frac{E_n n(-i)^n e^{in\varphi} + E_m m(-i)^m e^{im\varphi}}{E_n(-i)^n e^{in\varphi} + E_m(-i)^m e^{im\varphi}},$$ (3.25)

where Re is the real part of the number. It has been shown [55] that TC of Equation (3.25) takes a value of $TC = n$, if $E_n > E_m$ and $TC = m$ if $E_n < E_m$. If $E_n = E_m$, Equation (3.25) suggests that $TC = (n + m)/2$.

Indeed, the integral (3.25) can be written as an integral over the unit-radius circle in the complex plane (we suppose $n > m$, then make a substitution $(n - m)\varphi = \theta$ and then we substitute $Z = e^{i\theta}$):

$$TC = \frac{1}{2\pi} \text{Re} \oint_{|Z|=1} \frac{nZ + (E_m/E_n)m(-i)^{m-n}}{Z + (E_m/E_n)(-i)^{m-n}} \frac{dZ}{iZ}. \qquad (3.26)$$

The value of this integral depends on the ratio $|E_m/E_n|$, which defines the number of poles inside the circle $|Z| = 1$. If $|E_m| > |E_n|$, the integrand has the only pole $Z = 0$ and, according to the residue theorem, $TC = m$. If $|E_m| < |E_n|$, the integrand has two poles $Z = 0$ and $Z = -(-i)^{m-n}(E_m/E_n)$, and $TC = m + (n - m) = n$. If $|E_m| = |E_n|$, the integrand has one pole inside the circle $(Z = 0)$ and one pole right on the circle. Such a pole is typically bypassed along a half-circle (Figure 3.11) and therefore the corresponding residue is multiplied by 1/2. So, in this case, $TC = m + (n - m)/2 = (n + m)/2$.

Validity of Equations (3.24) and (3.25) for arbitrary distance z means that the half-integer TC of the superposition of two BG beams is conserved on propagation. Actually, at $E_n = E_m = E_0$ and $\alpha = \beta$, the sum of two BG beams in Equation (3.17) will be described by the amplitude (at $N = 2$ and large r):

$$E(r \to \infty, \varphi, z) \approx E_0 e^{ikz - i\pi/4} \sqrt{\frac{2}{\pi \alpha r q(z)}} \exp\left(-\frac{i\alpha^2 z}{2kq(z)} - \frac{r^2}{w^2 q(z)} + \frac{i\alpha r}{|q(z)|^2}\right)$$

$$\times \exp\left(\frac{\alpha r z}{z_0 |q(z)|^2} + \frac{i(n+m)}{2}\left(\varphi - \frac{\pi}{2}\right)\right) \cos\left[\frac{(n-m)}{2}\left(\varphi - \frac{\pi}{2}\right)\right]. \qquad (3.27)$$

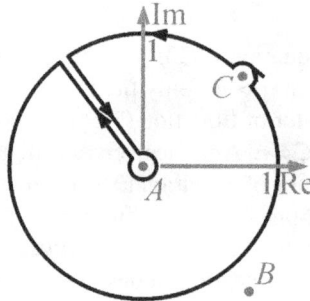

FIGURE 3.11 (a) Integration over a curve (circle) in a complex plane of a function with poles inside the circle (A), outside the circle (B), and right on the circle (C). The integral of the function is contributed by the residue in point A and by the half-residue in point C. Here, we show a contour for the integration by using the residues theorem.

From Equation (3.27), the sum is seen to produce an optical vortex that has $TC = (n + m)/2$ and $|n - m|$ intensity petals. It is worth noting in the linear combination of two similar Ovs with different TCs, OAM normalized to the beam power also equals $(n + m)/2$. If n and m are, respectively, even and odd (or vice versa) integers, TC of the linear combination in Equation (3.27) will be defined by a half-integer while the number of intensity petals will be an integer.

Near the center, at $r \to 0$, and at any values of E_m, E_n, α, and β, an OV with lesser TC $m < n$ will occur at the center, because the second term in the square brackets in the relationship below is much less in the absolute value than the first one:

$$E(r \to 0, \varphi, z) \approx \frac{1}{q(z)} \exp\left(ikz - \frac{r^2}{w^2 q(z)} \right)$$

$$\times e^{im\varphi} \left[\frac{E_m}{m!} \left(\frac{\beta r}{2q(z)} \right)^m \exp\left(-\frac{i\beta^2 z}{2kq(z)} \right) \right. \tag{3.28}$$

$$\left. + \frac{E_n}{n!} \left(\frac{\alpha r}{2q(z)} \right)^n \exp\left(-\frac{i\alpha^2 z}{2kq(z)} \right) e^{i(n-m)\varphi} \right].$$

Thus, the light field in Equation (3.27) with half-integer TC will have an isolated m-fold intensity null and $(n-m)$ null-intensity lines (edge dislocations) outgoing from the center to infinity along radii for which the polar angle is given by:

$$\varphi_p = \left(\frac{4p+1}{2|m-n|} \right) \pi, \quad p = 0, \pm 1, \pm 2, \ldots \tag{3.29}$$

When the scaling factors are different, $\alpha \neq \beta$, the beam in Equation (3.17) will rotate during propagation. At $N = 2$, the z-dependence of the intensity pattern rotation angle φ in Equation (3.17) takes the form:

$$\varphi = \frac{(\beta^2 - \alpha^2)z}{2k(n-m)} \left(1 + \frac{z^2}{z_0^2} \right)^{-1}, \tag{3.30}$$

or, equivalently:

$$\varphi = \frac{(\beta w)^2 - (\alpha w)^2}{4(n-m)} \frac{\operatorname{Im} q}{|q|}. \tag{3.31}$$

Thus, it is seen that the complex factor $q(z)$ in the argument of the Bessel functions in Equation (3.17) affects the rotation angle.

At $N = 2$, the paraxial beam in Equation (3.17) carries the power-normalized OAM, J_z, which is derived from [22,52]

$$J_z / W = \left(\int_0^\infty \int_0^{2\pi} |E(r,\varphi,z)|^2 \, rdrd\varphi \right)^{-1} \text{Im} \int_0^\infty \int_0^{2\pi} \bar{E}(r,\varphi,z) \frac{\partial E(r,\varphi,z)}{\partial \varphi} rdrd\varphi, \quad (3.32)$$

where $\bar{E}$ is the conjugate function. For a superposition of optical vortices, $E(r, \varphi , z) = \Sigma A_m(r, z)e^{im\varphi}$, Equation (3.32) leads to $J_z/W = \Sigma m W_m / \Sigma W_m$ with $W_m = \int \int |A_m(r, z)|^2 rdr$ being the total power of the vortex with TC of m. The power of each constituent BG beam in the superposition (3.17) can be obtained by using a reference integral (expression 2.12.39.3 in [157]). Thus, the normalized OAM of a superposition of two BG beams equals:

$$J_z / W = \frac{nE_n^2 e^{-\xi} I_n(\xi) + mE_m^2 e^{-\eta} I_m(\eta)}{E_n^2 e^{-\xi} I_n(\xi) + E_m^2 e^{-\eta} I_m(\eta)}, \quad (3.33)$$

where $\xi = (\alpha w / 2)^2, \eta = (\beta w / 2)^2$, $I_n(x)$ is a modified Bessel function. At $m < n$, from Equation (3.33) we find that the OAM is $m \leq J_z/W \leq n$ depending both on the constituent beam amplitudes $E_{n,m}$, and scaling factors α and β .

3.2.2 NUMERICAL SIMULATION

The complex amplitude of the superposition of two BG beams was calculated using Equation (3.17) at $N = 2$ and TC was calculated using Equation (3.20). The calculation was conducted for a wavelength of $\lambda = 532$ nm, Gaussian beam waist radius $w = 10\lambda$, TC of the first beam $n = 5$, TC of the second beam $m = 2$, the same scaling factors of the beams $\alpha = \beta = 1/\lambda \approx 1.88$ μm^{-1}, and propagation distance $z = z_0 \approx 167$ μm. The intensity and phase were computed in the regions $-R \leq x, y \leq R, R = 75$ μm, and the number of pixels was $N = 2048 \times 2048$. Certainly, the TC cannot be numerically computed strictly by Equation (3.20) since the circle of integration is of infinite radius. Instead, some large, but finite, radius should be used. Here we calculated the TC along a circle of radius $R_1 = 75$ μm dotted circle in Figure 3.12(b)), with the number of pixels along the circle being $N_1 = 10^7$. Such radius exceeds the waist radius of the Gaussian beam nearly 14 times and therefore the circle of integration encompasses the whole area of the notable intensity of the beam.

Figure 3.12 depicts (a) the intensity and (b) the phase of the sum of two BG beams with different amplitudes: $E_5 = 1.2, E_2 = 1.0$. Phase distributions in Figure 3.12(b) and in all other figures below are shown without the spherical component of the wavefront $\exp[ir^2 \text{Im } q/(w|q|)^2] = \exp[ikr^2/(2R)]$ with $R = z[1 + (z_0/z)^2]$ since this spherical component does not affect the positions of vortices.

Figure 3.13 depicts, for comparison, phase distributions, obtained in three different ways at the same parameters as in Figure 3.12, but at a smaller propagation distance $z = 30$ μm. The phase distribution in Figure 3.13(a) is obtained by

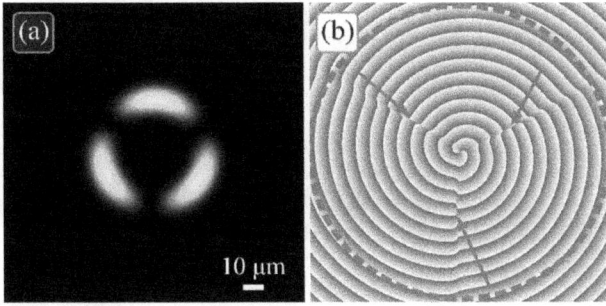

FIGURE 3.12 (a) Intensity and (b) phase of two BG beams with $E_5 = 1.2$, $E_2 = 1.0$ ($\alpha = \beta = 1/\lambda$). Arrows show the peripheral vortices

the integral Fresnel transform [158], which is a general solution of Equation (3.18). Figure 3.13(b) shows the same phase distribution but obtained by numerical solving the differential equation (Equation (3.18)) by the finite-difference beam propagation method (BPM), implemented in a commercial software BeamPROP (RSoft design, www.synopsys.com). Figure 3.13(c) shows the phase distribution computed directly by Equation (3.17).

According to Figure 3.13, all three phase distributions are almost coinciding and each of them contains four phase singularities marked by the circles. The summary *TC* of these singularities equals 5, as in Figure 3.12. The difference in phase distributions in Figure 3.13 and Figure 3.12(b) is that all three peripheral singularities in Figure 3.12(b) are located on a circle with a radius of nearly 30 μm, while in Figure 3.13 they are located on a smaller circle with a radius of 6 μm. The numerically calculated *TC* is 5.0. The maximum intensity on the ring of *TC* calculation is very low: 3.9e-039. The optical vortices with local *TC* +1 (under local *TC* we mean the number of 2π -jumps of the phase along a small contour around the vortex) are centered on rays defined by the angles of Equation (3.29). With decreasing amplitude of the beam with larger *TC*, three OVs indicated in Figure 3.12(b) with arrows start to move away from the center to the periphery (Figure 3.14). Figure 3.14 depicts (a) the intensity and (b) the phase of the sum of two BG beams with different amplitudes: $E_5 = 1.1$, $E_2 = 1.0$. The numerically calculated *TC* equals 5.0.

In Figure 3.14(b), three OVs with *TC* +1 are seen to be located farther from the center than in Figure 3.12. This behavior can be explained qualitatively by the properties of the Bessel functions of small arguments. If $\alpha = \beta$, we get the following approximate expression for the complex amplitude from Equation (3.17) at $N = 2$:

$$E(r,\varphi,z) \approx \frac{1}{q(z)} \exp\left[ikz - \frac{i\alpha^2 z}{2kq(z)} - \frac{r^2}{w^2 q(z)} \right]$$

$$\times \left[\frac{E_m}{m!}\left(\frac{\alpha \acute{r}}{2q(z)} \right)^m e^{im\varphi} + \frac{E_n}{n!}\left(\frac{\alpha r}{2q(z)} \right)^n e^{in\varphi} \right]. \quad (3.34)$$

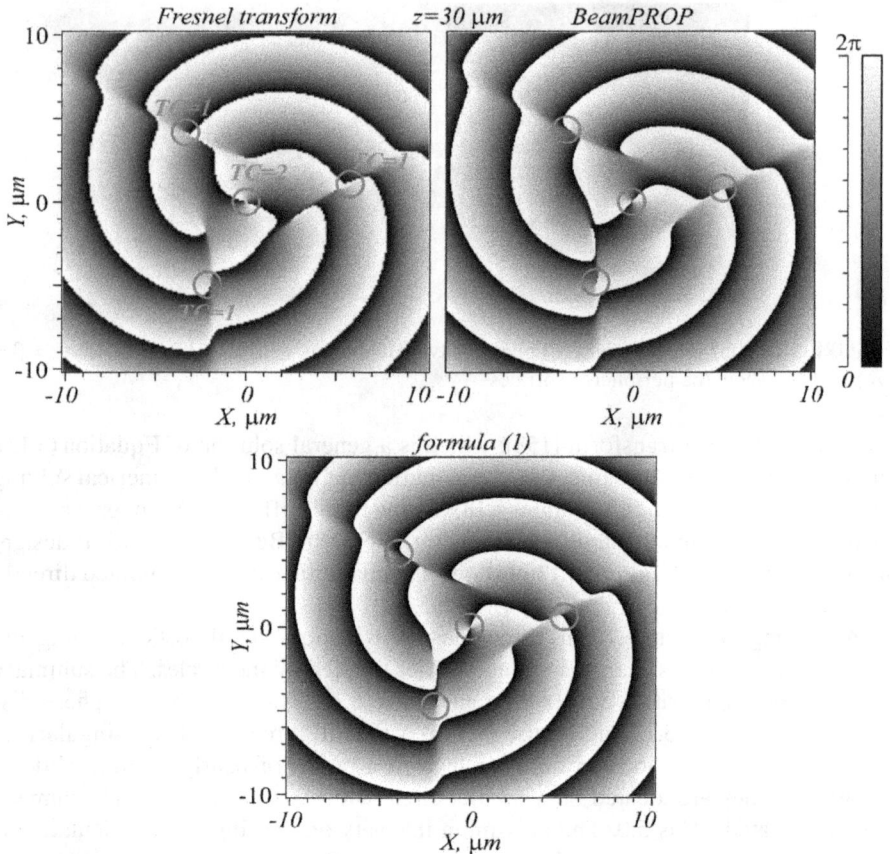

FIGURE 3.13 Phase distributions of the light beam from Figure 3.12 was computed at a lower propagation distance z by three different methods: Fresnel transform (a), beam propagation method (b), and direct computation by Equation (3.17).

Thus, the distance from the off-axis intensity nulls to the center are defined by the following equation (we suppose $n > m$): $[\alpha r/(2|q|)]^{n-m} = (n!/m!)(E_m/E_n)$. It is seen that the radius r grows with decreasing E_n. The maximum intensity on a circle drawn near the peripheral vortices (an dotted circle R_2 in Figure 3.14(b)), is also very low: 1.7e-019. Such an infinitesimally small intensity prevents the peripheral OVs in Figure 3.14(b) from being experimentally detected. According to Figure 3.14(b), changing E_5 affects the radius R_1 of the TC computation, needed to obtain the value $TC = 5$. This radius should exceed the radius R_2 of the circle with the three OVs with TC +1. Figure 3.14(c) depicts the radius R_2 against the amplitude ratio of the constituent BG beams, from which the radius is seen to increase as the amplitudes get closer to each other. If the amplitudes of the constituent BG beams are exactly the same, $E_5 = E_2 = 1.0$, the peripheral OVs are at infinity. Figure 3.15 is similar to Figure 3.14, except for $E_5 = E_2 = 1.0$. With no peripheral intensity nulls seen in Figure 3.15, just a central OV

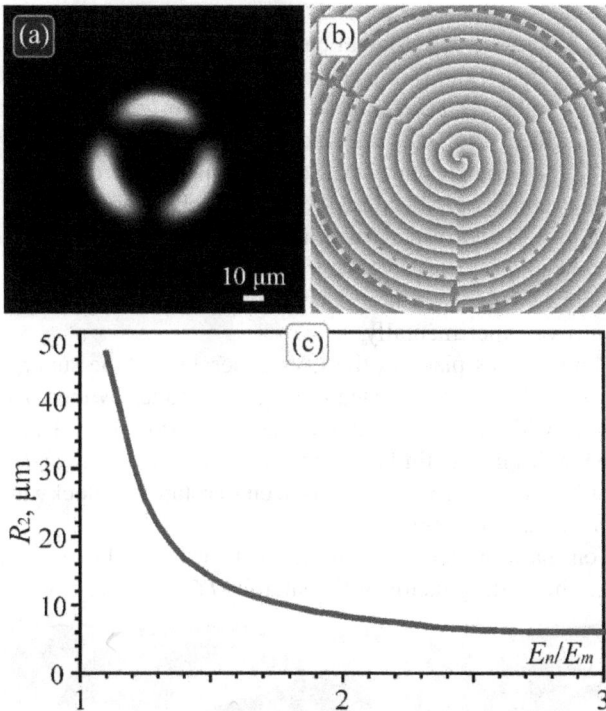

FIGURE 3.14 (a) Intensity and (b) phase of the superposition of two BG beams with $E_5 =$ 1.1, $E_2 = 1.0$ ($\alpha = \beta = 1/\lambda$). (c) Radius R_2 of the circle where TC jumps from 2 to 5 against the amplitude ratio.

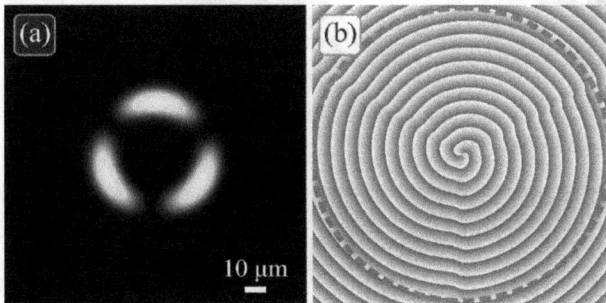

FIGURE 3.15 (a) Intensity and (b) phase of the superposition of two BG beams with $E_5 = E_2 = 1.0$.

with TC +2 is observed. The numerically calculated TC is 2.0. From a comparison of Figures 3.12 and 3.15, the intensity is seen to be almost the same although the BG beams have different TCs.

The case when the optical vortex moves to infinity has an analogy with the case when the vortex moves from the center to the border of an aperture of the radius

R. A light field beyond the aperture is $E(r, \varphi) = \text{circ}(r/R) (re^{i\varphi} - r_0)^n$ with r_0 being the vortex shift, n being its topological charge, and $\text{circ}()$ being the bounding function ($\text{circ}(\xi) = 1$ if $\xi \leq 1$ and 0 otherwise). When an nth-order vortex is inside the aperture (Figure 3.16(a)), its TC equals n. When the vortex moves to the edge of the aperture (Figure 3.16(b)), its TC is $n/2$ since the phase along a small contour around the vortex grows by πn, rather than $2\pi n$. So, when the aperture radius grows to infinity and the vortex remains on its edge, the TC of such a vortex should be two times lower than without the aperture. This confirms the idea about the half-integer TC given by Equations (3.25) and (3.28). Therefore, the TC of the beam shown in Figure 3.15 should be 3.5 ($2 + 3/2$), but it is derived analytically and cannot be confirmed either numerically or experimentally.

Similar evolution takes place in the OVs generated by the superposition of two BG if varying the scaling factors instead of the amplitude. By increasing the scaling factor of the beam with larger TC, e.g. putting $\alpha = 1.20/\lambda$, $\beta = 1/\lambda$, three peripheral OVs with TC +1 will once again be observed in the phase pattern (Figure 3.17(b)). From Figure 3.17(a), the intensity pattern is seen to rotate anticlockwise with a velocity derived from Equation (3.29).

The evolution pattern becomes different from the above-described pattern when increasing the scaling factor of the smaller-TC beam, e.g. by putting $\alpha = 1/\lambda$,

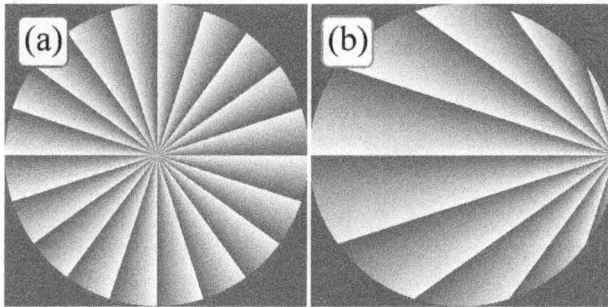

FIGURE 3.16 Phase of an optical vortex inside (a) and on the border (b) of an aperture (white color – 0, dark color – 2π).

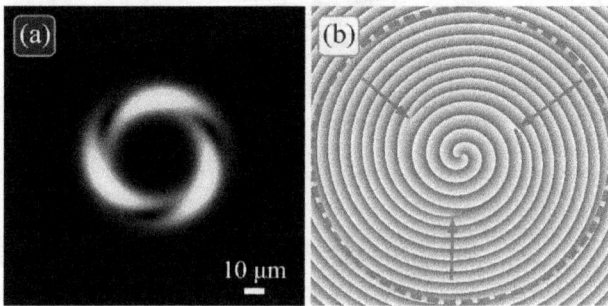

FIGURE 3.17 (a) Intensity and (b) phase of the superposition of two BG beams with $E_5 = E_2 = 1.0$ and $\alpha = 1.20/\lambda$, $\beta = 1/\lambda$.

$\beta = 1.20/\lambda$. In this case, it can be seen from Figure 3.18(b) that in addition to an OV with TC +2, three extra OVs with TC +1 will occur at the center and three OVs with TC −1 on the periphery. The resulting TC of the superposition will be equal to 2. From Figure 3.18(a), the intensity pattern is also seen to rotate clockwise in accordance with Equation (3.29).

Figure 3.19 demonstrates the anticlockwise rotation of the intensity pattern of the beam from Figure 3.18 in the near field ($z < z_0$).

Figure 3.20 demonstrates the propagation dynamics of two BG beams with more distinct scaling factors and at larger propagation distances, including the Fresnel diffraction zone and the far field.

At these distances, rotation is not so clearly seen, and in the far field the beam becomes rotationally symmetric and consists of two intensity rings corresponding to the two BG beams. The inner ring is for the 2nd-order BG beam, while the outer ring is for the fifth-order BG beam (since $\alpha > \beta$). That's why the TC of the whole superposition is 5.

To sum up, using the superposition of two BG beams (Equation (3.17)), we have shown that if both beams have equal scaling factors ($\alpha = \beta$) and equal amplitudes ($E_1 = E_2$), then the TC of the superposition (Equation (3.20)) is half-integer (Equation (3.25)) and it is conserved on beam propagation. However, some of the optical

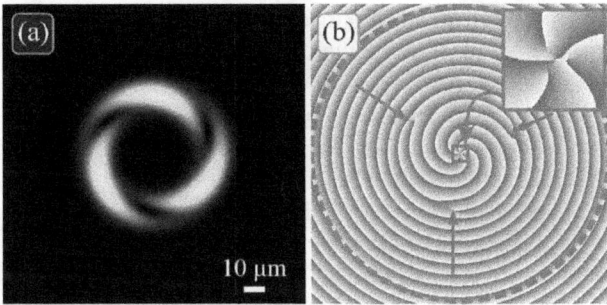

FIGURE 3.18   (a) Intensity and (b) phase of the superposition of two BG beams with $E_5 = E_2$ = 1.0 and $\alpha = 1/\lambda$, $\beta = 1.20/\lambda$.

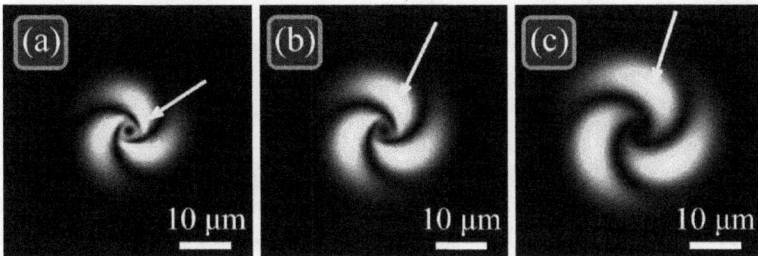

FIGURE 3.19   Intensity distributions of a superposition of two BG beams with $E_5 = E_2 = 1.0$ and $\alpha = 1/\lambda$, $\beta = 1.2/\lambda$ in three different transverse planes: $z = z_0/5$ (a), $z = z_0/4$ (b), $z = z_0/3$ (c). Arrows show the same intensity maximum.

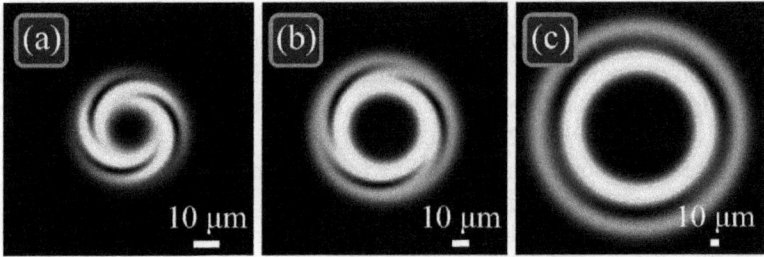

FIGURE 3.20 Intensity distributions of a superposition of two BG beams with $E_5 = E_2 = 1.0$ and $\alpha = 1.5/\lambda$, $\beta = 1/\lambda$ in three different planes: $z = z_0/2$ (near field) (a), $z = z_0$ (Fresnel diffraction) (b), $z = 3z_0$ (far field) (c).

vortices, the number of which equals the difference of TCs of the constituent beams, are located on the infinite-radius circle (Equation (3.27)) (Figure 3.14(c)). These optical vortices contribute to the fractional part of the TC [159].

3.3 TOPOLOGICAL CHARGE OF TWO PARALLEL LAGUERRE–GAUSSIAN BEAMS

Although they have been around in optics for over 40 years, optical vortices (OV) remain at the forefront of research as there are a number of aspects of their behavior that need to be elucidated. An important characteristic of an OV is the topological charge (TC), which can be calculated using two alternative approaches. The total TC of a vortex beam is derived as the sum of the TCs of all individual embedded OVs. Meanwhile, the TC of an individual OV (a screw dislocation) equals the integer number of 2π phase jumps when making a full circle around the center of the screw dislocation (phase singularity) [21]. This approach is rather cumbersome as it requires finding all centers of screw dislocations, or isolated intensity nulls, in the beam of interest. However, there is also a different, more fruitful approach to calculating the TC of the total vortex beam containing multiple screw dislocations [19], which is based on finding an integer number of 2π phase jumps when making a full circle of infinite radius, thus enabling all screw dislocations embedded into the beam to be accounted for. In this section, we utilize the latter approach.

When dealing with TC, one of the challenges stems from the fact that there is no proof of TC conservation for free-space vortex beam propagation. At the same time, in some publications, TC was shown not to be conserved [22,160,55,161,85]. Another challenge is associated with finding the TC of a superposition of several OVs. For instance, it has been well-known that a Laguerre–Gaussian beam whose amplitude is defined by a factor $\exp(in\varphi)$, where φ is the polar angle in the beam cross-section, has $TC = n$. Would it be possible to find the TC of an on-axis superposition of several Laguerre–Gaussian (LG) beams with different individual TCs? In such a superposition, TC was found to equal the largest constituent TC [55]. However, for an off-axis superposition of multiple LG beams, the total TC has not been derived yet. Another problem that remains to be addressed is as follows. For a Gaussian beam in which two identical screw dislocations with $TC = n$ are embedded at different points, the total TC of the beam is known to be $TC = 2n$. The question is: what will be the resulting TC of an off-axis superposition of two identical LG beams in which screw dislocations with the same $TC = n$ are embedded? Below, we show that in this case the total beam has $TC = n$.

In this section, we theoretically and numerically derive the TC of the sum of two off-set axisymmetric LG beams with the indices $(0,n)$ and $(0,m)$. We show that the superposition of two identical off-axis LG beams $(n = m)$ has $TC = n$. We also demonstrate that the superposition of two off-axis LG beams tilted to the axis so as to produce a structurally stable resulting beam has infinite TC, because the combination of such beams produces on a certain line an infinite number of screw dislocations with $TC = +1$. It is worth noting that the vortex beams with infinite TC were recently studied [162,163].

3.3.1   STRUCTURALLY STABLE SUPERPOSITION OF OFF-AXIS LG BEAMS

According to the theory of spiral beams [49,164], any function is given by:

$$E(x,y) = E_0 \exp\left(-\frac{x^2 + y^2}{w^2}\right) f(x \pm iy), \tag{3.35}$$

where E_0 is constant and $f(x)$ is any integer analytical function, describes a structurally stable beam with the Gaussian envelope. The beam has finite energy and retains the cross-section intensity pattern upon propagation while changing in scale and rotating. For the beam to retain its structure upon free-space propagation, the sign by the argument in Equation (3.35) needs always to remain either negative or positive. This may be illustrated by an example of the superposition of off-axis single-annulus LG beams, which is structurally stable:

$$E_1(x,y) = \sum_{n=0}^{N} E_n \exp\left(-\frac{(x - x_n)^2 + (y - y_n)^2}{w^2} + \frac{2i(x_n y - y_n x)}{w^2}\right)$$
$$\times (x - x_n + iy - iy_n)^{m_n}, \tag{3.36}$$

where (x_n, y_n) are the coordinates of the shifted beam center in the superposition, E_n are the weight coefficients (constant complex numbers), m_n is the TC of the n-th beam in the superposition. To make sure that the superposition of the off-axis LG beams in Equation (3.36) produces a structurally stable beam let us factor out the common Gaussian exponent from Equation (3.36):

$$E_1(x,y) = \exp\left(-\frac{x^2 + y^2}{w^2}\right)$$
$$\times \sum_{n=0}^{N} E_n \exp\left(-\frac{x_n^2 + y_n^2}{w^2} + \frac{2(x + iy)(x_n - iy_n)}{w^2}\right)(x + iy - x_n - iy_n)^{m_n}. \tag{3.37}$$

Equation (3.37) is seen to be similar to Equation (3.35) in form, containing a Gaussian exponent multiplied by a function of the argument $x + iy$. The function in Equation (3.37) can be rewritten in a more compact form:

$$E_1(x,y) = \exp\left(-\frac{x^2 + y^2}{w^2}\right) \sum_{n=0}^{N} \bar{E}_n \exp\left(\frac{2(x+iy)(x_n - iy_n)}{w^2}\right)(x + iy - x_n - iy_n)^{m_n} . \quad (3.38)$$

For simplicity, the superposition (3.38), whose TC we seek to determine is assumed to be composed of just two structurally stable beams, which are off-set from the optical axis along the horizontal axis by x_0. Then, the complex amplitude in the source plane ($z = 0$) is given by:

$$E_1(x,y) = A \exp\left(-\frac{(x-x_0)^2 + y^2}{w^2} + \frac{2ix_0 y}{w^2}\right)(x - x_0 + iy)^m$$

$$+ B \exp\left(-\frac{(x+x_0)^2 + y^2}{w^2} - \frac{2ix_0 y}{w^2}\right)(x + x_0 + iy)^n , \quad (3.39)$$

where A and B are constant, n and m are the TCs of each constituent beam in the superposition (positive integers), w is the Gaussian beam waist radius, and (x, y) are the transverse Cartesian coordinates. For the superposition in Equation (3.39), the TC can be found using Berry's formula [19]:

$$TC = \frac{1}{2\pi} \lim_{r \to \infty} \text{Im} \int_0^{2\pi} d\varphi \, \frac{\partial E(r,\varphi)/\partial \varphi}{E(r,\varphi)}, \quad (3.40)$$

where (r, φ) are the polar coordinates in the transverse plane and Im is the imaginary part of the number. Substituting Equation (3.39) into Equation (3.40) yields:

$$TC = \frac{1}{2\pi} \lim_{r \to \infty} \text{Im} \int_0^{2\pi} d\varphi \left\{ A \left[imre^{i\varphi}(re^{i\varphi} - x_0)^{m-1} + \frac{2ix_0 re^{i\varphi}(re^{i\varphi} - x_0)^m}{w^2} \right] \exp\left(\frac{2x_0 re^{i\varphi}}{w^2}\right) \right.$$

$$\left. + B \left[inre^{i\varphi}(re^{i\varphi} + x_0)^{n-1} - \frac{2ix_0 re^{i\varphi}(re^{i\varphi} + x_0)^n}{w^2} \right] \exp\left(-\frac{2x_0 re^{i\varphi}}{w^2}\right) \right\}$$

$$\times \left\{ A(re^{i\varphi} - x_0)^m \exp\left(\frac{2x_0 re^{i\varphi}}{w^2}\right) + B(re^{i\varphi} + x_0)^n \exp\left(-\frac{2x_0 re^{i\varphi}}{w^2}\right) \right\}^{-1} .$$

$$(3.41)$$

The integral (3.41) can be split into two integrals, $\int_0^{2\pi} = \int_{-\pi/2}^{\pi/2} + \int_{\pi/2}^{3\pi/2}$. In the first integral, appearing in Equation (3.41), $\cos \varphi > 0$. Hence, at $r \to \infty$, the terms in (3.41) containing a negative exponential factor will tend to zero and the first integral in Equation (3.41) will take the form:

$$\frac{1}{2\pi}\lim_{r\to\infty}\mathrm{Im}\int_{-\pi/2}^{\pi/2}d\varphi\left\{A\left[imre^{i\varphi}(re^{i\varphi}-x_0)^{m-1}+\frac{2ix_0re^{i\varphi}(re^{i\varphi}-x_0)^m}{w^2}\right]\exp\left(\frac{2x_0re^{i\varphi}}{w^2}\right)\right\}$$

$$\times\left\{A(re^{i\varphi}-x_0)^m\exp\left(\frac{2x_0re^{i\varphi}}{w^2}\right)\right\}^{-1}$$

$$=\frac{1}{2\pi}\mathrm{Im}\int_{-\pi/2}^{\pi/2}d\varphi\left\{iA\left[\left(m+\frac{2x_0re^{i\varphi}}{w^2}\right)r^me^{im\varphi}\right]\exp\left(\frac{2x_0re^{i\varphi}}{w^2}\right)\right\}\left\{A(re^{i\varphi})^m\right.$$

$$\left.\exp\left(\frac{2x_0re^{i\varphi}}{w^2}\right)\right\}^{-1}=\frac{m}{2}+\frac{2x_0}{\pi w^2}\lim_{r\to\infty}r.$$

$$(3.42)$$

The second integral in Equation (3.41) contains the negative-valued $\cos\varphi$ function. Hence, at $r\to\infty$ the terms in Equation (3.41) containing a positive exponential term will tend to zero. Then, in Equation (3.41) the second integral takes the form:

$$\frac{1}{2\pi}\lim_{r\to\infty}\mathrm{Im}\int_{\pi/2}^{3\pi/2}d\varphi\left\{B\left[inre^{i\varphi}(re^{i\varphi}+x_0)^{n-1}-\frac{2ix_0re^{i\varphi}(re^{i\varphi}+x_0)^n}{w^2}\right]\exp\left(-\frac{2x_0re^{i\varphi}}{w^2}\right)\right\}$$

$$\times\left\{B(re^{i\varphi}+x_0)^n\exp\left(-\frac{2x_0re^{i\varphi}}{w^2}\right)\right\}^{-1}$$

$$=\frac{1}{2\pi}\mathrm{Im}\int_{\pi/2}^{3\pi/2}d\varphi\left\{iB\left[(n-\frac{2x_0re^{i\varphi}}{w^2})r^ne^{in\varphi}\right]\exp\left(-\frac{2x_0re^{i\varphi}}{w^2}\right)\right\}$$

$$\left\{B(re^{i\varphi})^n\exp\left(-\frac{2x_0re^{i\varphi}}{w^2}\right)\right\}^{-1}=\frac{n}{2}+\frac{2x_0}{\pi w^2}\lim_{r\to\infty}r.$$

$$(3.43)$$

Summing up the values of the two integrals in Equations (3.42) and (3.43), we find that TC in Equation (3.41) is infinite:

$$TC=\frac{n+m}{2}+\frac{4x_0}{\pi w^2}\lim_{r\to\infty}r\to\infty.$$

$$(3.44)$$

The conclusion, that the *TC* of the superposition (3.39) is infinite, needs to be theoretically substantiated. Assume both terms in the sum in Equation (3.39) to be equal to each other in the absolute value for it is at points (x, y) of the same-value terms that phase singularities are evolving. Thus, we obtain:

$$\left|\frac{A}{B}\right| \exp\left(\frac{4xx_0}{w^2}\right) = \frac{\left[(x+x_0)^2 + y^2\right]^{n/2}}{\left[(x-x_0)^2 + y^2\right]^{m/2}}. \tag{3.45}$$

The analysis of Equation (3.45) in the general form is challenging. For simplicity, we put $m = n$ and $|A| = |B|$. Then, Equation (3.39) for the amplitude on the vertical axis ($x = 0$) reduces to:

$$E_1(0, y) = 2A\left(x_0^2 + y^2\right)^{\frac{n}{2}} \exp\left(-\frac{x_0^2 + y^2}{w^2}\right) \cos\left(\frac{2x_0 y}{w^2} - n \arctan\left(\frac{y}{x_0}\right)\right). \tag{3.46}$$

From Equation (3.46), amplitude zeros, or phase singularity centers (screw dislocations), each having $TC = +1$, are seen to be arranged on the vertical axis (n is even):

$$\frac{2x_0 y}{w^2} - n \arctan\left(\frac{y}{x_0}\right) = \frac{\pi}{2} + \pi p, \quad p = 0, 1, 2, \dots \tag{3.47}$$

Thus, given that $m = n$ and $|A| = |B|$, the infinite *TC* of superposition (3.39) is due to an infinite number of screw dislocations centered on the vertical axis at points of Equation (3.47). In the general case, from Equation (3.45) the screw dislocations are seen to lie on a curve, which will deflect toward either positive x (at $m > n$) or negative x ($n > m$) with y increasing in the absolute value. On the vertical axis, there will be just two singularity points, which follow from Equation (3.45) at $x = 0$.

Being structurally stable, the transverse intensity pattern of superposition in Equation (3.39) will rotate upon propagation as will the straight line where an infinite number of singularities are located. Hence, the conclusion that superposition in Equation (3.39) will have infinite *TC* at any z.

3.3.2 SUPERPOSITION OF TWO ON-AXIS LG BEAMS

Putting in (3.39) $x_0 = 0$, we derive a complex amplitude of two on-axis one-ring LG beams:

$$E(x, y) = \exp\left(-\frac{x^2 + y^2}{w^2}\right) \left[A\left(\frac{x+iy}{w}\right)^n + B\left(\frac{x+iy}{w}\right)^m\right]. \tag{3.48}$$

It can be shown [55] that superposition (3.48) has the $TC = \max\{m, n\}$. There is an on-axis OV with the least TC (say, m), with the remaining equidistant $(n - m)$ OVs lying on the radius:

$$r = w\,|B/A|^{1/(n-m)}. \tag{3.49}$$

If the values m and n are of opposite signs and close to each other (by absolute values), the corresponding interference patterns can be found in [165].

Hence, the total TC of the beam in Equation (3.48) is $TC = m + (n - m) = n$.

3.3.3 SUPERPOSITION OF TWO OFF-AXIS LG BEAMS

When combining two off-axis one-ring LG beams, which are structurally stable, their superposition is no more structurally stable. Its complex amplitude is:

$$E_2(x, y) = A\exp\left(-\frac{(x - x_0)^2 + y^2}{w^2}\right)\left(\frac{x - x_0 + iy}{w}\right)^m$$

$$+ B\exp\left(-\frac{(x + x_0)^2 + y^2}{w^2}\right)\left(\frac{x + x_0 + iy}{w}\right)^n. \tag{3.50}$$

Substituting Equation (3.50) into Equation (3.39) yields:

$$TC = \frac{1}{2\pi}\lim_{r\to\infty}\text{Im}\int_0^{2\pi} d\varphi\left\{A\left[imre^{i\varphi}(re^{i\varphi} - x_0)^{m-1} - \frac{2x_0 r\sin\varphi(re^{i\varphi} - x_0)^m}{w^2}\right]\right.$$

$$\exp\left(\frac{2x_0 r\cos\varphi}{w^2}\right) + B\left[inre^{i\varphi}(re^{i\varphi} + x_0)^{n-1} + \frac{2x_0 r\sin\varphi(re^{i\varphi} + x_0)^n}{w^2}\right]$$

$$\exp\left(-\frac{2x_0 r\cos\varphi}{w^2}\right)\right\} \times \left\{A(re^{i\varphi} - x_0)^m\exp\left(\frac{2x_0 r\cos\varphi}{w^2}\right)\right.$$

$$\left. + B(re^{i\varphi} + x_0)^n\exp\left(-\frac{2x_0 re^{i\varphi}}{w^2}\right)\right\}^{-1}. \tag{3.51}$$

As with the integral in Equation (3.41), the integral in Equation (3.51) can be split into two integrals, $\int_0^{2\pi} = \int_{-\pi/2}^{\pi/2} + \int_{\pi/2}^{3\pi/2}$. In the first integral, $\cos\varphi > 0$. Hence, at $r \to \infty$, the terms containing negative-power exponential factors tend to zero and the first integral is reduced to a finite value:

$$\frac{1}{2\pi} \lim_{r\to\infty} \operatorname{Im} \int_{-\pi/2}^{\pi/2} d\varphi \left\{ A \left[imre^{i\varphi}(re^{i\varphi} - x_0)^{m-1} - \frac{2x_0 r \sin\varphi (re^{i\varphi} - x_0)^m}{w^2} \right] \exp\left(\frac{2x_0 r \cos\varphi}{w^2} \right) \right\}$$

$$\times \left\{ A(re^{i\varphi} - x_0)^m \exp\left(\frac{2x_0 r \cos\varphi}{w^2} \right) \right\}^{-1}$$

$$= \frac{1}{2\pi} \operatorname{Im} \int_{-\pi/2}^{\pi/2} d\varphi \left\{ A \left[(im - \frac{2x_0 r \sin\varphi}{w^2}) r^m e^{mi\varphi} \right] \exp\left(\frac{2x_0 r \cos\varphi}{w^2} \right) \right\}$$

$$\times \left\{ A(re^{i\varphi})^m \exp\left(\frac{2x_0 r \cos\varphi}{w^2} \right) \right\}^{-1} = \frac{m}{2}.$$

$$(3.52)$$

In a similar way, for the second integral in Equation (3.51) we find:

$$\frac{1}{2\pi} \lim_{r\to\infty} \operatorname{Im} \int_{\pi/2}^{3\pi/2} \left\{ B \left[inre^{i\varphi}(re^{i\varphi} + x_0)^{n-1} + \frac{2x_0 r \sin\varphi (re^{i\varphi} + x_0)^n}{w^2} \right] \exp\left(-\frac{2x_0 r \cos\varphi}{w^2} \right) \right\}$$

$$\times \left\{ B(re^{i\varphi} + x_0)^n \exp\left(-\frac{2x_0 r \cos\varphi}{w^2} \right) \right\}^{-1} d\varphi$$

$$= \frac{1}{2\pi} \operatorname{Im} \int_{\pi/2}^{3\pi/2} d\varphi \left\{ B \left[(in + \frac{2x_0 r \sin\varphi}{w^2}) r^n e^{in\varphi} \right] \right.$$

$$\times \exp\left(-\frac{2x_0 r \cos\varphi}{w^2} \right) \right\} \left\{ B(re^{i\varphi})^n \exp\left(-\frac{2x_0 r \cos\varphi}{w^2} \right) \right\}^{-1} = \frac{n}{2}.$$

$$(3.53)$$

From summing up the results arrived at in Equations (3.52) and (3.53), we find that superposition in Equation (3.50) has the *TC* of:

$$TC = \frac{n+m}{2}. \qquad (3.54)$$

From Equation (3.54) follows a remarkable property of the sum of two identical off-axis one-ring LG: at $m = n$, the total *TC* of the superposition equals that of a single constituent LG beam, $TC = n$. Hence, the conclusion is that the sum of two identical screw dislocations separated in space has the same *TC* as the individual dislocation. We may hypothesize that this property can be extended to the case of several identical off-axis space-separated LG beams. In support of this hypothesis, let us remind that a superposition of several identical off-axis LG beams carries orbital angular momentum (OAM) equal to that of an individual constituent beam, which equals topological charge $TC = n$ [166].

If the off-axis shift is large enough ($x_0 \gg w$), the interference is near-negligible and the resulting TC may seem to equal the sum of all constituent beams, i.e. $TC = n + m$. However, counter-intuitively, this is not the case because considering that the superposition of a Gaussian OV and a plane wave has the amplitude:

$$E_2(r,\varphi) = A + r^n \exp\left(-r^2/w^2 + in\varphi\right), \tag{3.55}$$

where A is a constant of the on-axis plane wave, the total TC appears to be zero:

$$TC = \frac{1}{2\pi} \lim_{r \to \infty} \operatorname{Im} \int_0^{2\pi} d\varphi \left[inr^n \exp\left(-\frac{r^2}{w^2} + in\varphi\right) \right] \left[A + r^n \exp\left(-\frac{r^2}{w^2} + in\varphi\right) \right]^{-1} = 0. \tag{3.56}$$

Remarkably, TC in Equation (3.56) remains zero at any infinitesimally small value of constant A. It can be explained as follows: around the optical axis, on a circle of approximate radius:

$$r = |A|^{1/n}, \tag{3.57}$$

there will evolve n OVs with $TC = 1$, meanwhile on a larger circle whose radius meets the condition:

$$r^n \exp\left(-r^2/w^2\right) = |A|, \tag{3.58}$$

n more OVs with $TC = -1$ will be born. Thus, we infer that it is wrong to analyze two off-axis LG beams put aside at a large distance as being independent, summing up their respective TCs. At any value of the off-axis shift, the TC of the superposition should satisfy Equation (3.54).

3.3.4 NUMERICAL MODELING AND EXPERIMENT

Numerical modeling of structurally stable superposition of off-axis LG beams

As is theoretically predicted above, the beam from Equation (3.39), if $m = n$ and $|A| = |B|$, has an infinite number of screw dislocations centered on the vertical axis. Figure 3.21 confirms it. It illustrates the intensity and phase distributions of the beam in Equation (3.39) in the initial plane for three different values of the distance x_0 between the two constituent LG beams. The 4–6th columns of Figure 3.21 illustrate the same fields as in the 1–3rd columns, but in a more wide area, in order to demonstrate that the number of vortices on the y-axis grows.

It is seen in Figure 3.21 that, when the distance between the constituent beams decreases, two light rings (Figure 3.21(a, d)) start to interfere and generate a pattern

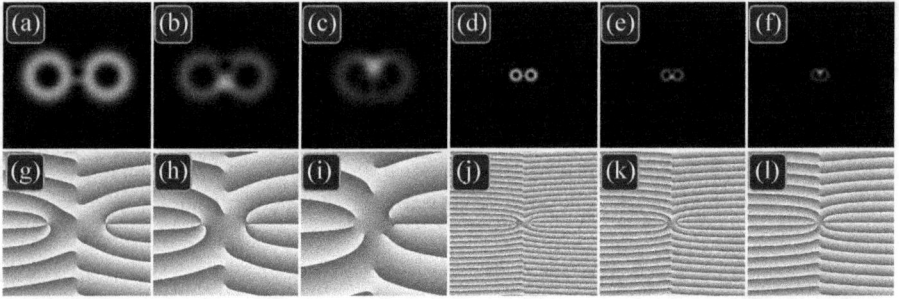

FIGURE 3.21 Intensities (a–f) and phases (g–l) in the initial plane of a structurally stable superposition in Equation (3.39) of two off-axis LG beams at different values of the off-axis shift. The following parameters were used: wavelength of light λ = 532 nm, waist radius w = 1 mm, TCs of the constituent LG beams m = n = 3, weight coefficients of the superposition A = 1, B = i, shifts of the LG beams from the optical axis x_0 = 2 mm (columns 1, 4), x_0 = 1.5 mm (columns 2, 5), x_0 = 1 mm (columns 3, 6), the depicted area is |x|, |y| ≤ R with R = 5 mm (a–c, g–i) and R = 20 mm (d–f, j–l).

with a shape of a "crown" (Figure 3.21(c, f)). As seen in the phase distributions, there are two off-axis vortices with the *TC* of 3 (Figure 3.21(g)), which split when x_0 decreases, and an infinite number of unit-charge vortices on the vertical axis.

Numerical modeling of a superposition of two off-axis LG beams

The numerical modeling aims to corroborate (or disprove) Equation (3.54). In the original plane, the complex amplitude of two off-axis one-ring LG beams is given by:

$$E(x,y,0) = A\left(\frac{e}{m}\right)^{m/2}\left[\frac{\sqrt{2}}{w}(x-x_0+iy)\right]^m \exp\left[-\frac{(x-x_0)^2+y^2}{w^2}\right]$$

$$+ B\left(\frac{e}{n}\right)^{n/2}\left[\frac{\sqrt{2}}{w}(x+x_0+iy)\right]^n \exp\left[-\frac{(x+x_0)^2+y^2}{w^2}\right].$$

(3.59)

where (x, y) are the Cartesian coordinates, w is the waist radius, x_0 is the off-axis shift of the beams (with the first beam shifted to the right and the second – to the left), m and n are the TCs of the beams, and A and B are the weight coefficients of the superposition. Constant factors by the weight coefficients are introduced to equalize the intensity rings in both of the beams.

Figure 3.22 shows the intensity and phase of the superposition in Equation (3.59) for λ = 532 nm, w_0 = 500 μm, m = 2, n = 5, A = B = 1, x_0/w = 0 (Figure 3.22(a,b)), x_0/w = 0.3 (Figure 3.22(c, d)), x_0/w = 0.6 (Figure 3.22(e, f)), x_0/w = 0.8 (Figure 3.22(g, h)). The computing domain is |x|, |y| ≤ R (R = 5 mm).

From Figure 3.22 it is seen that with no off-axis shift (x_0 = 0), superposition in Equation (3.59) has *TC* = 5, meaning that the superposition has the greater

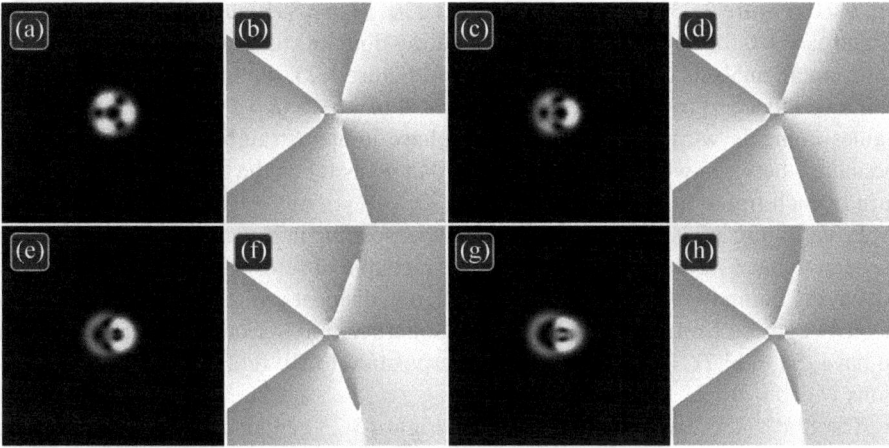

FIGURE 3.22 Intensities (a, c, e, g) and phases (b, d, f, h) in the source plane for superposition in Equation (3.59) at different values of the shift x_0/w_0: 0 (a, b), 0.3 (c, d), 0.6 (e, f), and 0.8 (g, h).

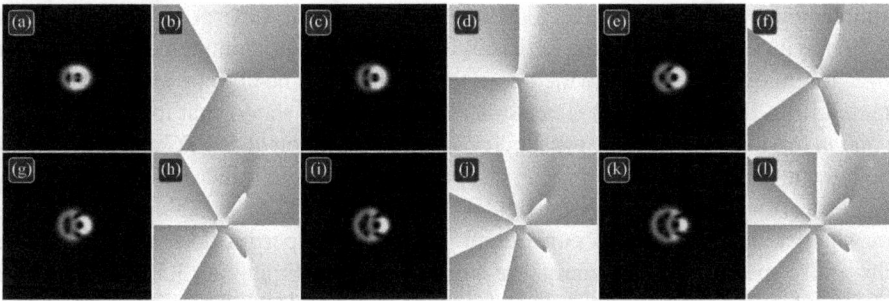

FIGURE 3.23 Intensity (a, c, e, g, i, k) and phase (b, d, f, h, j, l) patterns of superposition in Equation (3.59) for the following parameters: $\lambda = 532$ nm, w = 500 μm, m = 2, n = 3 (a, b), n = 4 (c, d), n = 5 (e, f), n = 6 (g, h), n = 7 (i, j), n = 8 (k, l), A = B = 1, $x_0/w = 0.5$ (a–l). Computing area $|x|, |y| \leq R$ (R = 5 mm).

TC of the two constituent beams ($m = 2$, $n = 5$). Following an off-axis shift, two screw dislocations out of the total five with $TC = +1$ get compensated by two screw dislocations with $TC = -1$ that instantaneously "arrive" from infinity (therefore, they are not seen yet in Figure 3.22(d)), with the *TC* of the superposition of the LG beams in Equation (3.59) becoming equal to $TC = 3$. This value presents an arithmetic mean of two TCs, 2 and 5, rounded off to the nearest smaller integer: $TC = (n + m - 1)/2 = (2 + 5 - 1)/2 = 3$. Remarkably, $TC = 3$ remains unchanged at any offset value x_0.

Below, we fix *m* while *n* is varied. Figure 3.23 shows intensity and phase patterns of superposition in Equation (3.59) for the following parameters: $\lambda = 532$ nm, w = 500 μm, $m = 2$, $n = 3$ (Figure 3.23(a, b)), $n = 4$ (Figure 3.23(c, d)), $n = 5$ (Figure 3.23(e, f)), $n = 6$ (Figure 3.23(g, h)), $n = 7$ (Figure 3.23(i, j)), $n = 8$ (Figure 3.23(k, l)), $A = B = 1$, and $x_0/w = 0.5$ (Figure 3.23(a–l)). Computing domain $|x|, |y| \leq R$ ($R = 5$ mm).

From an analysis of the phase patterns in Figure 3.23, the following values of TC can be deduced: 3 ($n = 3$), 4 ($n = 4$), 3 ($n = 5$), 4 ($n = 6$), 5 ($n = 7$), and 6 ($n = 8$). These values coincide with the number of 2π-phase jumps (straight lines that mark the boundary between white and black color) that go as far as the edge of the phase pattern in each frame in Figure 3.23.

From Figure 3.23, at $n = 5, 6, 7$, and 8, superposition in Equation (3.59) ($m = 2$) can be seen to have $TC = n - 2$, which is due to a couple of screw dislocations with $TC = +1$ having being compensated for by two screw dislocations with $TC = -1$, instantaneously coming from infinity. At $n = 3, 4$, superposition in Equation (3.59) turns out to have the total $TC = n$. In this case, superposition (3.59) retains all screw dislocations which it would have with no shift.

The findings graphically derived from Figure 3.23 can also be interpreted using Equation (3.54). Thus, we find that at $m = 2$ and $n = 3$, $TC = (n + m)/2 + 1/2 = 3$, at $m = 2$ and $n = 4 - TC = (n + m)/2 + 1 = 4$, at $m = 2$, $n = 5 - TC = (n + m)/2 - 1/2 = 3$, at $m = 2$ and $n = 6 - TC = (n + m)/2 = 4$; at $m = 2$ and $n = 7 - TC = (n + m)/2 + 1/2 = 5$, and at $m = 2$ and $n = 8 - TC = (n + m)/2 + 1 = 6$. Thus, we conclude that if the sum $m + n$ is even, then the superposition has a TC equal to the half-sum as in Equation (3.54) complemented by 0 or 1, with the odd sum $m + n$ leading to superposition in Equation (3.59) having a TC equal to the half-sum in Equation (3.54) complemented by 1/2 or −1/2. It remains to reveal in which case the above terms need to be added to or subtracted from Equation (3.54). The numerical simulation shows that for any modeled n and m (up to $m < 3$ and $n < 10$) superposition in Equation (3.59) TC can take one of the four integer numbers, defined by the relations shown in Table 3.1.

Hence, the choice of the particular relation for TC from Table 3.1 depends on whether the sum of two constituent TCs in the superposition is even or odd. Physically speaking, when analyzing superposition in Equation (3.50), the need to add or subtract 1/2 (for odd $n + m$) to/from the value of TC in Equation (3.54) stems from the fact that the light field cannot have fractional TC. The topological charge of the light field is supposed to be an integer, except for when we assume a fractional TC

TABLE 3.1

Relations that define TC of the superposition of two off-axis LG beams, as derived from the numerical simulation (Figures 3.22 and 3.23)

TC of the superposition

$TC_1 = (m + n)/2$

$TC_2 = (m + n + 2)/2$

$TC_3 = (m + n + 1)/2$

$TC_4 = (m + n - 1)/2$

---

in the original plane as a boundary condition. However, as soon as the fractional OV proceeds to propagate in space, TC becomes an integer. In our case, while having no boundary condition, we have a sum of two off-axis LG beams in the waist plane. In other words, with the beams of Equation (3.59) already propagating in the original plane ($z=0$), their TC should be an integer.

To deduce a rule for choosing a particular relation defining TC from Table 3.1, let us analyze Equation (3.45) in more detail, for its fulfillment may lead to the appearance of extra singularities (for simplicity, let $A = B$). At large positive x, Equation (3.45) can be rearranged as:

$$\exp\left(4xx_0/w^2\right) = \left[\left(x^2 + y^2\right)/w^2\right]^{(n-m)/2}. \tag{3.60}$$

The solution of Equation (3.60) can be points (x, y) whose y-coordinate is large enough ($y > w$) so that the left-hand side gets equal to the right-hand side (at $n > m$). Thus, in quadrants I and IV, there will arise an even number $2p$ of OVs with the coordinates ($x > 0$, $\pm y$) and $TC = -1$, which will compensate for $2p$ OVs with $TC = +1$. We note that "compensating" OVs are unable to arise in quadrants II and III of the Cartesian system because at large negative $x < 0$, Equation (3.60) has no solutions, with its left-hand side being near-zero and right-hand side being larger than 1 at any y. In a similar way, it can be shown that no "compensating" OVs can arise on the Cartesian axes. Actually, at $x = 0$, Equation (3.45) is replaced with the relation:

$$1 = \left[\left(x_0^2 + y^2\right)/w^2\right]^{(n-m)/2}, \tag{3.61}$$

which has no solutions at large y^2 ($y > w$). On the horizontal axis ($y = 0$), Equation (3.45) reads as:

$$\exp\left(\frac{4xx_0}{w^2}\right) = \frac{\left(x + x_0\right)^n}{\left(x - x_0\right)^m}. \tag{3.62}$$

At large positive x ($x > x_0$), Equation (3.62) has no solutions for the left-hand side is always larger than the right-hand side. Although there may be a solution at small $x > 0$, but "compensating" OVs need to arrive from infinity, meaning that they are unable to appear on the horizontal axis. Summing up, a qualitative analysis of Equation (3.45) has shown that following an x_0-shift, "compensating" OVs with $TC = -1$ arise just in quadrants I and IV ($x > 0$, $\pm y$), making TC of the superposition smaller by an even number $2p$, which equals the number of 2π-phase jumps that occur in superposition (3.59) with no shift ($x_0 = 0$):

$$TC = n - 2p = n - 2\left[(n-1)/4\right]_-, \quad n > 4, \tag{3.63}$$

where "$[]_-$" denotes rounding-off to the nearest smaller integer.

Now, assume that $m = 3$ and n is varied. Figure 3.24 depicts intensity and phase patterns for superposition in Equation (3.59) for the following parameters: $\lambda = 532$ nm, $w_0 = 500$ μm, $n = 1, 2, 5, 6, 7, 8, 9,$ and 10, $A = B = 1$, $x_0/w_0 = 0.5$. The computing domain is $|x|, |y| \leq R$ ($R = 5$ mm in Figure 3.24(a–o) and $R = 50$ mm in Figure 3.24(p)).

From the phase patterns in Figure 3.24, the following values of TC can be deduced: 3 ($n = 1$), 2 ($n = 2$), 5 ($n = 5$), 4 ($n = 6$), 5 ($n = 7$), 6 ($n = 8$), 7 ($n = 9$), and 6 ($n = 10$). These values coincide with the number of 2π-phase jumps (straight boundaries between white and dark color) that go as far as the edge of the phase pattern in each frame in Figure 3.24. Table 3.2 summarizes the TC values for the superposition of two off-axis LG beams, derived from the general Equation (3.40) and depicted in Figures 3.23 and 3.24.

Based on Equations (3.54) and (3.60)–(3.62) and on the numerical simulation (Figures 3.22–3.24), rules for the determination of TC of superposition in Equation (3.59) can be formulated as follows. Assume that in superposition in Equation (3.59) of two LG beams, constituent TCs are m and n ($n > m$). Then, with no off-axis shift ($x_0 = 0$), the superposition has a total $TC = n$. At any non-zero shift, total TC is defined by the relation TC_p, $p = 1,2,3,4$, from Table 3.1 Thus, at $m = n$, $TC_1 = n$. If $m + n$ is

FIGURE 3.24 Intensity (a,c,e,g,i,k,m) and phase (b,d,f,h,j,l,n) patterns for superposition in Equation (3.59) at $\lambda = 532$ nm, $w_0 = 500$ μm, n = 1 (a, b), 2 (c, d), 5 (e, f), 6 (g, h), 7 (i, j), 8 (k, l), 9 (m, n), and 10 (o, p), A = B = 1, $x_0/w_0 = 0.5$. Computing domain $|x|, |y| \leq R$ ($R = 5$ mm (a–o) and $R = 50$ mm (p)).

TABLE 3.2

TC of superposition in Equation (3.59) at different n and m

	$n = 0$	$n = 1$	$n = 2$	$n = 3$	$n = 4$	$n = 5$	$n = 6$	$n = 7$	$n = 8$	$n = 9$	$n = 10$
$m = 2$	2	1	2	3	4	3	4	5	6	5	6
$m = 3$	2	3	2	3	4	5	4	5	6	7	6

odd, we need to analyze how much the larger $TC = n$ has changed: $n - (n + m)/2 = (n - m)/2$. Considering that Equations (3.60)–(3.62) suggest that TC can only decrease by an even number, then at even $(n - m)/2$, the value of Equation (3.54) remains unchanged, with the total TC being defined by $TC_1 = (n + m)/2$. If; however, following the off-axis shift, the larger $TC = n$ changes to an odd number $(n - m)/2$, the value of Equation (3.54) needs to be increased by 1, since Equations (3.60)–(3.62) suggest that TC can decrease by an even number. So, we conclude that if $m + n$ is even and $(n - m)/2$ is odd, the total TC is defined by $TC_2 = (n + m + 2)/2$. If; however, $m + n$ is odd, from Equation (3.54) superposition appears to have half-integer TC, but it should be an integer. Hence, the TC of Equation (3.54) defined by $TC_1 = (n + m)/2$ needs to be complimented by $\pm 1/2$. With the total TC being able to change only by an even number following an off-axis shift, see Equations (3.60)–(3.62), the difference $n - (n + m - 1)/2 = (n - m + 1)/2$ needs to be analyzed. If the larger constituent $TC = n$ is decreased by an even number $(n - m + 1)/2$, the subtraction of $1/2$ gives a correct value for TC: $TC_4 = (n + m - 1)/2$. If, however, $m + n$ and $(n - m + 1)/2$ are both odd, the subtraction of $1/2$ from Equation (3.54) gives an incorrect result, and the value of $1/2$ needs to be added to Equation (3.54), meaning that the total TC is $TC_3 = (n + m + 1)/2$. All four feasible variants of the total TC are summarized in Table 3.3 The findings in Table 3.3 can be corroborated through a comparison with numerically simulated values from Table 3.2.

Tables 3.2 and 3.3 indicate that the superposition in Equation (3.59) has a different TC if the beams are shifted from the optical axis differently. For instance, it is seen in Table 3.2 that the TC of the superposition equals $TC = 2$ if $n = 2$ and $m = 3$, whereas $TC = 3$ if $n = 3$ and $m = 2$. In general, this means that if the LG beam $(0, m)$ is shifted from the axis by a distance $x_0 > 0$, and the LG beam $(0, n)$ is shifted from the axis by a distance $-x_0$, and if the sum $n + m$ is odd (e.g. $n = 2p$ and $m = 2q + 1$), then, according to Table 3.3, $(n - m + 1)/2 = p - q$. On the contrary, if the LG beam $(0, m)$ is shifted from the axis by a distance $-x_0$ and the LG beam $(0, n)$ is shifted from the axis by a distance $x_0 > 0$, then, according to Table 3.3, we should suppose $n = 2q + 1$, $m = 2p$ and, therefore, $(n - m + 1)/2 = q - p + 1$. For example, if $p - q$ is even, then $q - p + 1$ is odd, and, as follows from Table 3.3, the TC of the superposition is equal to $TC_4 = (m + n - 1)/2$ in the first case and $TC_3 = (m + n + 1)/2$ in the second case. Thus, if two LG beams with an odd sum of their orders are swapped in the superposition,

TABLE 3.3

Feasible variants of the total *TC* for the superposition of off-set LG beams and the selection rules

Rules for selecting the total *TC*	Total *TC* of the superposition
$n + m$ and $(n - m)/2$ are even	$TC_1 = (m + n)/2$
$n + m$ is even and $(n - m)/2$ is odd	$TC_2 = (m + n + 2)/2$
$n + m$ and $(n - m + 1)/2$ are both odd	$TC_3 = (m + n + 1)/2$
$n + m$ odd and $(n - m + 1)/2$ even	$TC_4 = (m + n - 1)/2$

the *TC* changes by 1. However, if $n + m$ is even (e.g. $n = 2p$, $m = 2q$), then, according to Table 3.3, when the beams are swapped in the superposition, the numbers $(n - m)/2 = p - q$ and $(n - m)/2 = q - p$ are of the same parity and, therefore, the *TC* is the same.

The above-explained interesting and nontrivial behavior occurs due to that the wrapped phase profile of an LG beam with an odd value of the *TC* is asymmetric with respect to the center of the beam, while for an LG beam having an even value of the *TC*, the wrapped phase is symmetric with respect to the beam's center. To show the differences more clearly, see the simulated interference patterns shown in Figure 3.25. These patterns show the superposition of different pairs of off-axis LG beams with a tilted plane wave for a lateral displacement of 0.85 mm of the LG beams. The calculated (according to Table 3.3) and observed *TC* for each pattern is the same and for (a) to (d) the TCs are 2, 1, 4, and 4, respectively.

Determining the topological charge by interferograms

Figure 3.26 depicts intensity (first row) and phase (second row) distributions of the superpositions in Equation (3.59) of two off-axis LG beams, as well as simulated interferograms, obtained with an inclined plane wave (third row). In the first column, the following parameters are used: wavelength $\lambda = 532$ nm, Gaussian beam waist radius $w_0 = 500$ μm, $A = B = 1$, and $x_0/w_0 = 2$ ($x_0 = 1$ mm), computation domain $|x|$, $|y| \leq R$ ($R = 5$ mm (a, g) and $R = 2.5$ mm (m)), $m = 3$, $n = 2$. For the first column, the theoretical *TC* value equals 2 ($TC = 2$), but counting the forks in the simulated interferogram yields $TC = 4$. This mismatch between the theory and the interferogram happens because two optical vortices, each with *TC* -1, are not seen in the interferogram, since they reside in an area of very low intensity. On the phase distribution in the first column of Figure 3.26, these two vortices are marked by solid white circles (Figure 3.26(g)).

In the second column of Figure 3.26, the TCs are opposite to those in the first column, i.e. the TCs of the constituent LG beams are equal to $m = 2$ and $n = 3$, while their shift is $x_0/w_0 = 0.2$ ($x_0 = 0.1$ mm). The computation domain is the same: $|x|$, $|y| \leq R$ ($R = 5$ mm (b, h) and $R = 2.5$ mm (n)). The theory predicts that the *TC*

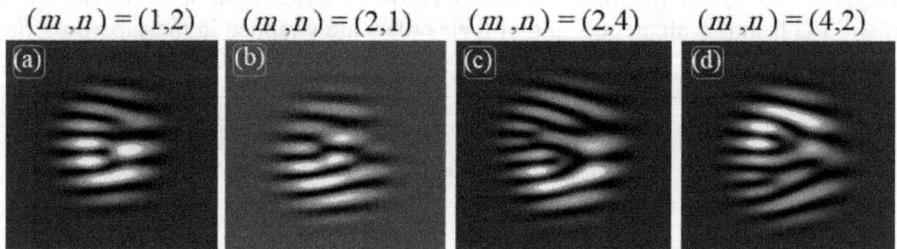

FIGURE 3.25 Simulation patterns for the interference of an inclined plane wave with a superposition of two off-axis LG beams having $\lambda = 532$ nm and $w \approx 1.7$ mm, $(m, n) = (1, 2)$ (a), $(m, n) = (2, 1)$ (b), $(m, n) = (2, 4)$ (c), and $(m, n) = (4, 2)$ (d). Calculated and the observed *TC* for each pattern is the same and for (a) to (d) the TCs are 2, 1, 4, and 4, respectively. Computing domain $|x|$, $|y| \leq R$ ($R \approx 3.3$ mm) and off-axis values for all of the patterns are $2x_0 = 0.85$ mm.

equals 3 ($TC=3$). Counting the forks on the computed interferogram also yields TC = 3. Comparison between the first and the second columns indicates that if two LG beams with an odd sum of their orders are swapped in the superposition, then the TCs of the superpositions are different.

In the third column of Figure 3.26, $m=3$, $n=5$. Other parameters are $x_0/w_0 = 1.2$ ($x_0 = 0.6$ mm), computation domain $|x|, |y| \leq R$ ($R = 2.5$ mm). According to the theory, the TC in the third column should equal 5 ($TC=5$). Counting the forks in the interferogram also yields $TC=5$.

In the fourth column of Figure 3.26, the following parameters are used: $m=5$, $n=2$, $x_0/w_0 = 1.6$ ($x_0 = 0.8$ mm), and the computation domain $|x|, |y| \leq R$ ($R = 5$ mm (d, j) and $R = 2.5$ mm (p)). According to the theory, the TC of the superposition equals 4 ($TC=4$). However, counting the forks in the interferogram yields a TC value equal to 6. This mismatch is because two off-axis optical vortices of the order -1 are not seen due to small intensity. On the phase distribution in the fourth column, these two vortices are marked by solid white circles (Figure 3.26(j)).

In the fifth column of Figure 3.26, we used the following parameters: $m=5$, $n=-2$, $x_0/w_0 = 1.6$ ($x_0 = 0.8$ mm), and the computation domain $|x|, |y| \leq R$ ($R = 5$ mm (e, k) and $R = 2.5$ mm (q)). The theory predicts the TC of the superposition is equal to 2 ($TC=2$). Counting the forks in the interferogram also yields the value $TC=2$.

In the sixth column of Figure 3.26, $m=3$, $n=0$, $x_0/w_0 = 0.8$ ($x_0 = 0.4$ mm), the computation domain is $|x|, |y| \leq R$ ($R = 5$ mm (f, l) and $R = 2.5$ mm(r)). The theoretical TC value is $TC=2$. Counting the forks in the interferogram also gives the value $TC=2$.

The above examples (Figure 3.26), comparing the theory and the simulation, demonstrate that the latter does not coincide with the theory that when some vortices in the superposition reside in a low-intensity area and do not manifest themselves as the forks in the interferogram.

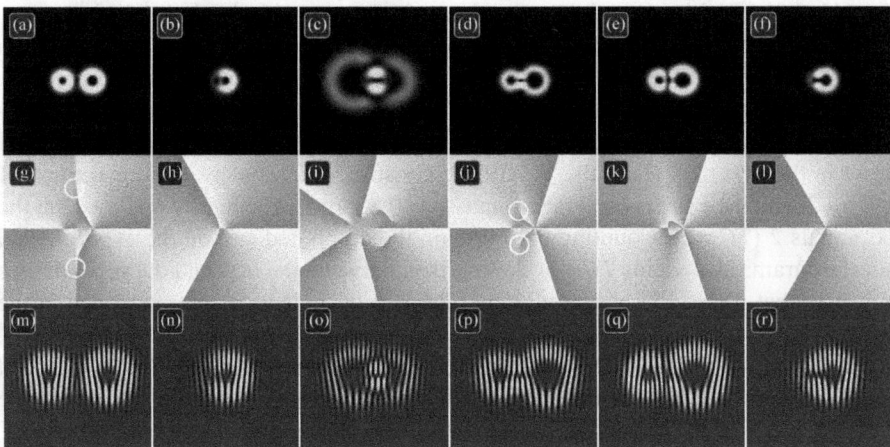

FIGURE 3.26 Distributions of intensity (first row) and phase (second row, dark color means 0, white color means 2π) of a superposition in Equation (3.59) of two off-axis LG beams, as well as computed interferograms (third row) with an inclined plane wave.

Comparison of analytical, simulated, and experimentally recorded results

Using the experimental setup shown in Figure 3.27 we investigate the interference of a plane wave with two off-axis LG beams having the TCs of n and m. The plane wave is selected from the central area of a Gaussian beam with the aid of an iris diaphragm. In Figure 3.27(a) by implementing a fork grating's structure on an SLM, two LG beams are generated over the different diffraction orders of the SLM and they permit to pass through Path 1 and Path 2, respectively. The plane wave propagates through Path 3. As is seen, a modified Mach–Zehnder interferometer is used at the end of the setup. In the interferometer, using two suitable obstacles over the two exiting sides of the BS2 we select the desired diffraction orders on each path, then overlap collinearly and laterally the selected LG beams. With the aid of two neutral density filters, DF1 and DF2, the amplitudes of the LG beams are balanced, especially when one of the beams is selected from 1st-order diffraction and the other one is selected from a higher order. A 2D positioner is used to displace the BS4, in which desired off-axis values can be implemented between the produced LG beams. By adjusting the mirror M2, the value of the angle between the propagation directions of the LG beams and the plane beam is adjusted.

Figure 3.28 shows interferograms produced by a simulation and experimentally superposition of two LG beams that have different TCs and different off-axis values. The propagation direction of the plane wave has a small angle with the propagation direction of the collinear and off-axis LG beams. The LG beams' parameters were $w \approx 1.7$ mm and $\lambda = 532$ nm. The computing domain of the simulated patterns and the illustrated area of the experimentally recorded images are $|x|$, $|y| \leq R$ ($R \approx 3.3$ mm). The two first columns from the left side show simulated (first column) and experimentally recorded (second column) patterns for $2x_0 = 0$ mm with equal TCs of $n = m = 1$ to 5. For better illustration and counting of the values of the TCs in the experimentally recorded patterns (second column), we added the lighter lines over the added forked fringes and we trace the main fringes with the darker lines.

The second pair of the columns in Figure 3.28 demonstrates the results for an off-axis value of 0.85 mm. In the first row, the following parameters are used $m = 2$, $n = 1$, and the theoretical TC value equals 1 ($TC = 1$). Counting the forks in the interferogram also yields the value $TC = 1$. In the second row, the TCs are the opposite of those in the first row. Here, the TCs of the constituent LG beams are equal to $m = 1$ and $n = 2$, and their shift is still 0.85 mm ($2x_0 = 0.85$ mm). The theory predicts that the TC equals 2 ($TC = 2$). Counting the forks on the computed and on the experimental interferograms also yields $TC = 2$. Comparison between the first and the second rows indicates that if two LG beams with an odd sum of their orders are swapped in the superposition, then the TCs of the superpositions are different (see also Figure 3.25). In the third row, $m = 4$, $n = 2$. According to the theory, the TC in the third row should equal 4 ($TC = 4$). Counting the forks in the interferogram also yields $TC = 4$. In the last row, we used the following parameters: $m = 6$, $n = 3$. The theory predicts the TC of the superposition is equal to 5 ($TC = 5$). Counting the forks in the interferogram also yields the value $TC = 5$.

In this section, we have proposed nontrivial rules for selecting the proper value of TC for a superposition of two symmetric off-axis LG beams with the indices (0,

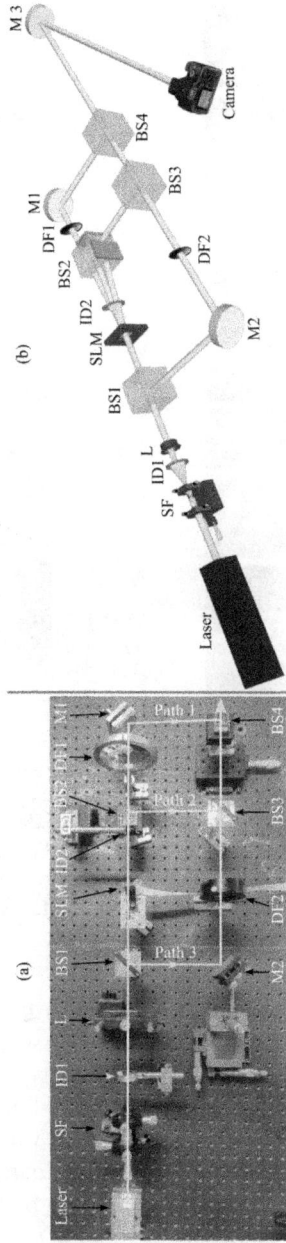

FIGURE 3.27 Used experimental setup (a), and the corresponding schematic arrangement (b). SF, ID, L, BS, SLM, DF, and M show spatial filter, iris diaphragm, lens, beam splitter, density filter, and mirror, respectively.

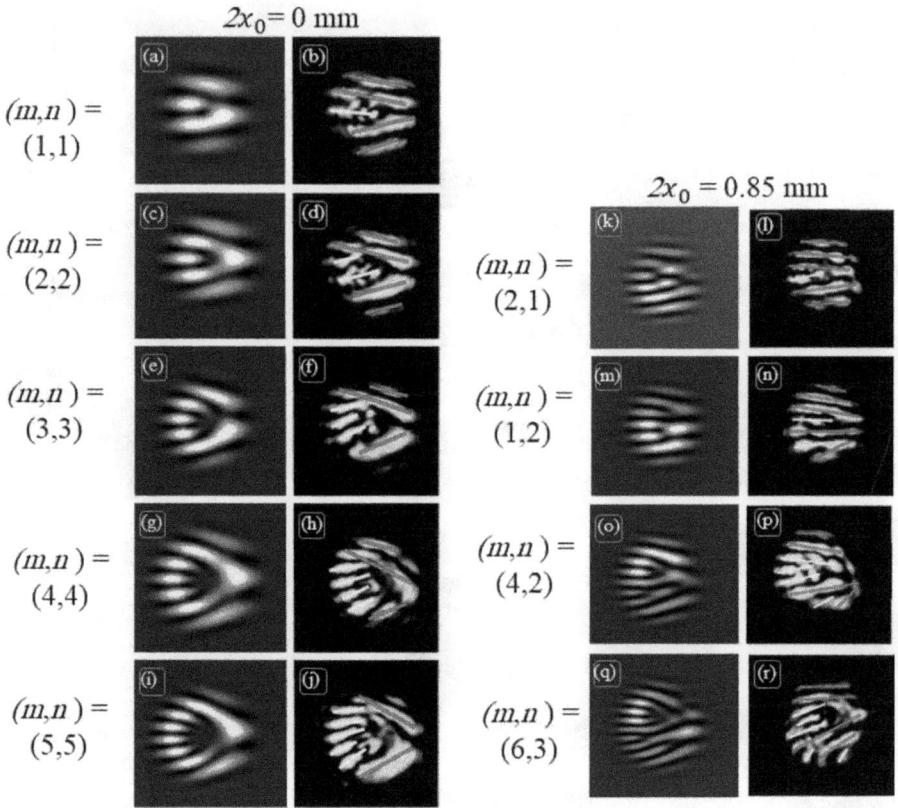

FIGURE 3.28 Simulation (a, c, e, g, i, k, m, o, q) and experimentally recorded (b, d, f, h, j, l, n, p, r) patterns for interference of an inclined plane wave with a superposition of two off-axis LG beams having $\lambda = 532$ nm and w ≈ 1.7 mm. For Figure 3.28(a–j) we have n = 1 to 5 and m = n with $2x0 = 0$ mm. In Figure 3.28(k–r) the LG beams' TCs are $(m, n) = (2, 1)$ (k, l), $(m, n) = (1, 2)$ (m, n), $(m, n) = (4, 2)$ (o, p) and $(m, n) = (6, 3)$ (q, r) with the same off-axis values of $2x0 = 0.85$ mm. The computing domain of the simulated patterns and illustrated area of the experimentally recorded images are |x|, |y| $\leq R$ ($R \approx 3.3$ mm).

n) and $(0, m)$ [167]. The straightforward use of Berry's formula for calculating TC of such a superposition gives arithmetic mean as the sought-for TC: $(n + m)/2$. It stands to reason that such a relation can take half-integer values, which are physically meaningless due to the continuous amplitude of a light field. Therefore, TC of the superposition can take four physically meaningful variants: $(n + m)/2$, $(n + m + 2)/2$, $(n + m + 1)/2$, and $(n + m - 1)/2$. When choosing a particular TC variant out of the four, one should take into account that in the nonshifted superposition, the total TC equals the larger constituent TC, e.g. being equal to n, if $n > m$. Using numerical modeling and a qualitative solution of transcendent equations, we have shown that following an off-axis shift of two LG beams, an even number of OVs with $TC = -1$ "arriving" from infinity will compensate for an equivalent number of OVs with TC

= +1 in the superposition. That is to say, when compared with the nonshifted super-position, TC of the superposition of off-axis LG beams either remains the same or decreases by an even number. Considering this fact, we have worked out rules for selecting the value of TC out of four feasible variants (see Table 3.3). We have also shown that when the LG beams of the orders $(0, n)$ and $(0, m)$ with $n + m$ being an odd number are swapped in the superposition, the TC of the superposition changes by 1. Notably, the TC of symmetric off-axis LG beams has been deduced for the first time. Finally, it was shown that the analytical, simulated, and experimentally recorded results confirm each other.

3.4 TOPOLOGICAL CHARGE OF POLYCHROMATIC OPTICAL VORTICES

Vortex beams, or optical vortices, are known in optics since the 1980s, but many fun-damental theoretical issues about these beams still have not been addressed. Some of these unsolved problems are related to such an important quantity of the optical vortices as their topological charge (TC) [168]. For instance, it was only recently discovered [167] that the TC of a superposition of two parallel Laguerre–Gaussian (LG) beams with their azimuthal indices of different parity can be different depend-ing on which of these two beams is on the left and which is on the right. If these beams are swapped in the superposition, their common TC changes by 1. However, while the TC of monochromatic optical vortices was discussed in some works, there are almost no works about the TC of "colored", or polychromatic optical vortices. Below we analyze briefly what is known about the TC of polychromatic vortices. In [169], a spiral phase plate (SPP) was illuminated by a white light beam and it was demonstrated that a rainbow is generated since the different wavelengths in the white light beam generate, after passing through the SPP, light rings of different radii. In [170], illuminating uniaxial crystals by polychromatic light, 1st- and 2nd-order optical vortices were generated. Their TC was measured experimentally by using a cylindrical lens. Using filters to select different spectral components out of the white vortex, it was shown that the TC of each color component in the beam is the same and equals either 1 or 2 [170]. In [171], white light illuminated a multisector spiral reflecting metasurface, fabricated on a surface of a gold film and composed of binary subwavelength gratings that rotate the polarization vector of the incident beam. As was demonstrated experimentally, when the beam is reflected from such a metasur-face, 1st-order optical vortices are generated in the whole visible range with almost equal effectiveness (nearly 60%). In [172], colored optical vortices were generated by a fork Bragg grating, generated in a cholesteric liquid crystal. Such a Bragg grating acts as a "thick" hologram, which has a significant angular selectivity. Changing the incidence angle of the white light onto a surface of the liquid crystal, colored (red and blue) 1st- and 2nd-order optical vortices were obtained at the output. In [173], white light from the LED was incident on a reflecting light modulator that imple-mented a spiral phase pattern embedded into a blazed grating. The grating reflects light of different wavelengths at different angles. Thus, colored (red, green, and blue) optical vortices with the topological charge of 3 were generated experimentally. The white vortex was obtained in [173] by adding a prism into the optical setup, which compensated for the dispersion of the grating in the modulator.

There are few works on the "colored" optical vortices (COV) and even fewer (or almost no) works, which derive the TC of a superposition of COVs. In this work, we study as an example a coaxial superposition of single-ringed (i.e. with zero radial index) LG beams with the same waist radius, but with different weight coefficients, TCs, and wavelengths. We show both theoretically and numerically that the topological competition is won by a more "red" LG beam, i.e. the common TC of the whole superposition is equal to the TC of the constituent LG beam with a larger wavelength.

3.4.1 TOPOLOGICAL CHARGE OF A TWO-COLOR SUPERPOSITION OF OPTICAL VORTICES

Under the colored optical vortices we mean here a coaxial superposition of single-ringed LG beams with the same waist radius, but with different topological charges (TCs) and with different wavelengths. The complex amplitude of such colored optical vortices is given by:

$$E\left(r,\varphi,z,t\right) = \sum_{s=1}^{N} c_s q_s^{-n_s-1}(z)\left(\frac{\sqrt{2}r}{w}\right)^{n_s} \exp\left(-\frac{r^2}{w^2 q_s(z)} + in_s\varphi + ik_s z - i\omega_s t\right), (3.64)$$

where (r, φ, z) is the cylindrical coordinates, t is time, w is the waist radius of the Gaussian beam, c_s are the weight coefficients, and n_s are the topological charges of each beam in the superposition:

$$q_s(z) = 1 + i\frac{z}{z_s}, \quad z_s = \frac{k_s w^2}{2}, \quad k_s = \frac{2\pi}{\lambda_s}, \quad (3.65)$$

z_s is the Rayleigh distance, k_s is the wavenumber, λ_s is the wavelength of monochromatic light, and ω_s is its frequency ($k_s = \omega_s/c$, with c being the speed of light in a vacuum). In practice, superposition in Equation (3.64) is generated if in the waist plane of a coaxial superposition of Gaussian beams with specific wavelengths a spiral phase plate (SPP) is placed with the order n and with a relief depth intended for the wavelength λ_0. At this, the wavelengths λ_s of the Gaussian beams should be such that the SPP order remained an integer number n_s: $\lambda_s n_s = \lambda_0 n$, $s = 1, 2, 3, ..., N$. Strictly speaking, after passing through the SPP, a Gaussian optical vortex [80] is generated, rather than the LG beam. It was proved, though, that the Gaussian vortex in the far field is almost coinciding with the LG beam [174,175].

Below we suppose for simplicity that there are only two terms in Equation (3.64). The topological charge is defined by Berry's formula [19]:

$$TC = \frac{1}{2\pi} \lim_{r\to\infty} \text{Im} \int_0^{2\pi} d\varphi \frac{\partial E\left(r,\varphi,z\right)/\partial\varphi}{E\left(r,\varphi,z\right)}. \quad (3.66)$$

It can be seen from Equation (3.66) that the right part depends on the propagation distance z. However, it will be shown below that the topological charge is z-independent. Substitution of Equation (3.64) with $N = 2$ into Equation (3.66) yields:

$$
\begin{aligned}
TC = \frac{1}{2\pi} \lim_{r \to \infty} \mathrm{Im} \int_0^{2\pi} & \left\{ \frac{inc_1}{q_1^{n+1}(z)} \left(\frac{\sqrt{2}r}{w} \right)^n \exp\left[-\frac{r^2}{w^2 q_1(z)} + in\varphi + ik_1 z - i\omega_1 t \right] \right. \\
& + \frac{imc_2}{q_2^{m+1}(z)} \left(\frac{\sqrt{2}r}{w} \right)^m \exp\left[-\frac{r^2}{w^2 q_2(z)} + im\varphi + ik_2 z - i\omega_2 t \right] \right\} \\
& \times \left\{ \frac{c_1}{q_1^{n+1}(z)} \left(\frac{\sqrt{2}r}{w} \right)^n \exp\left[-\frac{r^2}{w^2 q_1(z)} + in\varphi + ik_1 z - i\omega_1 t \right] \right. \\
& \left. + \frac{c_2}{q_2^{m+1}(z)} \left(\frac{\sqrt{2}r}{w} \right)^m \exp\left[-\frac{r^2}{w^2 q_2(z)} + im\varphi + ik_2 z - i\omega_2 t \right] \right\}^{-1} d\varphi.
\end{aligned}
$$

(3.67)

Supposing that both beams have nonzero amplitude: $c_1 \neq 0$, $c_2 \neq 0$, and replacing the variable $r = w\rho\,/2^{1/2}$, this expression can be rewritten in a more compact form:

$$
TC = \frac{1}{2\pi} \lim_{\rho \to \infty} \mathrm{Re} \int_0^{2\pi} \frac{nP(z)\rho^{n-m} e^{-Q(z)\rho^2} e^{i(k_1 - k_2)z} e^{-i(\omega_1 - \omega_2)t} e^{in\varphi} + m e^{im\varphi}}{P(z)\rho^{n-m} e^{-Q(z)\rho^2} e^{i(k_1 - k_2)z} e^{-i(\omega_1 - \omega_2)t} e^{in\varphi} + e^{im\varphi}} d\varphi, \quad (3.68)
$$

with:

$$
P(z) = \frac{c_1}{c_2} \frac{q_2^{m+1}(z)}{q_1^{n+1}(z)}, \quad Q(z) = \frac{1}{2}\left[\frac{1}{q_1(z)} - \frac{1}{q_2(z)} \right].
$$

Equation (3.68) indicates that if $|q_1(z)| < |q_2(z)|$ then the exponentials with ρ^2 have a coefficient with a negative real part (i.e. $\mathrm{Re}Q(z) > 0$) and, when ρ tends to infinity, the first terms in the integral both in the numerator and in the denominator tend to zero, regardless of the numbers n and m and regardless of the weight coefficients c_1 and c_2. The remained second terms, after simplifications, yield $TC = m$. On the contrary, if $|q_1(z)| > |q_2(z)|$, then the exponentials with ρ^2 have a coefficient with a positive real part and, when r tends to infinity, the first terms also tend to infinity and the second terms can be neglected. Then, simplifications yield $TC = n$. From Equation (3.68) follows that $|q_1(z)| < |q_2(z)|$ if $\lambda_1 < \lambda_2$ (and $TC = m$) and that $|q_1(z)| > |q_2(z)|$ if $\lambda_1 > \lambda_2$ (and $TC = n$). Thus, we can conclude that the TC of a superposition of two colored optical vortices is equal to the TC of the beam with a larger wavelength ("reds" win the "blues"). If we take into account a relation $\lambda_s n_s = \lambda_0 n$, $s = 1,2,...,N$, then the beam with a larger wavelength has a smaller topological charge.

We note that if there are more than two beams ($N > 2$) then the proof is bulkier, but quite similar, and is based on the overwhelming of one exponent over the others and therefore the *TC* of the whole superposition equals the *TC* of the beam with a larger wavelength.

We also note that if instead of the single-ringed LG beams in the superposition in Equation (3.64) we choose the LG beams with nonzero radial indices p_s:

$$E(r,\varphi,z,t) = \sum_{s=1}^{N} c_s q_s^{-n_s-1} \left(\frac{\sqrt{2}r}{w}\right)^{n_s} L_{p_s}^{n_s} \left[\frac{2r^2}{w^2|q_s(z)|^2}\right]$$

$$\times \exp\left[-\frac{r^2}{w^2 q_s(z)} + in_s\varphi - 2ip_s \arg q_s(z) + ik_s z - i\omega_s t\right], \qquad (3.69)$$

where $L_p^n(.)$ are the associated Laguerre polynomials, then similar derivation process will lead to the same *TC*, since it is defined by an overwhelming exponent (Gaussian envelope of the beam), rather than by the power growth of the radial polar coordinate.

If we suppose that all the beams in the superposition in Equation (3.64) have the same *TC* equal to n, then, instead of Equation (3.64), we get:

$$E(r,\varphi,z) = \left(\frac{\sqrt{2}r}{w}\right)^n e^{in\varphi} \sum_{s=1}^{N} c_s q_s^{-n-1}(z) \exp\left(-\frac{r^2}{w^2 q_s(z)} + ik_s z\right). \qquad (3.70)$$

Since the angular derivative of Equation (3.70) is $\partial E(r,\varphi,z)/\partial\varphi = inE(r,\varphi,z)$, then, according to Equation (3.66), the *TC* of the superposition in Equation (3.70) with arbitrary colors is equal to the *TC* of each beam: $TC = n$. This result is simple, but a practical generation of the superposition in Equation (3.70) is challenging since it requires that in the waist planes of each colored Gaussian beam different SPPs were placed, whose maximal relief depth h_s is matched with the wavelength λ_s of the incident light: $2\pi h_s(n_0 - 1) = n\lambda_s$ with n_0 being the refractive index of the SPP material (we suppose that there is no dispersion of the refractive index).

3.4.2 TOPOLOGICAL CHARGE OF A WHITE OPTICAL VORTEX

Here we consider a practically important case when in the waist plane of a white Gaussian beam a single SPP is placed, whose relief is matched with the wavelength λ_0. Then the complex amplitude of the superposition in the Fresnel diffraction zone is given by:

$$E(\rho,\theta,z,t) = \frac{-i}{z\lambda_0} \int_0^\infty \lambda^{-1} f(\lambda)d\lambda \int_0^\infty rdr \int_0^{2\pi} d\varphi$$

$$\times \exp\left\{i\frac{2\pi z}{\lambda} - i\frac{2\pi ct}{\lambda} - \frac{r^2}{w^2} + i\frac{n\lambda_0}{\lambda}\varphi + \frac{i\pi}{\lambda z}[r^2 + \rho^2 - 2r\rho\cos(\varphi-\theta)]\right\},$$

$$(3.71)$$

with $f(\lambda)$ being the envelope function of the spectrum of a white light source (e.g. LED). Since the *TC* of each monochromatic (single-color) vortex $\mu = n\lambda_0/\lambda$ is, in general, fractional, then the integral over φ in Equation (3.71) cannot be evaluated. Therefore, we expand the exponent $\exp(i\mu\,\varphi)$ into a series of optical vortices with integers TCs:

$$\exp\left(i\frac{n\lambda_0}{\lambda}\varphi\right) = \frac{e^{i\pi n\lambda_0/\lambda}\lambda\sin(\pi n\lambda_0/\lambda)}{\pi}\sum_{m=-\infty}^{\infty}\frac{e^{im\varphi}}{n\lambda_0 - m\lambda}. \tag{3.72}$$

Substituting Equation (3.72) into Equation (3.71), we get:

$$E(\rho,\theta,z,t) = \frac{-2i}{z\lambda_0}\sum_{m=-\infty}^{\infty}(-i)^m e^{im\theta}\int_0^{\infty}\frac{e^{i(\pi/\lambda)\left(n\lambda_0+\rho^2/z+2z-2ct\right)}\sin(\pi n\lambda_0/\lambda)f(\lambda)}{n\lambda_0 - m\lambda}d\lambda$$

$$\times \int_0^{\infty}\exp\left(-\frac{r^2}{w^2}+\frac{i\pi r^2}{\lambda z}\right)J_m\left(\frac{2\pi r\rho}{\lambda z}\right)r\,dr, \tag{3.73}$$

with $J_m(x)$ being the *m*th-order Bessel function of the first kind. The integral over the variable r in Equation (3.73) can be evaluated by using a reference integral from [176]:

$$\int_0^{\infty}r\,dr\exp\left(-\frac{r^2}{w^2}+\frac{i\pi r^2}{\lambda z}\right)J_m\left(\frac{2\pi r\rho}{\lambda z}\right)$$

$$= \frac{\pi^{3/2}\rho w^3(\lambda z)^{1/2}(\operatorname{sgn} m)^{|m|}}{4\left(\lambda z - i\pi w^2\right)^{3/2}}e^{-\xi}\left[I_{(|m|-1)/2}(\xi) - I_{(|m|+1)/2}(\xi)\right], \tag{3.74}$$

where:

$$\xi = \frac{(\pi\rho w)^2}{2\lambda z\left(\lambda z - i\pi w^2\right)}.$$

Substituting Equation (3.74) into Equation (3.73), we finally obtain:

$$E_n(\rho,\theta,z,t) = \frac{-i\pi^{3/2}\rho w^3}{2z^{1/2}\lambda_0}\sum_{m=-\infty}^{\infty}(-i)^m(\operatorname{sgn} m)^{|m|}e^{im\theta}D_{m,n}(\rho,z,t), \tag{3.75}$$

where:

$$D_{m,n}(\rho,z,t) = \int_0^{\infty}\frac{e^{-\xi}\left[I_{(|m|-1)/2}(\xi) - I_{(|m|+1)/2}(\xi)\right]}{\left(\lambda z - i\pi w^2\right)^{3/2}}$$

$$\times\frac{e^{i(\pi/\lambda)\left(n\lambda_0+\rho^2/z+2z-2ct\right)}\lambda^{1/2}\sin(\pi n\lambda_0/\lambda)f(\lambda)}{n\lambda_0 - m\lambda}d\lambda. \tag{3.76}$$

The integral in Equation (3.76) can hardly be reduced to a reference integral and therefore it cannot be evaluated in an explicit form, excepting a trivial case of monochromatic light, when $f(\lambda) = \delta (\lambda - \lambda_0)$. In this case, the expression (3.75) coincides with Equation (3.45) from [55]. However, even without evaluating Equation (3.76), some conclusions can be drawn from the obtained Equations (3.75) and (3.76). Namely, if there is a zero denominator of the integrand, then only some terms remain in the series in Equation (3.75), i.e. those terms, whose numbers m yield this zero denominator:

$$m = \frac{n\lambda_0}{\lambda}, \quad \lambda \in \left[\lambda_0 - \Delta\lambda, \lambda_0 + \Delta\lambda\right], \quad (3.77)$$

with $2\Delta\lambda$ being the width of the spectrum of the colored Gaussian beam. For instance, if $n = 10$, $\lambda_0 = 532$ nm, and $\Delta\lambda = 100$ nm, then only four terms remain in the sum (3.75) and their numbers are $m = 9$, 10, 11, and 12. Further, the expression in Equation (3.64) can be used. This expression indicates that the superposition contains effectively only four beams and the TC of the whole superposition is equal to the maximal TC, i.e. $TC = 12$. If the SPP order is increased two times, i.e. $n = 20$, then in the same example the sum in Equation (3.75) contains eight colored optical vortices with numbers m from 17 to 24. This means that the superposition (3.75) contains effectively eight optical vortices in Equation (3.64) and the TC of such a superposition is equal to $TC = 24$. On the contrary, if the SPP order is decreased, e.g. $n = 3$, then, as in [173], for a beam with an arbitrary wavelength from the range (432 nm and 632 nm) only one term in the Equation (3.75) becomes zero at $m = 3$. This explains the experimental results from [173], when after passing through a single SPP of the order $n = 3$ all colored vortices (white, blue, red, and green) had the TC equal to $m = 3$.

3.4.3 Numerical Simulation

Numerical simulation of two-color vortices

Here we consider a superposition of two different-color LG beams. According to Equation (3.64), such a light field is not stationary and is time-dependent. If in some point of space the complex amplitude of the first and second beam is respectively A and B, then the field in this point depends on time as follows: $E = A \exp(-i\omega_1 t) + B \exp(-i\omega_2 t)$, where ω_1 and ω_2 are the frequencies of the beams. In a simple case, when $A = B$, this field can be written as $E = 2A \cos[(\omega_1 - \omega_2)t/2] \exp[-i(\omega_1 + \omega_2)t/2]$. This simple example shows that if the frequencies ω_1 and ω_2 are not close to each other, then the field intensity oscillates from zero to maximum with a much higher frequency than a human eye can perceive. Therefore, this point is perceived as a point with constant (time-independent) intensity $|A|^2$ and with the average frequency $(\omega_1 + \omega_2)/2$. Consequently, if the beams are, for instance, of blue and red color, then some point with equal amplitudes of both beams looks purple and can be visualized as (255, 0, 255) (in RGB format). As to the phase, then, in contrast to the monochromatic waves, it cannot be determined, since in every point in space the time-dependence of the amplitude is not harmonic. Thus, only an instantaneous phase value can be visualized, i.e. the phase in some specific moment of time.

Figure 3.29 illustrates a superposition of two different-color LG beams. Intensity distributions (Figure 3.29(a, d, g)) are shown time-averaged, whereas the phase distributions (Figure 3.29(b, c, e, f, h, i)) are instantaneous (in two different time moments). The following parameters were chosen for computations: wavelengths λ_1 = 400 nm, λ_2 = 700 nm, waist radius of the Gaussian beam w = 500 μm, topological charges of the beams are respectively n_1 = 4 and n_2 = 3, radial indices $p_1 = p_2 = 0$ (single-ringed beams), weight coefficients of the superposition $c_1 = (p_1!/(n_1 + p_1)!)^{1/2}$ ≈ 0.2 and $c_2 = (p_2!/(n_2 + p_2)!)^{1/2} \approx 0.4$ (for such values, both beams are of the same

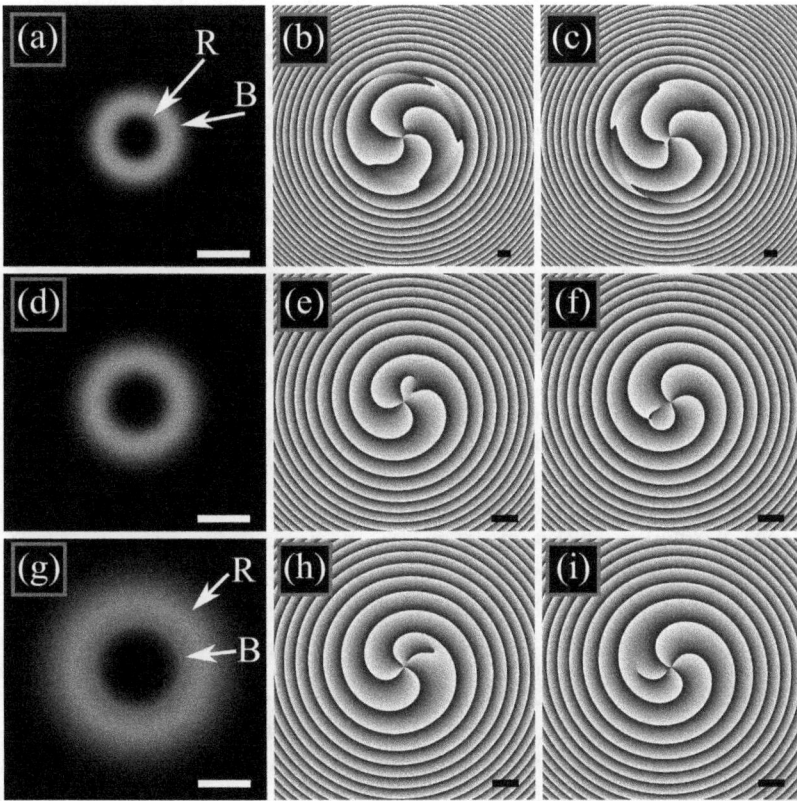

FIGURE 3.29 Distributions of time-averaged intensity (a, d, g) and of instantaneous phase (b, c, e, f, h, i) of a superposition of two single-ringed LG beams of different colors (λ_1 = 400 nm, λ_2 = 700 nm) at a distance $z = 0.2$ m (a–c), $z = 1$ m (d–f), $z = 2$ m (g–i) and at time moments $t_1 = 60$ s (b, e, h) and $t_2 = 61$ s (c, f, i). Other computation parameters are the following: Gaussian beam waist radius w = 500 μm, TCs n_1 = 4 and n_2 = 3, radial indices $p_1 = p_2 = 0$, superposition weight coefficients $c_1 \approx 0.2$ and $c_2 \approx 0.4$ (so that both beams are of the same power). The scaling mark in each figure shows 1 mm. Symbols "R" and "B" (a, g) show areas where, respectively, red and blue beam dominates.

power), propagation distance $z = 0.2$ m (Figure 3.29(a–c)), $z = 1$ m (Figure 3.29(d–f)), $z = 2$ m (Figure 3.29(g–i)), time of phase registration $t_1 = 60$ s (Figure 3.29(b, e, h)), and $t_2 = 61$ s (Figure 3.29(c, f, i)).

Since the TC of the blue beam exceeds that of the red beam, at a small propagation distance, the blue ring is of a larger radius than the red ring (Figure 3.29(a)). However, the red beam diverges stronger than the blue beam. Therefore, at a distance of nearly 1 m the radii of the rings are equal (one purple ring in Figure 3.29(d)), and, on further propagation, at a distance of nearly 2 m, the red ring becomes outer, while the blue ring becomes inner (Figure 3.29(g)).

According to the theory, a common topological charge of the whole superposition does not depend on the color of the outer ring and is determined solely by the topological charge of the beam with a larger wavelength. Then, for the beam from Figure 3.29, it should be equal to three ($TC = 3$). Computation of the topological charge using the instantaneous phase distributions from Figure 3.29 confirms this: computing by Berry's formula (3.66) along a circle with the radius $R_1 = 7.5$ mm yields the values 2.986 (Figure 3.29(b)), 2.982 (Figure 3.29 (c)), 2.874 (Figure 3.29 (e, f)), and 2.907 (Figure 3.29 (h, i)).

In the initial plane and in the initial time moment, optical vortices in the beam cross-section are distributed the following way. In the beam center on the optical axis, there is an optical vortex with the smallest TC of 3, whereas at some radius from the axis there is a vortex with a TC of +1 and at a greater radius there is a vortex with a TC of –1. After the evolution of such superposition in time and in space, on the optical axis, there is still a vortex with a TC of +3, while the optical "dipole" with TCs +1 and –1 on its ends approach to the optical axis. Thus, the TC of the superposition is equal to 3. This evolution can be seen in the instantaneous phase distributions in Figure 3.29. The vortices would behave differently, if, vice versa, the beam with a larger wavelength had the TC of 4, and the beam with a smaller wavelength had the TC of 3. Then, during the evolution of such a superposition, the vortex with TC of –1 would move away to infinity "almost immediately", and near the axis a vortex with TC of 4 would be generated, and the whole superposition would have the TC equal to 4.

The same holds for multiple-ringed LG beams. In Figure 3.30 we show a superposition of two LG beams of different colors with the same parameters as in Figure 3.29, but the radial indices are $p_1 = p_2 = 2$ (each beam has three rings). For equating the beam powers, the weight coefficients of the superposition are chosen equal to $c_1 = (p_1!/(n_1 + p_1)!)^{1/2} \approx 0.053$ and $c_2 = (p_2!/(n_2 + p_2)!)^{1/2} \approx 0.129$.

Since the TC of the blue beam is greater than that of the red beam, at a small propagation distance the radii of all three rings of the blue beam exceed the radii of the corresponding three rings of the red beam Figure 3.30(a)). Since the red beam diverges stronger than the blue beam, at a distance of nearly 1 m the radii of the inner (most bright) rings become equal and a single purple ring appears (Figure 3.30 (d)), whereas the radii of two outer red rings already exceed those of the blue outer rings. On further propagation, at a distance of nearly 2 m, all three red rings become outer with respect the to three blue rings (Figure 3.30(g)).

According to the theory, the presence of several rings and their radii do not affect the common topological charge of the whole superposition and, as in Figure 3.29, it

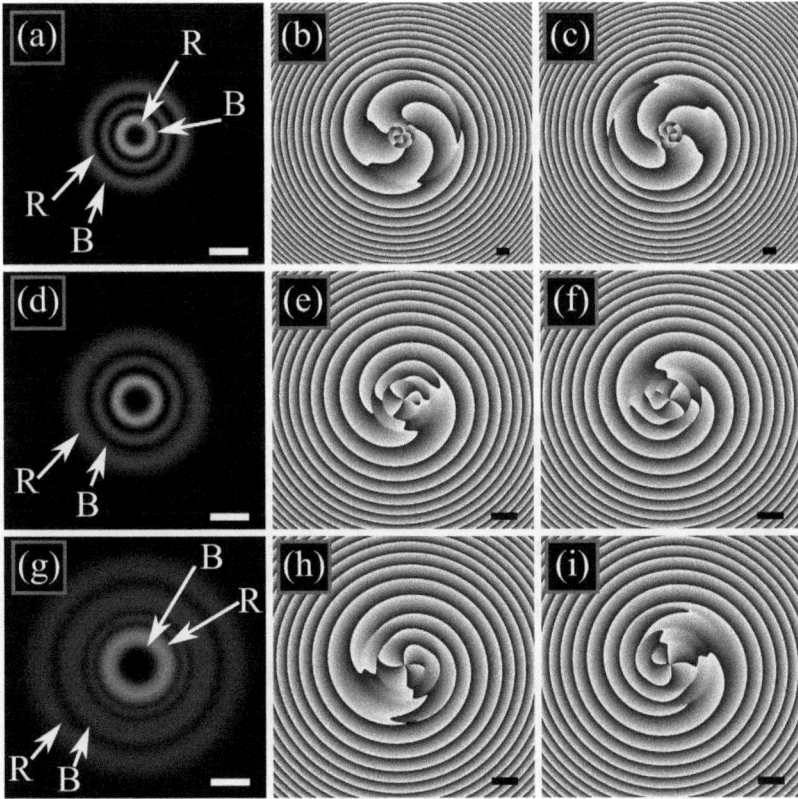

FIGURE 3.30 Distributions of time-averaged intensity (a, d, g) and of instantaneous phase (b, c, e, f, h, i) of a superposition of two three-ringed LG beams of different colors ($\lambda_1 = 400$ nm, $\lambda_2 = 700$ nm) at a distance $z = 0.2$ m (a–c), $z = 1$ m (d–f), $z = 2$ m (g–i) and at time moments $t_1 = 60$ s (b, e, h) and $t_2 = 61$ s (c, f, i). Other computation parameters are the following: Gaussian beam waist radius $w = 500$ μm, TCs $n_1 = 4$ and $n_2 = 3$, radial indices $p_1 = p_2 = 2$, superposition weight coefficients $c_1 \approx 0.053$ and $c_2 \approx 0.129$ (so that both beams are of the same power). The scaling mark in each figure shows 1 mm. Symbols "R" and "B" (a, g) show areas where, respectively, red and blue beam dominates.

should be equal to three ($TC = 3$). Computation of the topological charge by Berry's formula using the instantaneous phase distributions from Figure 3.30 confirms this. Numerically obtained values are 2.987 (Figure 3.30 (b)), 2.982 (Figure 3.30 (c)), 2.874 (Figure 3.30 (e, f)), and 2.907 (Figure 3.30 (h, i)).

Numerical simulation of a three-color vortex

Here we try in a similar way to construct a superposition of three Laguerre–Gaussian beams of three different colors that are spatially separated in the initial plane, then add up in a single ring and form an optical vortex that looks white or gray (superposition of red, green, and blue beams), and then separated again. At a distance z

from the initial plane, the radius of the maximal-intensity ring of each of these three vortices is equal to:

$$r_{\text{max},s} = w \sqrt{\frac{n_s}{2}} \sqrt{1 + \left(\frac{z}{z_s} \right)^2}.$$

(3.78)

If these radii are almost equal for all three vortices, the following condition should be fulfilled:

$$n_1 \left[1 + \left(\frac{z}{z_1} \right)^2 \right] \approx n_2 \left[1 + \left(\frac{z}{z_2} \right)^2 \right] \approx n_3 \left[1 + \left(\frac{z}{z_3} \right)^2 \right].$$

(3.79)

It is seen that three vortices cannot add up at small propagation distances ($z \ll z_1, z_2, z_3$), since in this case, their topological charges should be nearly equal, and therefore in the initial plane these three rings cannot be spatially separated. Now we suppose that the beams add up in the Fresnel diffraction zone and the propagation distance is equal, for instance, to the average Rayleigh distance of the three beams. Let $\lambda_1 < \lambda_2 < \lambda_3$. Then, since the waist radius is the same for all three beams, putting $z = z_2$ in Equation (3.79), we get the following condition for the topological charges:

$$\frac{n_2}{n_1} \approx \frac{1}{2} \left[1 + \left(\frac{\lambda_1}{\lambda_2} \right)^2 \right], \quad \frac{n_2}{n_3} \approx \frac{1}{2} \left[1 + \left(\frac{\lambda_3}{\lambda_2} \right)^2 \right].$$

(3.80)

For example, if the wavelengths of the interfering beams are equal to $\lambda_1 = 400$ nm, $\lambda_2 = 550$ nm, $\lambda_3 = 700$ nm, then $n_2/n_1 \approx 0.76 \approx 3/4$ and $n_2/n_3 \approx 1.31 \approx 4/3$. For example, we can choose $n_1 = 16$, $n_2 = 12$, and $n_3 = 9$. However, these topological charges are small and, according to Equation (3.78), the radii of the maximal-intensity rings in the initial plane are close to each other: $r_{\text{max},1} \approx 2.83w$, $r_{\text{max},2} \approx 2.45w$, and $r_{\text{max},3} \approx 2.12w$. Since each ring has a thickness comparable to the waist radius w, these three rings are not spatially separated in the initial plane. Therefore, for computations, we increase the order of each beam four times: $n_1 = 64$, $n_2 = 48$, and $n_3 = 36$.

Shown in Figure 3.31 are time-averaged intensity distributions and instantaneous phase distributions of a superposition of three single-ringed LG beams of different colors ($\lambda_1 = 400$ nm, $\lambda_2 = 550$ nm, and $\lambda_3 = 700$ nm) at a distance $z = 0$ m, $z = z_3/2 \approx 0.56$ m, $z = z_2 \approx 1.43$ m, and $z = 2z_1 \approx 3.93$ m and at a time moment $t = 60$ s. Other computation parameters are the following: Gaussian beam waist radius $w = 500$ µm, topological charges $n_1 = 64$, $n_2 = 48$, and $n_3 = 36$, radial indices $p_1 = p_2 = p_3 = 0$, superposition weight coefficients $c_s = 1/(n_s!)^{1/2}$, $s = 1,2,3$ (so that all the beams are of the same power).

Since for the chosen parameters the topological charge of the beams decreases with the wavelength, then in the initial plane and at a small distance from it the red ring is within the green ring, and the green ring is within the blue ring (Figure 3.31(a, c)). It is also seen that, due to the divergence in propagation, the borders between the beams (low-intensity rings) in Figure 3.31(c) become less distinct than in

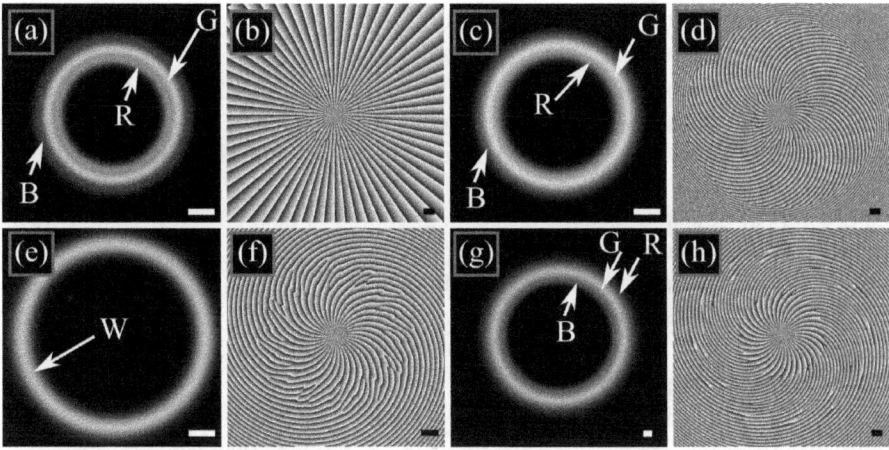

FIGURE 3.31 Distributions of time-averaged intensity (a, c, e, g) and of instantaneous phase (b, d, f, h) of a superposition of three single-ringed LG beams of different colors ($\lambda_1 =$ 400 nm, $\lambda_2 = 550$ nm, $\lambda_3 = 700$ nm) at a distance $z = 0$ m (a, b), $z = z_3/2 \approx 0.56$ m (c, d), $z = z_2 \approx 1.43$ m (e, f), and $z = 2z_1 \approx 3.93$ m (g, h) and at a time moment t = 60 s. Other computation parameters are the following: Gaussian beam waist radius $w = 500$ μm, topological charges $n_1 = 64$, $n_2 = 48$, and $n_3 = 36$, radial indices $p_1 = p_2 = p_3 = 0$, superposition weight coefficients cs $= 1/(ns!)^{1/2}$, s = 1,2,3 (so that all the beams are of the same power). The scaling mark in each figure shows 1 mm. Symbols "R", "G", and "B" (a, c, g) show areas where, respectively, red, green, and blue beam dominates. Symbol "W" shows a white (light-gray) ring, generated by the addition of all three beams.

Figure 3.31(a). Since the divergence of the beams increases with the wavelength, the red beam diverges most strongly and the blue beam – most weakly. Choosing the topological charges according to the condition in Equation (3.80), at the Rayleigh distance of the green beam ($z = z_2 \approx 1.43$ m), all three beams have the light ring of the same radius and therefore the diffraction pattern looks like a single white (light-gray) ring (Figure 3.31(e)). On further propagation, due to the different divergence, the rings separate again, but now the blue ring becomes inner and the red ring becomes outer (Figure 3.31(g)).

According to the theory, a common topological charge of the whole superposition is equal to the topological charge of the beam with a larger wavelength. For the chosen parameters, it should be equal to $TC = 36$. Numerical computation by using the instantaneous phase distributions from Figure 3.31 by Berry's formula in Equation (3.66) over a circle with the radius $R_1 \approx 9.3$ mm yields the values 63.944 (Figure 3.31(b)), 35.847 (Figure 3.31(d)), 35.775 (Figure 3.31(f)), and 35.934 (Figure 3.31(h)), i.e. computation confirms the theoretical value $TC = 36$ in all transverse planes excepting the initial one. In the initial plane, the value $TC = 64$ was obtained since the peripheral vortices of the order –1 are outside the computation area, but even at $z = z_3/2$ (half of the minimum of the three Rayleigh distances, for the red beam), there are 28 vortices of the order –1 on the phase distribution (Figure 3.31(d) and therefore the net topological charge is 36 (64 –28).

Numerical simulation of three-color vortex with different permutations of light rings colors

In all the considered examples (Figures 3.29–3.31), the topological charge of the beams decreases with the wavelength. Thus, it can seem that the topological charge of the whole superposition is equal to the minimal topological charge, rather than to the topological charge of the beam with the largest wavelength. In order to confirm the theoretical outcome that the topological charge is defined solely by the wavelength, below we consider all possible permutations of three colors and three topological charges. There are six such permutations.

Figure 3.32 depicts time-averaged intensity distributions and instantaneous phase distributions of these six different superpositions of three single-ringed LG beams of different colors ($\lambda_1 = 400$ nm, $\lambda_2 = 550$ nm, and $\lambda_3 = 700$ nm) at a distance of $z = z_2/2 \approx 0.71$ m and at a time moment $t = 60$ s. Other computation parameters are the following: Gaussian beam waist radius $w = 500$ μm, topological charges $n_1 = 1$, $n_2 = 2$, and $n_3 = 4$ (a, b), $n_1 = 1$, $n_2 = 4$, and $n_3 = 2$ (c, d), $n_1 = 2$, $n_2 = 1$, and $n_3 = 4$ (e, f), $n_1 = 2$, $n_2 = 4$, and $n_3 = 1$ (g, h), $n_1 = 4$, $n_2 = 1$, and $n_3 = 2$ (I,j), $n_1 = 4$, $n_2 = 2$, and $n_3 = 1$ (k, l), radial indices $p_1 = p_2 = p_3 = 0$, superposition weight coefficients $c_s = 1/(n_s!)^{1/2}$, $s = 1, 2, 3$ (to make the beam powers equal to each other).

According to the theory, the topological charge should be equal to 4, 2, and 1 on those figures where the red ring is respectively outer (Figure 3.32 (a, b, e, f)), middle (Figure 3.32 (c, d, i, j)), and inner (Figure 3.32 (g, h, k, l)). Numerical computation by using the instantaneous phase distributions from Figure 3.32 by Berry's formula in Equation (3.66) over a circle with the radius $R_1 \approx 5$ mm yields the value 3.996 (Figure 3.32 (a, b, e, f)), 1.998 (Figure 3.32 (c, d, i, j)), and 0.999 (Figure 3.32 (g, h, k, l)), i.e. the computations confirm the corresponding theoretical values. In addition, the topological charge can be determined on the phase distributions visually. Each of Figure 3.32 (b, f) contains four optical vortices of the order +1 and thus the net total topological charge is $TC = 4$. In Figure 3.32 (d, j), in addition to the four vortices of the order +1 there are two vortices of the order –1, i.e. the net topological charge equals $TC = 2$. In Figure 3.32 (h, l), there are already three such vortices of the order –1 and therefore the net topological charge is $TC = 1$.

Thus, in this section, we have investigated different variants of a coaxial superposition of the Laguerre–Gaussian beams with different wavelengths. Using Berry's well-known formula, we derived the topological charge of a coaxial superposition of two Laguerre–Gaussian beams of different "colors", each with its own wavelength and its own TC. It turned out that the TC of such a superposition equals the TC of the LG beam with a larger wavelength (more red), regardless of the weight coefficient of this beam in the superposition and regardless of the TC of this beam. This TC derivation can be generalized to a superposition of an arbitrary finite number of the LG beams with different wavelengths, both single-ringed and multiple-ringed. At this, the TC of the whole superposition is equal to the TC of the constituent LG beam with a larger wavelength. This result was confirmed numerically for a superposition of three single-ringed beams and for a superposition of two three-ringed beams. Since the phase velocities of the beams are different, the transverse intensity section of the beam changes on propagation with a velocity proportional to the

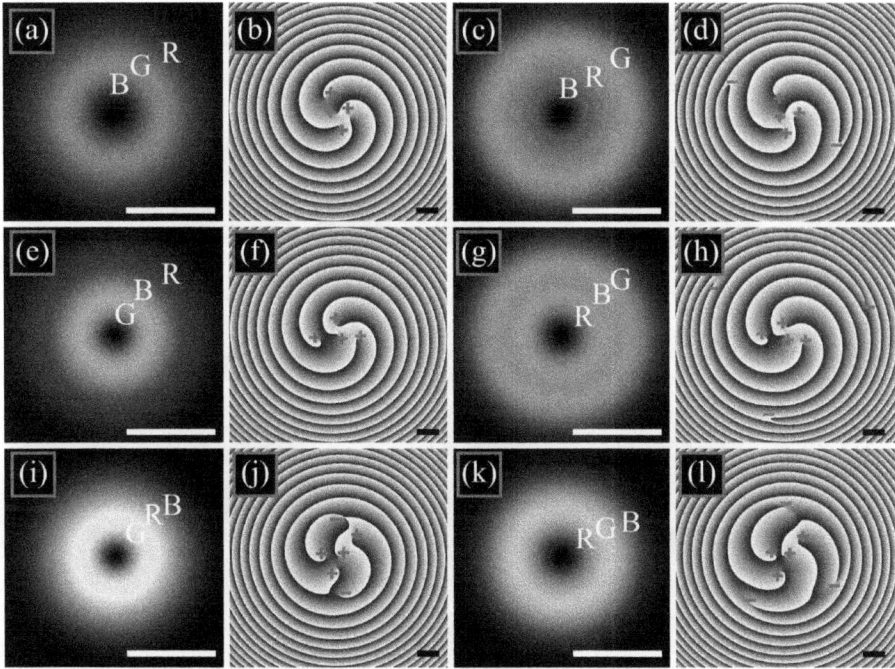

FIGURE 3.32 Distributions of time-averaged intensity (a, c, e, g, i, k) and of instantaneous phase (b, d, f, h, j, l) of six different superpositions of three single-ringed LG beams of different colors ($\lambda_1 = 400$ nm, $\lambda_2 = 550$ nm, and $\lambda_3 = 700$ nm) at a distance $z = z_2/2 \approx 0.71$ m (half of the Rayleigh distance of the beam with the average wavelength) and at a time moment t = 60 s. Other computation parameters are the following: Gaussian beam waist radius $w = 500$ μm, topological charges $n_1 = 1$, $n_2 = 2$, and $n_3 = 4$ (a, b), $n_1 = 1$, $n_2 = 4$, and $n_3 = 2$ (c, d), $n_1 = 2$, $n_2 = 1$, and $n_3 = 4$ (e, f), $n_1 = 2$, $n_2 = 4$, and $n_3 = 1$ (g, h), $n_1 = 4$, $n_2 = 1$, and $n_3 = 2$ (i, j), $n_1 = 4$, $n_2 = 2$, and $n_3 = 1$ (k, l), radial indices $p_1 = p_2 = p_3 = 0$, superposition weight coefficients $c_s = 1/(ns!)^{1/2}$, s = 1,2,3 (so that all the beams are of the same power). The scaling mark in each figure shows 1 mm. Symbols "R", "G", and "B" (a, c, g) show areas where, respectively, red, green, and blue beam dominates. Red symbols "+" and "–" in the phase distributions show the optical vortices of the orders +1 and –1.

difference between the maximal and minimal wavelengths. At the same time, the instantaneous *TC* of such a superposition is conserved, while the intensity distribution (time-averaged) of the "colored" optical vortex changes its light "gamut". For example, for a two-color superposition, if in the near field the colors of the light rings (rainbow) arrange along a radius according to their TCs in the superposition from lower to greater, then, on space propagation (in the far field) the colors of the light rings in the rainbow are arranged in the reverse order, from the greater *TC* to the lower *TC*. It was also demonstrated that by appropriately choosing the wavelengths (blue, green, and red) in a three-color superposition of single-ringed LG beams it is possible to generate at some distance a time-averaged light ring of the white color.

4 Optical Vortex Beams with an Infinite Topological Charge

4.1 PROPAGATION-INVARIANT LASER BEAMS WITH AN ARRAY OF PHASE SINGULARITIES

In 1992, L. Allen [33] has shown that light beams with helical wavefronts (optical vortices) carry the orbital angular momentum (OAM). Such light fields are characterized also by another quantity, the topological charge (TC). In optical communications, data can be encoded into different OAM states, and, theoretically, the number of these states is unlimited. In practice; however, the generation of large-topological-charge vortex beams is not easy. This motivated researchers to study light beams with large values of TC or OAM, methods of obtaining, detecting, and converting such beams, as well as their applications. In 1997, it was shown that achieving large OAM does not require the optical vortex at all. It can be done for a Gaussian beam by a simple cylindrical lens [17]. Later, in 2012, high-order optical vortices were generated by spiral phase mirrors [177], produced by direct machining with a diamond tool. Thus, high-quality optical vortices were produced with TCs ranging from 1 to upwards of 100 at a wavelength of 532 nm. In 2015, optical vortices were created by using a digital micromirror device (DMD) [178]. Adopting either the binary Lee hologram or the superpixel binary encoding technique, optical vortices were generated with a topological charge as large as 90. In a recent work [179], the cross-phase was applied to experimentally generate the Laguerre–Gaussian beams and to measure their topological charges up to 200. In [180], cylindrical vector beams were generated with TC up to 14, but in a compact solid-state Yb:YAG microchip laser. As to measuring large TC values, we can mention works [15,181,182]. In [15], a cylindrical lens was used to split the vortex beam into an array of light spots, which allowed determining TC up to 100. In [181], based on the Mach–Zehnder configuration, a self-referenced interferometric method was developed to estimate the magnitude and sign of an incoming optical vortex. Topological charges of up to 130 were measured. In [182], a method of TC detection by using gradually changing-period spiral spoke grating was developed. The detection of TCs up to 160 were demonstrated. In [183], a method was developed to enhance the modal purity (to reduce unwanted radial modes) of classical OAM beams and in the quantum detection of OAM photons. High-purity OAM modes of the orders as high as 100 were obtained.

DOI: 10.1201/9781003326304-4

In [184], hollow ring core silica photonic crystal fiber was proposed that can support 101 OAM-modes, maintaining a high mode quality without phase distortion, a low confinement loss, and a large effective index difference between the adjacent modes. In [185], a spiral phase mirror was used, which generated photons with more than 10,000 quanta of OAM. As to applications, in [186], OAM states up to 15 were used for data encoding in a free-space one-to-many multicasting link. Finally, in [187], multi-vortex laser beams with TC greater than 100 were created at the laser source (in the cavity mirror) for enabling spatial and temporal encoding in optical data transmission.

So, the question arises: how large the TC can be, at least theoretically? Can it even be infinite? For light beams with optical vortices, but free of circular symmetry, the quantities TC and OAM are different. Whereas the OAM depends both on the intensity and phase distributions in the transverse plane [17,188], TC depends only on the phase distribution [19] and it is equal to the sum of orders of all vortices in the beam. Gaussian beams with multiple optical vortices were theoretically investigated by G. Indebetouw [49] and, in more general form, structurally stable (form-invariant) Gaussian beams were studied in [164,189].

In this section, adopting the approach from [164,189], we construct theoretically four examples of scalar light fields with an infinite topological charge. The first example is the Gaussian beam with a vortex-argument cosine function, whereas the second example is the Gaussian beam with a vortex-argument Bessel function. The third example is the Gaussian beam with the cosine function of the squared vortex-argument and, finally, the fourth example is the Gaussian beam with the vortex-argument cosine function raised to power. For the first, third, and fourth beams, we study analytically their intensity distributions and their OAMs, which are shown to be finite. All fields are form-invariant. At certain parameters, the shape of the first, second, and fourth beams contains two bright light spots accompanied by pale (invisible) local low-intensity maxima. Therefore, these beams are visually similar to propagation-invariant Hermite–Gaussian beams of the order (1,0) [4] and to the astigmatic Fourier-invariant Gaussian beams [190]. However, Hermite–Gaussian beams have zero TC, and astigmatic Fourier-invariant Gaussian beams have finite TC, while the TC of the beams studied here is infinite.

4.1.1 FORM-INVARIANT GAUSSIAN BEAMS

It has been shown in [164,189] (expression (6.1) in [189]) that any function given by:

$$E_{\pm}\left(x,y,z\right) = \frac{1}{q}\exp\left(-\frac{x^2+y^2}{qw_0^2}\right)f\left(\frac{x\pm iy}{qw_0}\right), \qquad (4.1)$$

where (x, y, z) are the Cartesian coordinates, w_0 is the waist radius of the Gaussian beam, $q = 1 + iz/z_0$ ($z_0 = kw_0^2/2$ is the Rayleigh range, $k = 2\pi/\lambda$ is the wavenumber),

$f(x \pm iy)$ is an arbitrary entire analytical function, describes a solution of the scalar paraxial Schrödinger-type Helmholtz equation:

$$2ik \frac{\partial E}{\partial z} + \frac{\partial^2 E}{\partial x^2} + \frac{\partial^2 E}{\partial y^2} = 0, \tag{4.2}$$

where E is a transverse component of a linearly polarized field. In this section, we limit our study only to scalar paraxial fields since, in other cases, spin-orbit effects can modify the fine structure of the beams under consideration.

Expression in Equation (4.1) describes form-invariant light fields, whose transverse intensity distribution is conserved on space propagation (up to scale and rotation around the optical axis).

In addition, expression in Equation (4.1) allows an analytical description of light fields with optical vortices located in the initial plane ($z = 0$) in arbitrary points with the Cartesian coordinates (a_j, b_j) ($j = 1, ..., m$) [49]:

$$E(x,y,z) = \frac{1}{q} \exp\left(-\frac{x^2 + y^2}{qw_0^2}\right) \prod_{j=1}^{m} \left(\frac{x+iy}{q} - a_j - ib_j\right). \tag{4.3}$$

The light field in Equation (4.3) contains a finite number m of optical vortices. Below we investigate four light fields with an infinite array of optical vortices. More of it, all these vortices are of the same sign and therefore the topological charge of such fields is infinite.

4.1.2 GAUSSIAN BEAM WITH A VORTEX-ARGUMENT COSINE FUNCTION

Theory

If we choose the following function $f(Z)$ (Z is an arbitrary complex argument) in the complex amplitude in Equation (4.1):

$$f(Z) = \cos\left(\frac{w_0}{\alpha_0} Z\right), \tag{4.4}$$

then the complex amplitude of the light field in the initial plane reads as:

$$E(x,y,0) = \cos\left(\frac{x+iy}{\alpha_0}\right) \exp\left(-\frac{x^2 + y^2}{w_0^2}\right), \tag{4.5}$$

while in an arbitrary transverse plane at a distance z it is given by:

$$E(x, y, z) = \frac{1}{q} \cos\left(\frac{x+iy}{\alpha_0 q}\right) \exp\left(-\frac{x^2+y^2}{qw_0^2}\right). \tag{4.6}$$

The quantity α_0 defines the distance between the optical vortices in the initial field and is inverse proportional to the linear optical vortices density $\rho_0 = 1/(\pi \alpha_0)$, i.e. number of vortices per unit length.

Rotating the coordinates by an angle equal to the Gouy phase [4]:

$$\zeta(z) = \arctan\left(\frac{z}{z_0}\right), \tag{4.7}$$

we introduce new Cartesian coordinates:

$$\begin{pmatrix} x' \\ y' \end{pmatrix} = \begin{pmatrix} \cos\zeta & \sin\zeta \\ -\sin\zeta & \cos\zeta \end{pmatrix}\begin{pmatrix} x \\ y \end{pmatrix} = \frac{1}{\sqrt{1+z^2/z_0^2}}\begin{pmatrix} x+yz/z_0 \\ y-xz/z_0 \end{pmatrix}, \tag{4.8}$$

and the complex amplitude in these coordinates reads as:

$$E(x', y', z) = \frac{w_0}{w(z)} \cos\left[\frac{x'+iy'}{\alpha(z)}\right]$$

$$\times \exp\left[-\frac{x'^2+y'^2}{w^2(z)} + \frac{ik(x'^2+y'^2)}{2R(z)} - i\zeta(z)\right], \tag{4.9}$$

where

$$w(z) = w_0\sqrt{1+\frac{z^2}{z_0^2}}, \quad \alpha(z) = \alpha_0\sqrt{1+\frac{z^2}{z_0^2}}, \quad R(z) = z\left(1+\frac{z_0^2}{z^2}\right). \tag{4.10}$$

Below, the argument z of the functions ζ, w, α, R from Equations (4.7) and (4.10) is omitted for brevity.

Equation (4.10) indicates that the waist radius of the Gaussian beam $w(z)$ and the period of the cosine function $2\pi \alpha(z)$ are scaled by the same coefficient. Therefore, the field in Equation (4.9) conserves its intensity shape on propagation. Its intensity distribution only changes in scale and rotates. The phase distribution also scales and rotates, but acquires a spherical wavefront, which does not affect the positions of optical vortices though.

According to Equation (4.9), intensity nulls are located uniformly on the line $y' = 0$ in the points $x' = \alpha(\pi/2 + \pi p)$ (p are integer numbers).

To determine the topological charge, we introduce local polar coordinates (δ, θ) in the vicinity of an arbitrary intensity null (δ, θ). Thus, near the intensity null, in a point ($x' = \alpha\,(\pi\,/2 + \pi p) + \delta\,\cos\theta$, $y' = \delta\,\sin\theta$) the phase of the vortex factor in Equation (4.9) is given by:

$$\arg\cos\left(\frac{x'+iy'}{\alpha}\right) = \arg\cos\left[\frac{\alpha\left(\pi/2+\pi p\right)+\delta e^{i\theta}}{\alpha}\right]$$

$$= \arg\cos\left[\pi/2+\pi p+\frac{\delta}{\alpha}e^{i\theta}\right] \tag{4.11}$$

$$= \arg\left\{(-1)^{p+1}\sin\left[\frac{\delta}{\alpha}e^{i\theta}\right]\right\} \approx \arg\left\{(-1)^{p+1}e^{i\theta}\right\} = \theta + \text{const.}$$

The Gaussian envelope does not affect the phase distribution and therefore, according to Equation (4.11), the topological charge of an optical vortex around each intensity null (screw dislocation) is unitary. Since the number of intensity nulls is infinite, the total topological charge of the light field in Equation (4.9) is also infinite.

Now we determine the intensity shape of the light field in Equation (4.9). The intensity distribution reads as:

$$I\left(x',y',z\right) = \frac{1}{2}\frac{w_0^2}{w^2}\exp\left(-2\frac{x'^2+y'^2}{w^2}\right)$$

$$\times\left[\cos\left(\frac{2x'}{\alpha}\right)+\cosh\left(\frac{2y'}{\alpha}\right)\right]. \tag{4.12}$$

Obtaining the partial derivatives $\partial\,I/\partial\,x'$ and $\partial\,I/\partial\,y'$, we get the following equations to determine the intensity maxima:

$$\begin{cases} -\sin\left(\dfrac{2x'}{\alpha}\right) = \left[\cos\left(\dfrac{2x'}{\alpha}\right)+\cosh\left(\dfrac{2y'}{\alpha}\right)\right]\dfrac{2\alpha x'}{w^2}, \\[4mm] \sinh\left(\dfrac{2y'}{\alpha}\right) = \left[\cos\left(\dfrac{2x'}{\alpha}\right)+\cosh\left(\dfrac{2y'}{\alpha}\right)\right]\dfrac{2\alpha y'}{w^2}. \end{cases} \tag{4.13}$$

One obvious solution of these equations is the point $(x', y') = (0, 0)$. The intensity at this point is $I(0, 0, z) = (w_0/w)^2$.

If $x' \neq 0$ and $y' \neq 0$, the system in Equation (4.13) is reduced to an equation $-(1/x')\sin(2x'/a) = (1/y')\sinh(2y'/a)$. However, the equation $-\sin u/u = \sinh v/v$ does not have real solutions (it can be easily seen on the plots of these functions that the right part is equal to or greater than 1, whereas the left part is less than 1). Therefore, all the local intensity maxima can reside only on the Cartesian axes x' and y'.

If $y' = 0$ and $x' \neq 0$, the system in Equation (4.13) leads to an equation $[\sin(x'/\alpha)+\cos(x'/\alpha)(2\alpha x'/w^2)]\cos(x'/\alpha) = 0$. The points x' with $\cos(x'/a) = 0$ are the local intensity minima, whereas the intensity maxima reside in points given by an equation $\tan(x'/\alpha) = -2\alpha x'/w^2$. The number of these maxima is infinite and, according to Equation (4.12), the intensity of each such maximum is equal to $I(x',0,z) = (w_0/w)^2 \cos^2(x'/\alpha)\exp(-2x'^2/w^2)$ and does not exceed the value $(w_0/w)^2$.

Similarly, if $x' = 0$ and $y' \neq 0$, the system in Equation (4.13) reduces for the following equation for the local intensity maxima: $\tanh(y'/\alpha) = 2\alpha y'/w^2$. This equation can have either one root ($y' = 0$) or two additional roots. If $\alpha \ll w$ (i.e. $\alpha_0 \ll w_0$), two nonzero roots are approximately equal to $y' \approx w^2/(2\alpha)$ and the intensity in these points is $I(0,y',z) = [w_0/(2w)]^2 \exp[w^2/(2\alpha^2)]$ and can greatly exceed the intensity of the central maximum $(w_0/w)^2$. Therefore, if $\alpha_0 \ll w_0$, the intensity distribution contains two symmetric global intensity maxima on the axis $x' = 0$ and an infinite number of local maxima on the axis $y' = 0$, between which there are isolated intensity nulls surrounded by optical vortices with the topological charge +1.

With increasing α (or decreasing w), the two intensity maxima become closer and at the condition $\alpha_0 \geq w_0/\sqrt{2}$, the equation $\tanh(y'/\alpha) = 2\alpha y'/w^2$ has only one root $y' = 0$. In this case, instead of two light spots, the diffraction pattern contains a single spot in the center accompanied by secondary low-intensity maxima.

Light beams in Equation (4.9) can be generated experimentally since their initial complex amplitude in Equation (4.5) can be rewritten as follows:

$$
E(x,y,0) = \frac{1}{2}\exp\left\{-\frac{x^2 + (y+y_c)^2}{w_0^2} + ikx\sin\theta - \frac{w_0^2}{4\alpha_0^2}\right\}
$$

$$
+ \frac{1}{2}\exp\left\{-\frac{x^2 + (y-y_c)^2}{w_0^2} - ikx\sin\theta - \frac{w_0^2}{4\alpha_0^2}\right\},
$$

(4.14)

with $y_c = w_0^2/(2\alpha_0)$ and $\sin\theta = 1/(k\alpha_0)$. It is seen that this field is a superposition of two symmetric off-axis Gaussian beams with a tilted wavefront. The shift of the beam centers is along the coordinate y, whereas the tilt of the wavefront is along the coordinate x. Two such Gaussian beams can be combined interferometrically, as was done in [191], although it is impossible to see the infinite number of vortices experimentally.

The increasing distance from the optical axis to the two light spots ($y' \approx w^2/(2\alpha)$) with increasing density of optical vortices (i.e. when α decreases) has an analogy with the classical rotationally symmetric optical vortex beams with a single on-axis phase singularity, for example, with the Laguerre–Gaussian beams (Figure 4.1). When their topological charge (vortex order) grows, the light ring widens, i.e. its radius grows as well (Figure 4.1(b, e)). Thus, the stronger phase gradients in the center (Figure 4.1 (a, d)) push the light ring out. So, the optical vortex affects the intensity distribution, even though the vortex itself is in the low-intensity central area. In this area, it is difficult (if even possible) to measure the phase directly and determine the topological charge. However, it can be done indirectly, on the light ring, when the beam interferes with a plane wave (Figure 4.1 (c, f)).

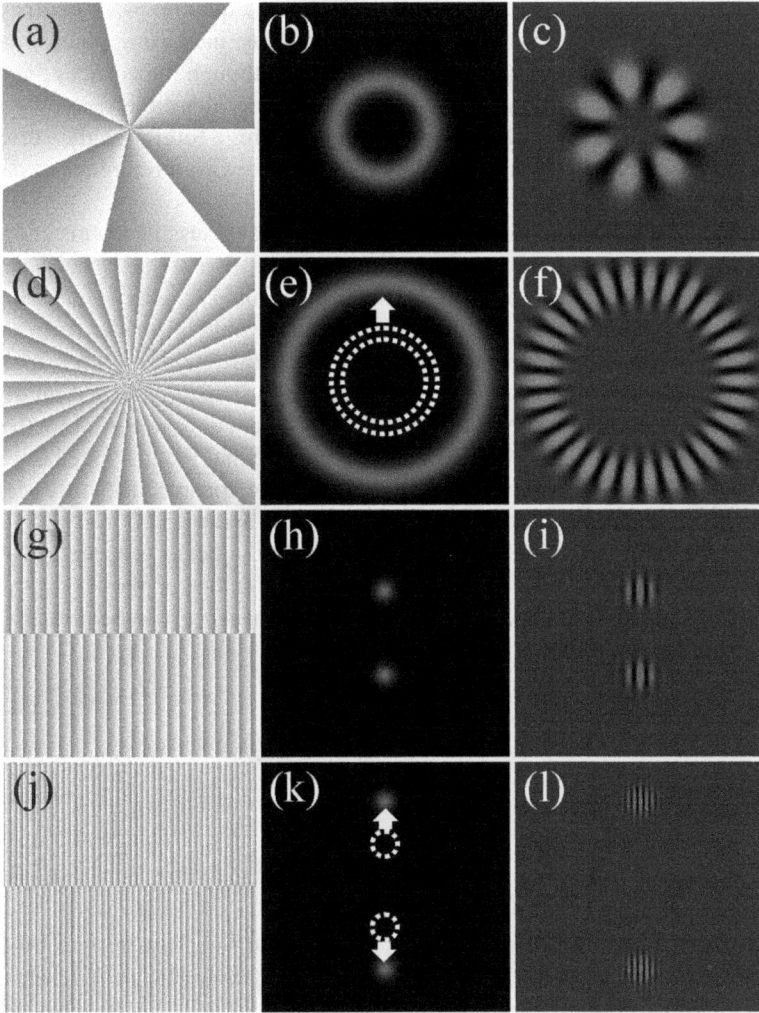

FIGURE 4.1 Single-ringed Laguerre–Gaussian beams (a–f) versus the infinite-topologi-cal charge beams given by Equation (4.9) (g–l): distributions of phase (a, d, g, j), intensity (b, e, h, k), and intensity with a reference plane wave (c, f, i, l). Increasing the topological charge of the Laguerre–Gaussian beams (a, d) pushes the light ring out and its radius grows (b, e). Increasing the density of optical vortices (g, j) pushes out the two light spots (h, k). Interference with a plane wave allows indirect determination of the topological charge of the Laguerre–Gaussian beam (c, f) and the optical vortices density of the beam from Equation (4.9) (i, l). Dashed contours (e, k) show the light ring or spots before increasing the vortex order (b) or density of the vortices (h).

The same effect takes place for the light beam in Equation (4.9), i.e. increasing density (Figure 4.1 (g, j)) of optical vortices (and, thus, increasing phase gradients) pushes out the two light spots (Figure 4.1 (h, k)), even though the vortices themselves are also in the low-intensity area (the area is; however, along the whole horizontal straight line instead of the dark area within the light ring). Similarly, the vortex density can be determined also indirectly by interference with a reference plane wave (Figure 4.1 (i, l)).

We note that widening the light ring of the Laguerre–Gaussian beams and retreating the two light spots of the beams in Equation (4.9) is not solely because of the optical vortices. If some beam would be the same as the Laguerre–Gaussian beam, but without the optical vortex, its complex amplitude would be written as $(r/w)^n \exp(-r^2/w^2)$ and its light ring would also grow with n. However, the vortex makes the Laguerre–Gaussian beam a mode with its transverse shape conserved on propagation. Similarly, without the vortex in the cosine argument in Equation (4.9), the intensity distribution in Equation (4.12) would still indicate that the distance between the two intensity maxima is proportional to $(1/\alpha_0)$, but the vortex makes this beam also a mode, which, on propagation, changes only in scale and rotates.

Despite the infinite number of optical vortices in the beam in Equation (4.9), its OAM is finite. This happens because only a small number of vortices reside near high-intensity areas. Below we derive the OAM and power of the beam in Equation (4.9) by using the well-known expressions [17,188]:

$$J_z = \text{Im} \int_{-\infty}^{\infty} \int_{-\infty}^{\infty} E^*(x,y)\left(x\frac{\partial}{\partial y} - y\frac{\partial}{\partial x}\right)E(x,y)\,dxdy, \tag{4.15}$$

$$W = \int_{-\infty}^{\infty} \int_{-\infty}^{\infty} E^*(x,y)E(x,y)\,dxdy, \tag{4.16}$$

Representing the cosine function from Equation (4.9) as a sum of two exponents, using the reference integrals 2.4.15.3 and 2.5.36.5 from [157], and substituting Equation (4.9) into Equations (4.15) and (4.16), we get:

$$J_z = \frac{\pi w_0^4}{4\alpha_0^2}\sinh\left(\frac{w_0^2}{2\alpha_0^2}\right), \tag{4.17}$$

$$W = \frac{\pi w_0^2}{2}\cosh\left(\frac{w_0^2}{2\alpha_0^2}\right). \tag{4.18}$$

Then the normalized OAM (OAM divided by power) reads as:

$$\frac{J_z}{W} = \left(\frac{w_0^2}{2\alpha_0^2}\right)\tanh\left(\frac{w_0^2}{2\alpha_0^2}\right), \tag{4.19}$$

and, if $\alpha_0 \ll w_0$, it can be estimated by a simple expression $J_z/W \approx (w_0/\alpha_0)^2/2$.

Expression (4.19) defines the OAM with respect to the center of the intensity distribution (a middle point between the two light spots). It is an intrinsic OAM and it does not depend on the chosen coordinate system and is minimal among the OAMs derived with respect to other points. It depends solely on the ratio between the Gaussian beam waist radius and the period of the optical vortices (w_0/α_0).

The linear density of the optical vortices in the beam in Equation (4.9) is defined above as the number of vortices per unit length:

$$\rho_0 = \frac{1}{\pi\alpha_0}. \tag{4.20}$$

Thus, if the density of vortices is high ($\alpha_0 \ll w_0$), the OAM is approximately equal to $J_z/W = (\pi w_0\rho_0)^2/2$, i.e. grows with the density parabolically.

Numerical simulation

According to the theory, the intensity distribution consists of either two bright light spots (if $\alpha_0 \ll w_0$), or one central bright spot (if $\alpha_0 \geq w_0/\sqrt{2}$). In this subsection, simulation results are given for both these cases.

Figure 4.2 illustrates the intensity and phase distributions of the beam from Equation (4.9) in the initial plane $z=0$ (Figure 4.2(a, b)) and after propagation in space (Figure 4.2(c–f)). Initial distributions (Figure 4.2(a, b)) are computed by Equation (4.5), while at a distance z they are obtained by using Equation (4.9) (Figure 4.2(c, d)) and by the Fresnel transform implemented numerically as a convolution with using the fast Fourier transform (Figure 4.2(e, f)). The following parameters were used for computation: wavelength $\lambda = 532$ nm, waist radius $w_0 = 0.5$ mm, scaling factor of the vortex-argument cosine-function $\alpha_0 = w_0/4$, propagation distance $z = 1$ m, computation domain $-R \leq x, y \leq R$ with $R = 2$ mm (Figure 4.2(a, b)) and $R = 3$ mm (Figure 4.2(c–f)).

Figure 4.2 demonstrates the coincidence of the distributions obtained theoretically by using Equation (4.9) and numerically by using the Fresnel transform. Figure 4.2 also confirms that all optical vortices reside on a single straight line, which rotates counter-clockwise on propagation, and that, as theory predicts, there are indeed two light spots on the intensity pattern, located on a line orthogonal to the line with the vortices. For the parameters chosen, the normalized OAM of the light field from Figure 4.2 should be equal to $J_z/W = 8$. Numerically obtained values are 7.990329 (Figure 4.2(c, d)) and 7.988885 (Figure 4.2(e, f)). To demonstrate the infinite topological charge, it should be computed along several circles. It can be done by using the phase distribution from Figure 4.2(d), but obtained for a wider area $-R \leq x$, $y \leq R$ with $R = 10$ mm (Figure 4.2). Computation of the topological charge over the circumferences with the radii $R_1 = 3$ mm, 6 mm and 9 mm (shown by dashed curves in Figure 4.3) yields the values TC = 12, 26, 38 (11.98, 25.87 and 37.60), respectively. Integer values are obtained by manually counting the phase singularities within a circumference of the radius R_1 in Figure 4.3, while the approximate fractional values are computed by using Berry's formula [19]:

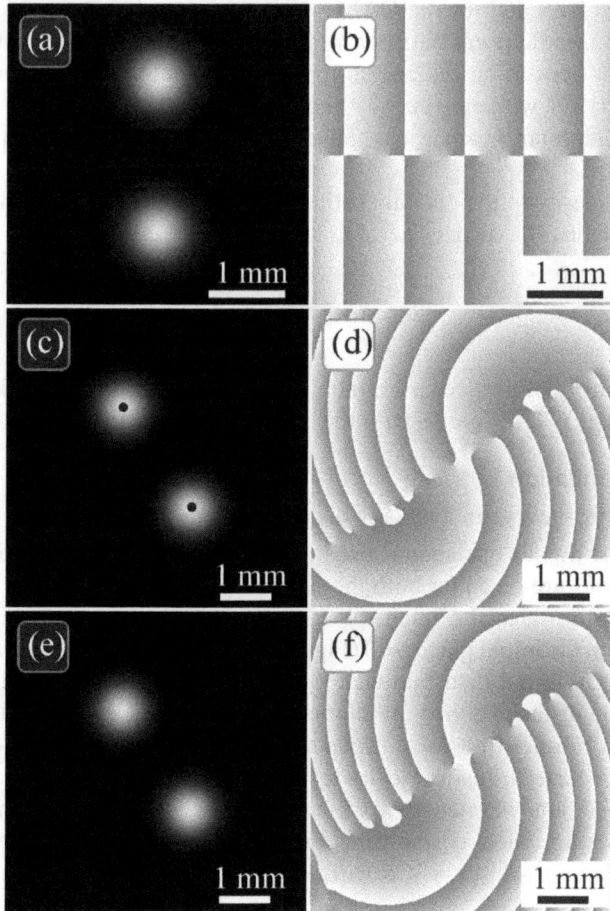

FIGURE 4.2 Distributions of intensity (a, c, e) and phase (b, d, f) of the beam from Equation (4.9) in the initial plane and after propagation in space. Distributions in the initial plane (a, b) are obtained by Equation (4.5), while at a distance from it they are computed by using Equation (4.9) (c, d) and by numerical Fresnel transform (e, f). Computation parameters: wavelength $\lambda = 532$ nm, waist radius $w_0 = 0.5$ mm, scaling factor of the vortex argument cosine-function $\alpha_0 = w_0/4$, propagation distance $z = 1$ m, computation domain $-R \leq x, y \leq R$ with $R = 2$ mm (a, b), and $R = 3$ mm (c–f). Black dots (c) show positions of the intensity maxima obtained by the approximate formula $y' \approx w^2/(2\alpha)$.

$$TC = \frac{1}{2\pi} \lim_{r \to \infty} \int_0^{2\pi} \frac{\partial}{\partial \varphi} \arg E\left(r, \varphi\right) d\varphi, \qquad (4.21)$$

with an infinite-radius circle replaced by a circle of finite radius R_1.

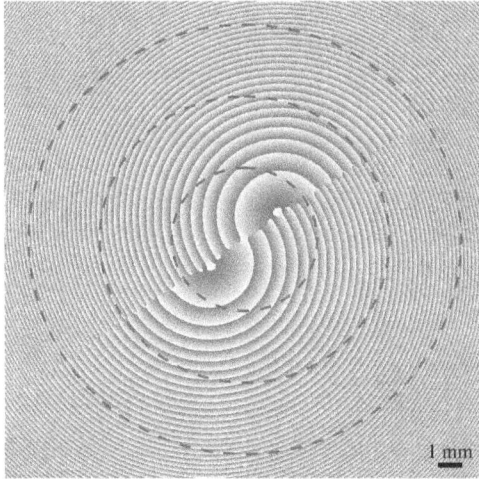

FIGURE 4.3 Phase distribution of the beam from Figure 4.2 in a wider area, obtained by using Equation (4.9).

As is predicted above by theory, increasing α should lead to the merge of the two light spots into one central spot. Figure 4.4 depicts the intensity and phase distributions of the beam from Equation (4.9) in the initial plane and after propagation in space. Initial distributions (Figure 4.4(a, b)) are computed by Equation (4.5), while at a distance they are obtained by using Equation (4.9) (Figure 4.4(c, d)) and by the numerical Fresnel transform (Figure 4.4(e, f)). Computation parameters are the same as in Figure 4.2 but the scaling factor of the vortex-argument cosine-function is $\alpha_0 = w_0$.

Figure 4.4 also demonstrates the coincidence of the distributions obtained theoretically by using Equation (4.9) and numerically by using the Fresnel transform. Figure 4.4(c–f) also show four optical vortices, residing on a single straight line, and one elliptic light spot on the intensity pattern. For the parameters chosen, the normalized OAM of the light field from Figure 4.4 should be equal to $J_z/W = 0.5$ tanh(0.5) $\approx$ 0.231. Numerically obtained values are 0.231 (Figure 4.4(c, d)) and 0.228 (Figure 4.4(e, f)). To demonstrate the infinite topological charge, we also obtain the phase distribution from Figure 4.4(d) in a wider area $-R \leq x, y \leq R$ with $R = 10$ mm (Figure 4.5) and then we compute the topological charge over several circles.

In contrast to Figure 4.3, there are not so many phase singularities in Figure 4.5. Therefore, we marked them with small circles. Computation of the topological charge over the circumferences with the radii $R_1 = 3$ mm, 6 mm, and 9 mm (shown by dashed curves in Figure 4.5) yields the values TC = 4, 6, 10 (3.99, 5.97 and 9.89), respectively. This means that TC grows with increasing radius of the circumference, over which the TC is evaluated.

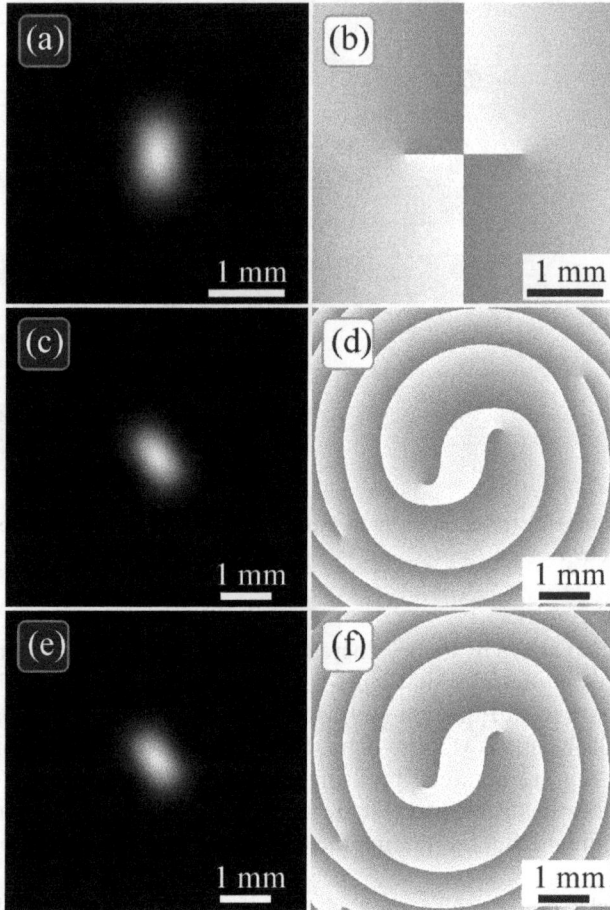

FIGURE 4.4 Distributions of intensity (a, c, e) and phase (b, d, f) of the beam (4.9) in the initial plane and after propagation in space. Distributions in the initial plane (a, b) are obtained by Equation (4.5), while at a distance from it they are computed by using Equation (4.9) (c, d) and by numerical Fresnel transform (e, f). Computation parameters: wavelength λ = 532 nm, waist radius w_0 = 0.5 mm, scaling factor of the vortex argument cosine-function $\alpha_0 = w_0$, propagation distance $z = 1$ m, computation domain $-R \leq x, y \leq R$ with $R = 2$ mm (a, b) and $R = 3$ mm (c–f).

4.1.3 GAUSSIAN BEAM WITH A VORTEX-ARGUMENT BESSEL FUNCTION

Theory

Now we choose another function $f(Z)$ (Z is an arbitrary complex argument) in the complex amplitude in Equation (4.1):

$$f(Z) = J_m\left(\frac{w_0}{\alpha_0} Z\right), \tag{4.22}$$

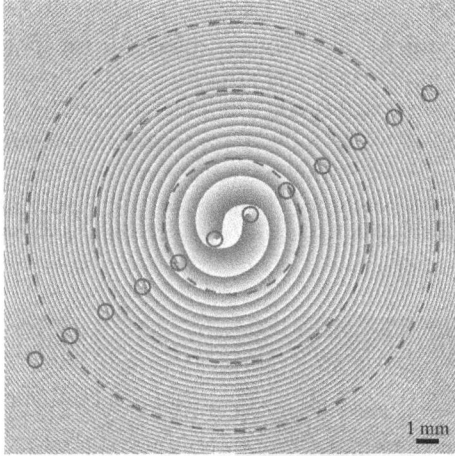

FIGURE 4.5 Phase distribution of the beam from Figure 4.4 in a wider area, obtained by using Equation (4.9). Small circles show the phase singularities.

with $J_m(x)$ being the mth-order Bessel function of the first kind. Then, the complex amplitude in an arbitrary transverse plane is given by:

$$E(x,y,z) = \frac{1}{q} \exp\left(-\frac{x^2+y^2}{qw_0^2}\right) J_m\left(\frac{x+iy}{\alpha_0 q}\right). \tag{4.23}$$

Using the rotated coordinates in Equation (4.8), complex amplitude in Equation (4.23) can be rewritten as:

$$E(x',y',z) = \frac{w_0}{w} J_m\left(\frac{x'+iy'}{\alpha}\right)$$
$$\times \exp\left[-\frac{x'^2+y'^2}{w^2} + \frac{ik}{2R}\left(x'^2+y'^2\right) - i\zeta\right], \tag{4.24}$$

with the functions w, α, R, ζ depending on the propagation distance z according to Equations (4.7) and (4.10).

It is known that all zeros of the Bessel function are real. Then, from Equation (4.24), all intensity nulls reside on a single straight line, similar to the beam from Equation (4.9). However, in contrast to the beam in Equation (4.9), the nulls are not uniform and are located in the points $(x', y') = (\pm\alpha\,\gamma_{m,p}, 0)$ with $\gamma_{m,p}$ being pth zero of the mth-order Bessel function. In addition, the central intensity null is of mth order since the zero root of the mth-order Bessel function has mth order as well. Similarly to the beam in Equation (4.9), it can be shown that all the other intensity nulls have the order +1 and, therefore, TC of the whole beam is infinite.

Numerical simulation

Shown in Figure 4.6 are the intensity and phase distributions of the beam from Equation (4.24) in the initial plane and after propagation in space. The distributions are computed by Equation (4.24) (Figure 4.6(a–d)) and by the numerical Fresnel transform (Figure 4.6(e, f)). The following parameters were used for computation: wavelength $\lambda = 532$ nm, waist radius $w_0 = 0.5$ mm, order and scaling factor of the vortex-argument Bessel function $m = 3$ and $\alpha_0 = w_0/4$, respectively, propagation distance $z = 1$ m, computation domain $-R \leq x, y \leq R$ with $R = 2$ mm (Figure 4.6(a, b)) and $R = 3$ mm (Figure 4.6(c–f)).

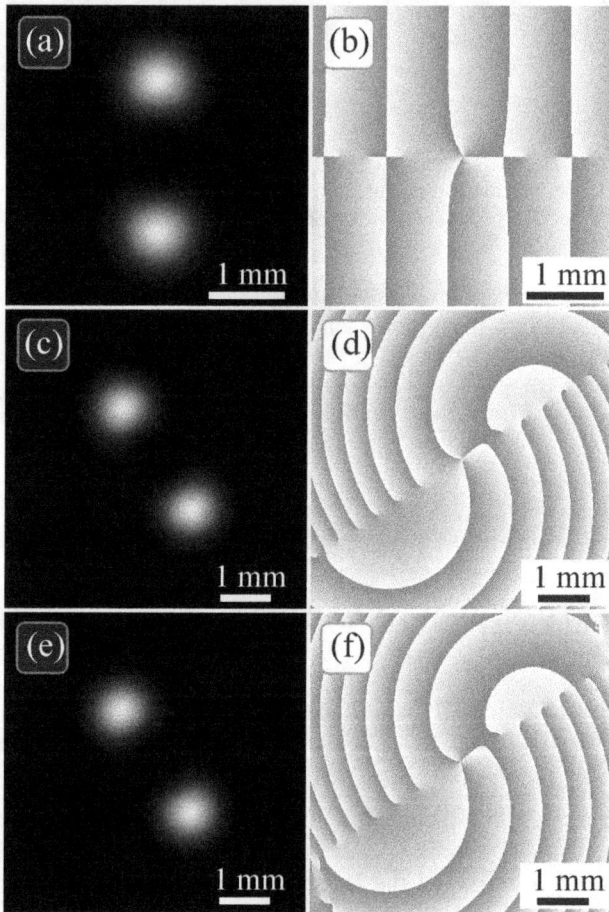

FIGURE 4.6   Distributions of intensity (a, c, e) and phase (b, d, f) of the beam (4.24) in the initial plane and after propagation in space. Distributions are obtained by Equation (4.24) (a–d) and by the numerical Fresnel transform (e, f). Computation parameters: wavelength λ = 532 nm, waist radius $w_0 = 0.5$ mm, order and scaling factor of the vortex-argument Bessel function $m = 3$ and $\alpha_0 = w_0/4$, propagation distance $z = 1$ m, computation domain $-R \leq x, y \leq R$ with $R = 2$ mm (a, b) and $R = 3$ mm (c–f).

As seen in Figure 4.6, distributions obtained theoretically by Equation (4.24) coincide with those obtained numerically by the Fresnel transform (excepting the areas near the corners). It is also seen that all optical vortices reside on a single straight line, rotating counter-clockwise on propagation. The intensity distribution has the same shape as that of the beam from Equation (4.9) with the vortex-argument cosine function, but the intensity nulls reside on the straight line not uniformly, and the central null is of the third order, rather than the first. Computation of the TC over three different circumferences (with the radii $R_1 = 3$ mm, 6 mm, and 9 mm) on the phase distribution (Figure 4.7) yields the values increasing with the radius: 13, 25, 37 (12.98, 24.88, 36.62), i.e. the TC of the beam from Equation (4.24) is infinite. The OAM of the beam from Equation (4.24) is; however, finite and equals approximately 8.14.

4.1.4 GAUSSIAN BEAM WITH THE COSINE FUNCTION OF THE SQUARED VORTEX-ARGUMENT

Theory

Here we study another form-invariant beam with the infinite TC. It is similar to the beam from Equation (4.6), but the vortex argument of the cosine function is squared:

$$E(x,y,z) = \frac{1}{q}\cos\left(\frac{x+iy}{\alpha_0 q}\right)^2 \exp\left(-\frac{x^2+y^2}{qw_0^2}\right). \tag{4.25}$$

In the rotated coordinates in Equation (4.8), the complex amplitude of the beam from Equation (4.25) reads as:

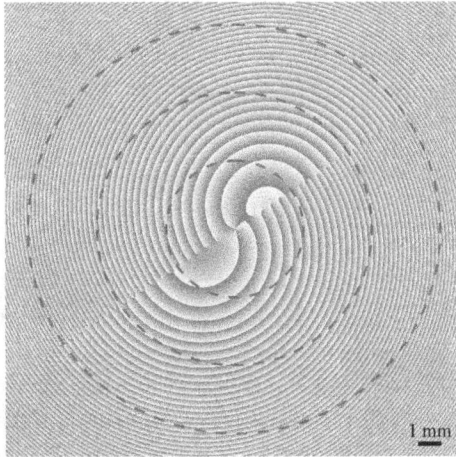

FIGURE 4.7 Phase distribution of the beam from Figure 4.6 in a wider area $-R \leq x, y \leq R$ ($R = 10$ mm), obtained by Equation (4.24).

$$E(x',y',z) = \frac{w_0}{w(z)} \cos\left[\frac{x'+iy'}{\alpha(z)}\right]^2$$

$$\times \exp\left[-\frac{x'^2+y'^2}{w^2(z)} + \frac{ik(x'^2+y'^2)}{2R(z)} - i\zeta(z)\right].$$

(4.26)

Squaring the argument of the cosine function in Equation (4.26), we get:

$$E(x',y',z) = \frac{w_0}{w(z)} \cos\left[\frac{x'^2-y'^2+i2x'y'}{\alpha^2(z)}\right]$$

$$\times \exp\left[-\frac{x'^2+y'^2}{w^2(z)} + \frac{ik(x'^2+y'^2)}{2R(z)} - i\zeta(z)\right].$$

(4.27)

From Equation (4.27), the argument of the cosine function is real if either $x' = 0$ or $y' = 0$. Therefore, intensity nulls reside on the two axes x' and y' and have the following coordinates:

$$\begin{cases} x' = \pm\alpha\sqrt{\pi/2 + \pi p}, & y' = 0, \quad p = 0,1,2,..., \\ y' = \pm\alpha\sqrt{\pi/2 + \pi p}, & x' = 0, \quad p = 0,1,2,.... \end{cases}$$

(4.28)

According to Equation (4.28), the density of the isolated intensity nulls of the light field in Equation (4.27) grows with the increasing number p, i.e. the distance between the neighbor nulls decreases parabolically. Around each such null, there is an optical vortex with TC +1. The total number of vortices is countable and therefore TC of the beam from Equation (4.24) is infinite.

The intensity of the beam in Equation (4.26) is given by:

$$I(x',y',z) = \frac{1}{2}\frac{w_0^2}{w^2} \exp\left(-2\frac{x'^2+y'^2}{w^2}\right)$$

$$\times \left[\cos\left(\frac{2(x'^2-y'^2)}{\alpha^2}\right) + \cosh\left(\frac{4x'y'}{\alpha^2}\right)\right].$$

(4.29)

It can be shown that the intensity distribution in Equation (4.29) contains only one maximum located in the center ($x' = y' = 0$). However, this maximum occurs only at a certain condition. As seen in Equation (4.29), the intensity at large distances $x' = \pm y' \to \pm\infty$ is proportional to the Gaussian exponential function:

$$I(x' = y' \to \infty, z) \sim \exp\left(-4x'^2\left(\frac{1}{w^2} - \frac{1}{\alpha^2}\right)\right).$$

(4.30)

Therefore, the light field in Equation (4.25) is of finite energy and has one central maximum only if $w < \alpha$.

Substituting Equation (4.25) into Equations (4.15) and (4.16), we get the OAM and the power of the beam from Equation (4.25):

$$J_z = \left(\frac{\pi \alpha_0^2 w_0^6}{4} \right) \left[\left(\alpha_0^4 - w_0^4 \right)^{-3/2} - \left(\alpha_0^4 + w_0^4 \right)^{-3/2} \right], \tag{4.31}$$

$$W = \left(\frac{\pi \alpha_0^2 w_0^2}{4} \right) \left[\left(\alpha_0^4 - w_0^4 \right)^{-1/2} + \left(\alpha_0^4 + w_0^4 \right)^{-1/2} \right]. \tag{4.32}$$

It is seen in Equations (4.31) and (4.32) that if $w = \alpha$ then the OAM and the power of the beam in Equation (4.25) become infinite, whereas for $w > \alpha$ these values become imaginary. The normalized OAM is equal to

$$\frac{J_z}{W} = \frac{(\chi - 1)^{-3/2} - (\chi + 1)^{-3/2}}{(\chi - 1)^{-1/2} + (\chi + 1)^{-1/2}} \tag{4.33}$$

with $\chi = (\alpha_0/w_0)^4$.

Expression in Equation (4.33) defines the OAM with respect to the center of the intensity distribution (the central intensity maximum) and does not depend on the chosen coordinate system. As the OAM from Equation (4.19), it also depends solely on the ratio between the period of the optical vortices and the Gaussian beam waist radius (α_0/w_0).

As seen in Equation (4.33), if $\alpha_0 > w_0$ (it is the condition of the finite beam energy) then the OAM can have only positive values. In addition, the OAM can have notable values at $\alpha_0 \approx w_0$, whereas for large values $\alpha_0 >> w_0$ the normalized OAM of the beam from Equation (4.25) tends to zero. It is explainable since at $\alpha_0 >> w_0$ all the intensity nulls (all optical vortices) are in the beam periphery where the intensity is almost zero. At the same time, the intensity shape in Equation (4.29) is defined by the Gaussian function with the center in the origin. Thus, despite the infinite TC of the beam from Equation (4.25) and despite the countable number of the optical vortices, residing on two orthogonal straight lines with their density increasing to periphery, the normalized OAM of such beam is not large and tends to zero if $\alpha_0 >> w_0$.

Numerical simulation

Figure 4.8 illustrates the intensity and phase distributions of the beam from Equation (4.25) in the initial plane and after propagation in space. All distributions are computed by Equation (4.25) for the following parameters: wavelength $\lambda = 532$ nm, waist radius $w_0 = 0.5$ mm, scaling factor of the cosine function with the squared vortex argument $\alpha_0 = 1.1w_0$, propagation distance $z = 1$ m, computation domain $|x|,|y| \leq R$ with $R = 1.5$ mm (Figure 4.8(a)), $R = 5$ mm (Figure 4.8(b, d)), and $R = 3$ mm (Figure 4.8(c)).

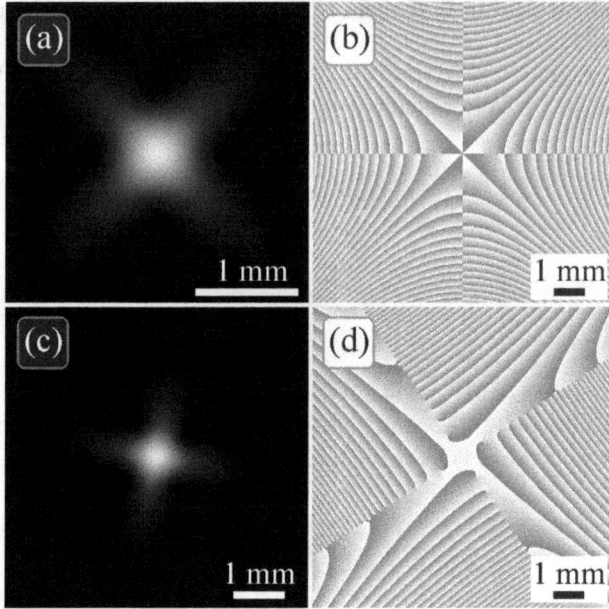

FIGURE 4.8 Distributions of intensity (a, c) and phase (b, d) of the beam from Equation (4.25) in the initial plane (a, b) and after propagation in space (c, d), computed by Equation (4.25) for the following parameters: wavelength $\lambda = 532$ nm, waist radius $w_0 = 0.5$ mm, scaling factor of the cosine function with the squared vortex argument $\alpha_0 = 1.1w_0$, propagation distance $z = 1$ m, computation domain $|x|, |y| \leq R$ with $R = 1.5$ mm (a), $R = 5$ mm (b, d), and $R = 3$ mm (c).

Figure 4.8 confirms that all optical vortices reside on two orthogonal straight lines $x = 0$ and $y = 0$, which rotate on propagation. The intensity distribution has a shape of a square with sharpened corners (i.e. a four-pointed star) and has one maximum in the center. The intensity nulls are located on the Cartesian axes and their density grows from the center to the periphery. Since all these nulls are of the order +1, the total TC of the beam from Equation (4.25) is infinite. Computation of the normalized OAM on the distributions from Figure 4.8(c, d) with using Equation (4.33) yields nearly the same values of 1.380 and 1.379.

If $\alpha_0 = w_0(1 + \delta)$ with $\delta \ll 1$, the normalized OAM can have large values and, according to Equation (4.33), can be estimated by the following formula:

$$\frac{J_z}{W} \approx \frac{1}{4\delta}. \tag{4.34}$$

Figure 4.9 depicts the intensity and phase distributions of the beam from Equation (4.25) in the initial plane and after propagation in space for the same parameters as in Figure 4.8, but with $\alpha_0 = 1.01w_0$. Since $\alpha_0 \approx w_0$, the intensity along the lines $y = \pm x$ decays much slower than in Figure 4.8, thus making the transverse section of the beam looking like a cross. According to Equation (4.34), the normalized OAM

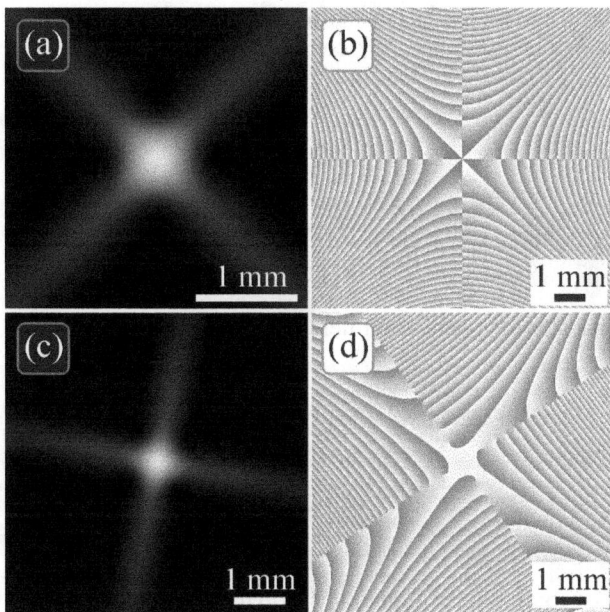

FIGURE 4.9 Distributions of intensity (a, c) and phase (b, d) of the beam from Equation (4.25) in the initial plane (a, b) and after propagation in space (c, d). Computation parameters are the same as in Figure 4.8, but with $\alpha_0 = 1.01w_0$.

should be equal approximately to $J_z/W \approx 25$. Computation by the distributions from Figure 4.9 with using Equations (4.15) and (4.16) yields the value 21.42.

4.1.5 Higher-Order Cosine Vortex Optical Beam

Theory

Here we study yet another form-invariant beam with an infinite TC, which is similar to the beam from Equation (4.6), but with the vortex-argument cosine function raised to some power:

$$E\left(x,y,z\right) = \frac{1}{q}\cos^m\left(\frac{x+iy}{\alpha_0 q}\right)\exp\left(-\frac{x^2+y^2}{qw_0^2}\right). \qquad (4.35)$$

It is seen in Equation (4.35) that the cosine function becomes real at $y = 0$ (in the initial plane), and the intensity nulls are located on the horizontal axis (in the rotated coordinates (4.8)) in the points $x_p = \alpha\ (\pi/2 + \pi p)$ with p being the integer numbers. Near these isolated nulls, the cosine function can be written approximately as $(x - x_p + iy)^m$. Therefore, there are optical vortices around the intensity nulls, which have the TC m. We note that for residing the intensity nulls (or phase singularities) on the vertical axis, the argument of the cosine function should be equal to $ix-y$, rather than $x+iy$. In the rotated coordinates (4.8), complex amplitude (4.35) is given by:

$$E\left(x',y',z\right)=\frac{w_0}{w\left(z\right)}\cos^m\left[\frac{x'+iy'}{\alpha\left(z\right)}\right]$$

$$\times\exp\left[-\frac{x'^2+y'^2}{w^2\left(z\right)}+\frac{ik\left(x'^2+y'^2\right)}{2R\left(z\right)}-i\zeta\left(z\right)\right].$$

(4.36)

The intensity distribution of the light field from Equation (4.36) is as follows:

$$I\left(x',y',z\right)=2^{-m}\frac{w_0^2}{w^2}\exp\left(-2\frac{x'^2+y'^2}{w^2}\right)$$

$$\times\left[\cos\left(\frac{2x'}{\alpha}\right)+\cosh\left(\frac{2y'}{\alpha}\right)\right]^m.$$

(4.37)

The intensity distribution from Equation (4.37) has one local maximum in the origin $(x', y') = (0, 0)$, where the intensity value equals $I(0, 0, z) = 2^{-m}(w_0/w)^2$. Two other main local maxima reside on the axis y' (i.e. $x' = 0$). Indeed, it is seen in Equation (4.37) that the intensity initially grows along the axis y' as a hyperbolic cosine and then decays because of the Gaussian envelope. These local maxima can be found from the Equation (4.37) for intensity at $x' = 0$:

$$I\left(x'=0,y'\gg1,z\right)=2^{-2m}\frac{w_0^2}{w^2}\exp\left(2\frac{my'}{\alpha}-2\frac{y'^2}{w^2}\right).$$

(4.38)

Differentiation of the intensity (4.38) with respect to y' allows obtaining the coordinates of the intensity maxima (the two maxima are located symmetrically): $y'_{max}=\pm mw^2/(2\alpha)$. It is seen from this expression that increasing the power m leads to increasing the distance between the intensity maxima on the axis y'. The maximal intensity value is:

$$I_{max}\left(x'=0,y'_{max},z\right)=2^{-2m}\frac{w_0^2}{w^2}\exp\left(\frac{m^2w^2}{2\alpha^2}\right).$$

(4.39)

So, the intensity distribution of the optical vortex beam from Equation (4.35) has two main circular light spots residing on a straight line orthogonal to the straight line with the countable number of the intensity nulls (phase singularities). Around each such singular point, there is an optical vortex with the topological charge m. On propagation in space, the beam widens and rotates. At the Rayleigh distance and in the far field, the beam is rotated respectively by 45 and 90 degrees.

Despite the infinite TC of the beams from Equation (4.35), their OAM is finite and can be shown to be equal to:

$$\frac{J_z}{W} = m \frac{\sum_{s=0}^{m} \binom{m}{s} \xi_s \cosh^m \xi_s \tanh \xi_s}{\sum_{s=0}^{m} \binom{m}{s} \cosh^m \xi_s}. \tag{4.40}$$

with $\xi_s = (w_0/\alpha_0)^2(s - m/2)$. Equation (4.40) indicates that for the high density of the phase singularities (i.e. $\alpha_0 << w_0$), the normalized OAM can be obtained by an approximate formula:

$$\frac{J_z}{W} = \frac{m^2}{2} \frac{w_0^2}{\alpha_0^2}. \tag{4.41}$$

Similar to Equation (4.19), expressions from Equations (4.40) and (4.41) define the OAM with respect to the middle point between the two light spots (center of the intensity distribution) and do not depend on the chosen coordinate system. They depend solely on the topological charge of the optical vortices on the straight line and on the product of the Gaussian beam waist radius with the optical vortices density (w_0/α_0).

Equation (4.41) indicates that increasing the Gaussian beam waist radius leads to a parabolic growth of the normalized OAM of the beams from Equation (4.35). It is worth noting that for conventional rotationally symmetric optical vortices (for example, Laguerre–Gaussian or Bessel–Gaussian beams), the Gaussian beam waist radius does not affect the normalized OAM.

The parabolic dependence of the OAM on the order m has a physical explanation. Above, based on the intensity distribution from Equation (4.37), we derive the coordinates of two main intensity maxima: $x'_{max} = 0$ and $y'_{max} = \pm mw^2/(2\alpha)$, revealing that the distance from the light spot to the center is proportional to m. In addition, according to Equation (4.35), the period of the cosine function decreases m times. Thus, the phase gradient in the vicinity of the intensity maximum increases m times. As a result, a microscopic dielectric particle, occurring in the mth-order light field from Equation (4.35), is being trapped by the light spot, located m times further from the center, while the azimuthal component of the force (Poynting vector), that is acting onto the particle, is also increasing m times. When both the force and the leverage increase m times, the angular momentum increases m^2 times.

Numerical simulation

Shown in Figure 4.10 are the intensity and phase distributions of the beam from Equation (4.35) in the initial plane and after propagation in space. All distributions are computed by Equation (4.35) for the following parameters: wavelength $\lambda = 532$ nm, waist radius $w_0 = 0.5$ mm, scaling factor of the cosine function $\alpha_0 = w_0/4$, the

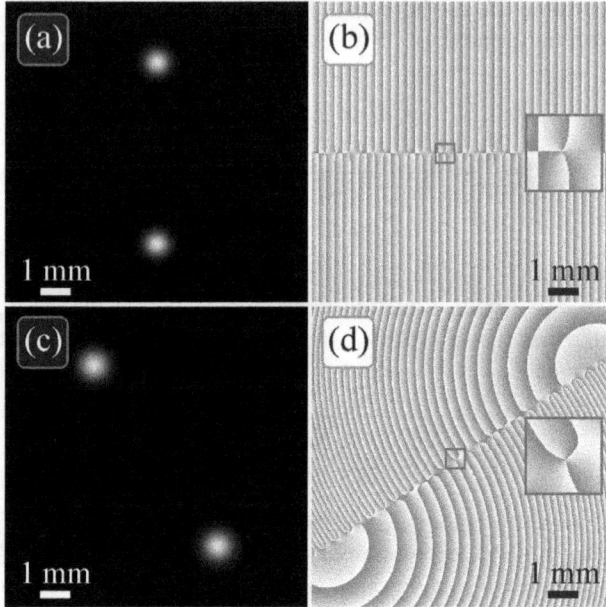

FIGURE 4.10 Intensity (a, c) and phase (b, d) distributions of the beam from Equation (4.35) in the initial plane (a, b) and after propagation in space (c, d). Calculation parameters: wavelength $\lambda = 532$ nm, waist radius $w_0 = 0.5$ mm, scaling factor and power (order) of the cosine function $\alpha_0 = w_0/4$ and $m = 3$, propagation distance $z = 1$ m, computation domain $|x|$, $|y|$ $\leq R$ with $R = 5$ mm. The insets (b, d) show the 4× magnified area from the box near the center.

cosine function is raised to the power $m = 3$, propagation distance $z = 1$ m, computation domain $|x|$, $|y| \leq R$ with $R = 5$ mm.

As seen in Figure 4.10, all optical vortices reside uniformly on a single straight line (horizontal in the initial plane), which rotates on propagation. However, these vortices are of the third order ($m = 3$), rather than first. Intensity distribution contains two light spots residing on a straight line orthogonal to the line with the intensity nulls (phase singularities). Since all the nulls are of the order +3, the summary TC of the beam from Equation (4.35) is infinite. For the chosen parameters, the normalized OAM, is computed by using Equations (4.15) and (4.16) over the intensity and phase distributions from Figure 4.10(a, b), which is equal to 71.76. This numerical value is consistent with our theoretical estimation given by Equation (4.41): $J_z/W = 8m^2 = 72$.

In this section, we have investigated optical vortex beams with an infinite TC [163]. Such beams have a countable number of phase singularities (isolated intensity nulls), typically with the unitary TC, which reside uniformly (or not uniformly) on a straight line (or on two orthogonal straight lines) in the beam transverse cross-section. Such vortex beams are form-invariant and on propagation in space they only widen and rotate. The investigated beam families have the transverse intensity section in a form of two symmetric light spots, one elliptic spot, or a four-pointed star. We derived the intrinsic orbital angular momentum (OAM) of such beams, which is

TABLE 4.1

Equations for the intensity and OAM of the studied infinite-topological-charge light beams

Beam type	Intensity distribution	OAM
Gaussian beam with a vortex-argument cosine function	Equation (4.12)	Equation (4.19)
Gaussian beam with the cosine function of the squared vortex-argument	Equation (4.29)	Equation (4.33)
Higher-order Gaussian beam with a vortex-argument cosine-function	Equation (4.37)	Equation (4.40)

finite, since only a finite number of screw dislocations is within the area of a notable intensity of the Gaussian beam. All the other phase singularities are in the periphery (and in the infinity), where the intensity is almost zero. Table 4.1 lists the equations for the intensity distributions and OAMs of the studied beams:

In practice, such beams can be generated by using the SLM, but with using the algorithms for encoding the amplitude-phase functions $\cos(x + iy)$, $J_m(x + iy)$, $\cos(x + iy)^2$, and $\cos^m(x + iy)$ into pure-phase function. Such rotating light beams (Figures 4.2, 4.6, and 4.10) can be used for increasing the longitudinal resolution in optical single-molecules microscopy [192].

4.2 ORBITAL ANGULAR MOMENTUM OF GENERALIZED COSINE GAUSSIAN BEAMS WITH AN INFINITE NUMBER OF SCREW DISLOCATIONS

Topological charge (TC) is one of the most important quantities characterizing the optical vortices [19,21]. For simple optical vortices (screw dislocations) with a single n-fold degenerate phase singularity, or isolated intensity null, (n is an integer number), TC equals the integer number of full periods of phase variation (one period is 2π) along an arbitrarily closed contour around the intensity null. Optical vortices with a single central screw dislocation can have large, but finite, TC, for instance, 10010 [185]. It is clear that if a light field has several screw dislocations related to several intensity nulls located at different points of the beam transverse cross-section, the TC of the entire field equals the sum of the TCs of each separate optical vortex. An optical vortex with a single intensity null and with a single n-fold screw dislocation can be split into n separate optical vortices with the unitary TC by simple interference of the optical vortex with either a Gaussian beam or a plane wave of arbitrary low amplitude [109]. Currently, there is a plenitude of known ways of obtaining the multiple-singularity optical vortices. Based on the principle of splitting a single optical vortex with TC n into n ($TC = 1$) simple optical vortices, many methods of TC measurement are developed [132,193,194,14,195,15,196,197,182]. In [193,194,14], the TC of an optical vortex was determined due to diffraction by a triangular aperture.

It is also possible to measure the TC by using a tilted convex spherical lens [195], cylindrical lens [15], astigmatic phase mask [196], sectorial half-wave plates [197], and a gradually changing-period spiral spoke grating [182]. Currently, novel variants of the traditional interferometric methods for TC measurement are also adopted [198,199,116,200]. It is not difficult to determine the TC of separate symmetric optical vortices (for instance, Laguerre–Gaussian or Bessel–Gaussian beams) and of optical fields with multiple singularities. However, there are light fields, for which obtaining the TC is not easy. These include light fields with a fractional initial TC [19] and superpositions of the optical vortices [14]. In a general case, TC of composite optical vortices can be derived by using a formula given by Berry [19]. Light fields with a fractional initial TC acquire an integer TC on propagation in space and its value can change on propagation [160,55,161,85]. Superposition of the optical vortices [22] can also change its TC on propagation. Moreover a superposition of optical vortices can even have undetermined TC. For example, a coaxial superposition of two different diffraction-free Bessel beams has a complex amplitude defined as a sum of two different Bessel functions of the orders n and m, and TC can be equal to either n or m depending on the radius of the circumference, along which the TC is determined. As another example of a light field, whose TC changes on propagation, we can mention the optical vortices generated by spiral slits in an opaque screen (Archimedes or Fermat spirals) [74]. Optical vortices with a controllable TC can be generated by using microlasers [201], metasurfaces [202], and other nanophotonic components [132].

In this section, we study a family of form-invariant mth-order cosine Gaussian optical vortices with an infinite topological charge. These beams are shown to be a superposition of the Laguerre–Gaussian modes of the orders $(2p, 0)$ with p being nonnegative integer numbers. An expression for the OAM of such beams is obtained, which is proportional to the squared order (power) m, to which the cosine function is raised. These beams contain an infinite (but countable) [203] number of phase singularities (screw dislocations), residing on a straight line in a plane transverse to a propagation direction. All these screw dislocations have the topological charge m.

4.2.1 Complex Amplitude, Space Propagation, and Intensity Distribution of the Generalized Cosine Gaussian Beams with an Infinite Number of Screw Dislocations

It has been shown in [164] that any function given by:

$$E_{\pm}(x,y,z) = \frac{1}{q}\exp\left(-\frac{x^2+y^2}{qw_0^2}\right)f\left(\frac{x\pm iy}{qw_0}\right), \qquad (4.42)$$

where (x, y, z) are the Cartesian coordinates, w_0 is the waist radius of the Gaussian beam, $q = 1 + iz/z_0$, $z_0 = kw_0^2/2$ is the Rayleigh range, $k = 2\pi/\lambda$ is the wavenumber,

$f(x \pm iy)$ is an arbitrary entire analytical function, describes a solution of the paraxial Schrödinger-type Helmholtz equation:

$$2ik\frac{\partial E}{\partial z} + \frac{\partial^2 E}{\partial x^2} + \frac{\partial^2 E}{\partial y^2} = 0. \tag{4.43}$$

Formula in Equation (4.42) describes form-invariant light fields with their intensity distribution conserved on space propagation (up to scale and rotation around the optical axis). Here we study a family of form-invariant beams from Equation (4.42) with an infinite topological charge (TC):

$$E(x, y, z) = \frac{1}{q}\cos^m\left(\frac{x+iy}{\alpha_0 q}\right)\exp\left(-\frac{x^2+y^2}{qw_0^2}\right), \tag{4.44}$$

with m being an integer positive power, to which the cosine function is raised, and α_0 being a real constant, defining the period of the cosine function. In the rotated Cartesian coordinates:

$$\begin{pmatrix} x' \\ y' \end{pmatrix} = \begin{pmatrix} \cos\zeta & \sin\zeta \\ -\sin\zeta & \cos\zeta \end{pmatrix}\begin{pmatrix} x \\ y \end{pmatrix} = \frac{z_0}{\sqrt{z^2+z_0^2}}\begin{pmatrix} x+yz/z_0 \\ y-xz/z_0 \end{pmatrix}, \tag{4.45}$$

with the rotation angle being equal to the Gouy phase [22]: $\zeta(z) = \arctan(z/z_0)$, the complex amplitude of the beam (4.44) is given by:

$$E(x', y', z) = \frac{w_0}{w(z)}\cos^m\left[\frac{x'+iy'}{\alpha(z)}\right]\exp\left[-\frac{x'^2+y'^2}{w^2(z)} + \frac{ik(x'^2+y'^2)}{2R(z)} - i\zeta(z)\right], \tag{4.46}$$

with $w(z) = w_0[1+(z/z_0)^2]^{1/2}$, $\alpha(z) = \alpha_0[1+(z/z_0)^2]^{1/2}$, $R(z) = z[1+(z_0/z)^2]$. According to Equation (4.46), intensity nulls are located on the axis $y' = 0$ in the points $x_p' = \alpha(z)$ $(\pi/2 + \pi p)$ (p are integer numbers). Near these isolated nulls, the cosine function can be replaced by the amplitude $(x'-x_p'+iy')^m$. Therefore, around these intensity nulls, there are optical vortices with TC m. We note that for residing the intensity nulls (points of phase singularity) on the vertical axis y' ($x' = 0$), the argument of the cosine function should be equal to $ix'-y'$, rather than $x'+iy'$.

The intensity of the light field (4.46) is:

$$I(x', y', z) = 2^{-2m}\frac{w_0^2}{w^2}\exp\left(-2\frac{x'^2+y'^2}{w^2}\right)\left[\cos\left(\frac{2x'}{\alpha}\right) + \cosh\left(\frac{2y'}{\alpha}\right)\right]^m. \tag{4.47}$$

The intensity distribution in Equation (4.47) has one local maximum in the origin $(x', y') = (0,0)$, where the intensity value equals $I(0,0,z) = (w_0/w)^2$. Two other main local maxima reside on the axis y' (i.e. $x' = 0$): $y'_{max} = \pm m w^2/(2\alpha)$. It is seen from this expression that increasing the number m leads to increasing the distance between the intensity maxima on the axis y'. The intensity at these two maxima is:

$$I_{max}\left(x' = 0, y'_{max}, z\right) = 2^{-2m} \frac{w_0^2}{w^2} \exp\left(\frac{m^2 w^2}{2\alpha^2}\right). \tag{4.48}$$

It is seen in Equation (4.48) that the maximal intensity grows with m exponentially. So, the intensity distribution of the optical vortex beam from Equation (4.44) has two main circular light spots residing on a straight line orthogonal to the straight line with the infinite (but countable [203]) number of the intensity nulls (points of phase singularity). Around each such singular point, there is an optical vortex with the topological charge m. On propagation in space, the beam widens and rotates. At the Rayleigh distance and in the far field, the beam is rotated respectively by 45 and 90 degrees.

4.2.2   OAM Spectrum, Energy, and Normalized-to-Power OAM of the Generalized Cosine Gaussian Beams

To obtain the OAM-spectrum of the beam (4.44), we rewrite the field from Equation (4.44) in the initial plane in the polar coordinates (r, φ) and expand it into a series of the OAM-harmonics:

$$E\left(r,\varphi,0\right) = \cos^m\left(\frac{r e^{i\varphi}}{\alpha_0}\right)\exp\left(-\frac{r^2}{w_0^2}\right) = \sum_{n=-\infty}^{\infty} E_n\left(r\right) e^{in\varphi}. \tag{4.49}$$

We decompose the light field from Equation (4.44) into the OAM-harmonics, since the complex amplitude contains the terms $(x + iy)$ and $(x^2 + y^2)$, revealing the vortex nature of the field and its Gaussian envelope. So, using the basis of vortex beams is preferable compared to other sets of orthogonal functions (e.g. Hermite–Gaussian beams).

Using the Taylor expansion of the cosine function ($\cos \xi = 1 - \xi^2/2! + \xi^4/4! - \xi^6/6! + \ldots$), it can be shown that the OAM-harmonics in Equation (4.49) is zero for the negative orders $n < 0$, whereas for positive orders $n \geq 0$ the weight functions in Equation (4.49) read as:

$$E_n\left(r\right) = \frac{i^n C_{nm}}{2^m n!}\left(\frac{r}{\alpha_0}\right)^n \exp\left(-\frac{r^2}{w_0^2}\right), \tag{4.50}$$

where:

$$C_{nm} = \sum_{s=0}^{m}\binom{m}{s}\left(2s - m\right)^n.$$

For odd orders n, the OAM-harmonics from Equation (4.49) can also be shown to vanish, since the sum C_{nm} consists of the opposite terms (e.g. $C_{3m} = (-m)^3 + m(2-m)^3 + \ldots + m(m-2)^3 + m^3$). Therefore, the OAM-spectrum in Equation (4.49) contains only even positive terms, which describe the Laguerre–Gaussian modes of the orders $(0, n = 2p)$ in the initial plane. Thus, the light field from Equation (4.44) is a superposition of an infinite number of the Laguerre–Gaussian modes with zero radial index and with even nonnegative topological charges. The energy of each mode (OAM-spectrum of the beam in Equation (4.49)) is given by:

$$W_n = 2\pi \int_0^\infty |E_n(r)|^2 \, r dr = \frac{\pi w_0^2 C_{nm}^2}{2^{2m+1} n!} \left(\frac{w_0^2}{2\alpha_0^2} \right)^n. \tag{4.51}$$

The energy of the whole beam from Equation (4.44) is the sum of energies of all the modes:

$$W = \sum_{n=-\infty}^{\infty} W_n = \frac{\pi w_0^2}{2^{2m+1}} \sum_{n=0}^{\infty} \frac{C_{nm}^2}{n!} \left(\frac{w_0^2}{2\alpha_0^2} \right)^n = \frac{\pi w_0^2}{2^{m+1}} \sum_{s=0}^{m} \binom{m}{s} \cosh^m \xi_s, \tag{4.52}$$

with $\xi_s = (w_0/\alpha_0)^2 (s - m/2)$. The OAM of the beam from Equation (4.44) can be obtained by the known formula [55]:

$$J_z = \mathrm{Im} \int_{-\infty}^{\infty} \int_{-\infty}^{\infty} E^*(x,y) \left(x \frac{\partial}{\partial y} - y \frac{\partial}{\partial x} \right) E(x,y) \, dxdy, \tag{4.53}$$

Substituting Equation (4.44) into Equation (4.53), we get an expression for the OAM:

$$J_z = \sum_{n=-\infty}^{\infty} n W_n = m \frac{\pi w_0^2}{2^{m+1}} \sum_{s=0}^{m} \binom{m}{s} \xi_s \cosh^m \xi_s \tanh \xi_s. \tag{4.54}$$

The OAM normalized to the beam power is:

$$\frac{J_z}{W} = m \frac{\sum_{s=0}^{m} \binom{m}{s} \xi_s \cosh^m \xi_s \tanh \xi_s}{\sum_{s=0}^{m} \binom{m}{s} \cosh^m \xi_s}. \tag{4.55}$$

Equation (4.55) leads to simple expressions for the normalized OAM when the orders m are small. For example, if $m = 1$, the OAM is $J_z/W = \xi \tanh \xi$ with $\xi = (w_0/\alpha_0)^2/2$, whereas if $m = 2$, then the OAM is $J_z/W = \xi \sinh(2\xi)/[1 + \cosh^2(\xi)]$ with $\xi = (w_0/\alpha_0)^2$. According to Equation (4.14), if the cosine function in Equation (4.44) has a large period ($\alpha_0 \gg w_0$), the OAM (4.55) tends to zero as $J_z/W \sim (w_0/\alpha_0)^4$. If, on the contrary, $\alpha_0 \ll w_0$, then Equation (4.55) leads to a simple approximate formula for the normalized OAM, valid for an arbitrary order m:

$$\frac{J_z}{W} = \frac{m^2}{2} \frac{w_0^2}{\alpha_0^2}. \tag{4.56}$$

The parabolic dependence of the OAM on the order m has a physical explanation. Based on the intensity distribution from Equation (4.47), we derive the coordinates of two main intensity maxima: $x'_{max} = 0$ and $y'_{max} = \pm mw^2/(2\alpha)$, revealing that the distance from the light spot to the center is proportional to m. In addition, according to Equation (4.47), the period of the cosine function decreases m times. Thus, the phase gradient in the vicinity of the intensity maximum increases m times. As a result, a microscopic dielectric particle, occurring in the mth-order light field from Equation (4.46), is being trapped by the light spot, located m times further from the center, while the azimuthal component of the force (Poynting vector), that is acting onto the particle, is also increasing m times. When both the force and the leverage increase m times, the angular momentum increases m^2 times.

4.2.3 NUMERICAL SIMULATION

Shown in Figure 4.11 are intensity and phase distributions of the beam from Equation (4.44) in the initial plane and after propagation in space. All distributions are computed for the following parameters: wavelength $\lambda = 532$ nm, waist radius $w_0 = 0.5$ mm, scaling factor of the cosine function $\alpha_0 = w_0/4$, the order (power) of the cosine function $m = 3$ (Figure 4.11 (a–d)) and $m = 5$ (Figure 4.11 (e–h)), propagation distance $z = 1$ m, computation domain $|x|, |y| \leq R$ with $R = 5$ mm (Figure 4.11 (a–d)) and $R = 7.5$ mm (Figure 4.11 (e–h)).

   As seen in Figure 4.11, all optical vortices reside uniformly on a single straight line (horizontal in the initial plane), which rotates on propagation. Intensity distribution contains two light spots residing on a straight line orthogonal to the line with the intensity nulls (points of phase singularity). All the nulls are of the order $m = 3$ (Figure 4.11 (a–d)) or $m = 5$ (Figure 4.11 (e–h)), and the summary TC of the beam (4.44) is infinite. According to Equation (4.55), for the parameters chosen, the normalized OAM should be equal to $J_z/W = 8m^2 = 72$ for $m = 3$ and $J_z/W = 200$ for $m = 5$. Numerical computation (by using Equation (4.56)) over the intensity and phase distributions from Figure 4.11 (a, b) yields the value 71.76, while for Figure 4.11 (e, f) the obtained value is 197.13. Figure 4.12 depicts the OAM-spectrum (4.51) of the light beam (4.44) with $m = 2$, 3, and 5 (other parameters are the same as in Figure 4.11). As seen in Figure 4.12, it consists of single-ringed even-order Laguerre–Gaussian modes. Although the OAM spectrum widens with increasing m (at $m = 2$, 3, and 5, there are 7, 10, and 17 modes of the energy, exceeding half-maximal value), only the modes of some range of their topological charges have the notable energy. The maximal energy has the modes of 32th (Figure 4.12 (a)), 72th (Figure 4.12 (b)), and 200th

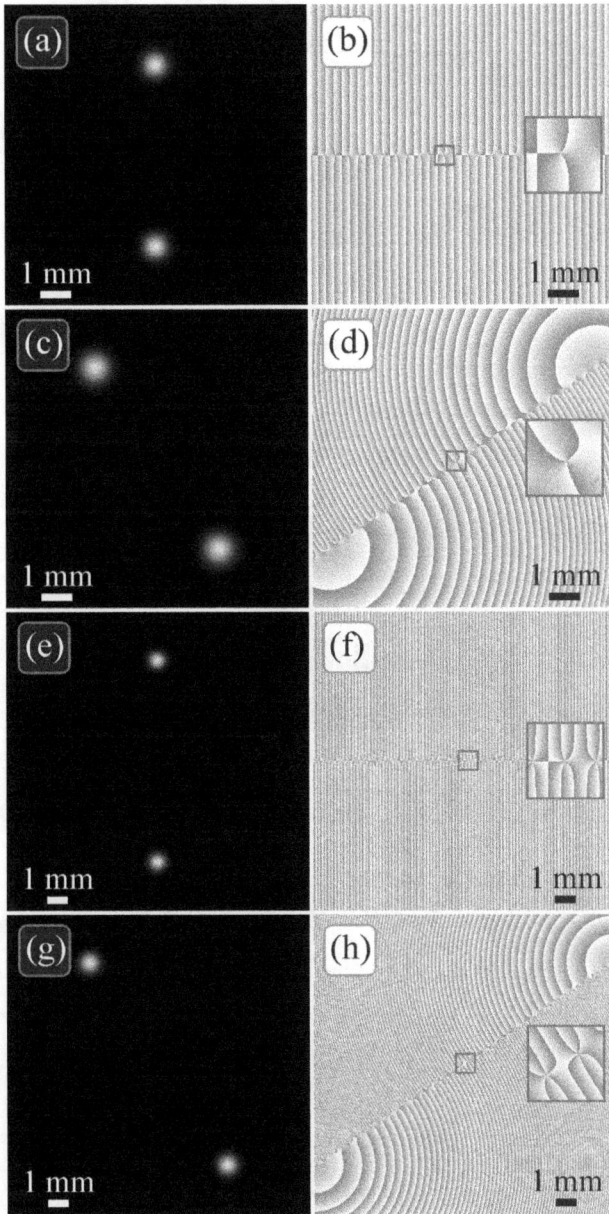

FIGURE 4.11 Intensity (a, c, e, g) and phase (b, d, f, h) distributions of the beam (4.44) in the initial plane (a, b, e, f) and after propagation in space (c, d, g, h), computed by Equation (4.44) for $m = 3$ (a–d) and $m = 5$ (e–h). The insets (b, d, f, h) show the 4× magnified area from the box near the center.

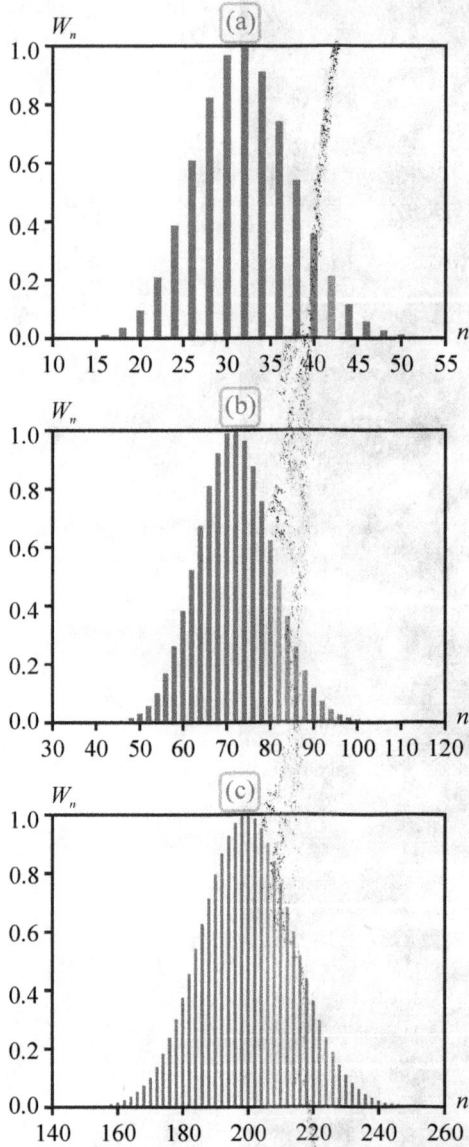

FIGURE 4.12 Normalized-to-maximum OAM-spectrum in Equation (4.10) of the beam from Equation (4.44) for $m = 2$ (a), $m = 3$ (b), and $m = 5$ (c). Other parameters are the same as in Figure 4.11. The vertical axis means the energy of the OAM-harmonics, while the horizontal axis means their orders (topological charges).

(Figure 4.12 (c)) order. Thus, the order (topological charge) of the central OAM-harmonic in this spectrum (Figure 4.12) is approximately equal to the normalized OAM of the beam.

The approximate coincidence of the order (topological charge) of the central OAM-harmonic of the wide OAM-spectrum (Figure 4.12) with the OAM value of the whole beam follows from the symmetry of the OAM-spectrum in Figure 4.12. Indeed, if the OAM-spectrum is symmetric, then the OAM can be written as:

$$J_z = \sum_{p=-n}^{n} (n-p)W_{n-p} = nW_n + \sum_{p=1}^{n} \left[(n-p)W_{n-p} + (n+p)W_{n+p}\right]$$

$$= nW_n + 2\sum_{p=1}^{N} nW_{n-p} = nW_n + 2n\sum_{p=1}^{N} W_{n-p} = nW,$$

(4.57)

where n is the average topological charge (order) of the optical vortices in the spectrum (Figure 4.12), W_{n-p} is the weight coefficients of each component of the OAM-spectrum, and W is the power of the whole beam. Then, for a symmetric OAM-spectrum, $J_z/W = n$.

In conclusion, we investigated in this section a family of multi-vortex high-order cosine Gaussian light fields [204]. They have a countable number of phase singularities (screw dislocations) and, thus, an infinite topological charge. These beams are described by the Gaussian function multiplied by the cosine function, raised to an arbitrary power, and having the vortex argument $x + iy$. There is a countable number of optical vortices in such beams, residing uniformly on one Cartesian axis in the waist plane, and each vortex has the topological charge equal to the power (order), to which the cosine function is raised. The intensity distribution of such beam contains two light spots, located on a straight line orthogonal to the straight line with the optical vortices. The orbital angular momentum of such beams is proportional to the squared topological charge of each vortex. The OAM-spectrum of such beams is almost symmetric with respect to the central OAM-harmonic. It is interesting that the order of this central OAM-harmonic is almost equal to the normalized OAM of the whole beam. Such rotating two-petal light beams can be used for increasing the longitudinal resolution in optical single-molecules microscopy [192].

4.3 ASTIGMATIC TRANSFORM OF A GAUSSIAN BEAM WITH AN INFINITE NUMBER OF EDGE DISLOCATIONS

Now we consider yet another example of an OV with the infinite TC, which can be obtained by the astigmatic transform. In the initial plane, the complex amplitude of such a beam is given by:

$$E(x,y,z=0) = \cos\left(\frac{x}{\alpha_0}\right)\exp\left[-\frac{x^2+y^2}{w_0^2} - \frac{ik}{4f}(x+y)^2\right].$$

(4.58)

The light field in Equation (4.58) is the Gaussian beam in its waist, along with a grating with the transmission $\cos(x/\alpha_0)$, and with a cylindrical lens with the focal length f, rotated by 45 degrees from the Cartesian axes. Field from Equation (4.58) has an infinite number of zero-intensity lines (edge dislocations), parallel to the y-axis and separated by a distance $\pi\,\alpha_0$. At the double focal length from the initial plane ($z = 2f$), field from Equation (4.58) has the following complex amplitude ($\gamma = z_0/(2f)$):

$$E\left(\xi,\eta,z=2f\right) = \frac{-i\gamma}{\sqrt{1+\gamma^2}}\cosh\left[\frac{\gamma\left(\xi-i\gamma\eta\right)}{\alpha_0\left(1+\gamma^2\right)}\right]$$

$$\times\exp\left[-\frac{w_0^2}{4\alpha_0^2\left(1+\gamma^2\right)} - \frac{\gamma^2\left(\xi^2+\eta^2\right)}{w_0^2\left(1+\gamma^2\right)}\right] \qquad (4.59)$$

$$\times\exp\left[\frac{ik}{4f}\left(\xi^2+\eta^2+\frac{2\gamma^2\xi\eta}{1+\gamma^2}\right)\right].$$

The elliptic optical vortex $\xi-i\gamma\,\eta$ in Equation (4.59) is in the argument of the hyperbolic cosine. At $\xi = 0$, the hyperbolic cosine turns into the trigonometric cosine and we obtain an equation for the intensity nulls (optical vortices):

$$\cos\left(\frac{\gamma^2\eta}{\alpha_0\left(1+\gamma^2\right)}\right) = 0. \qquad (4.60)$$

Equation (4.60) indicates that vertical zero-intensity lines in the initial plane transform into an infinite array of equidistant optical vortices with the period $T = (\pi\,\alpha_0)(1 + 1/\gamma^2)$ and with the TC of -1, residing on the vertical axis η in the transverse plane at the distance $z = 2f$. Therefore, the TC of the light field from Equation (4.59) is infinite. Figure 4.13 shows the intensity (Figure 4.13(a, c)) and phase (Figure 4.13(b, d)) distributions of the cosine Gaussian beam (Equations (4.58) and (4.59)) in the initial plane and at the double focal distance of the cylindrical lens. The following parameters were used: wavelength $\lambda = 532$ nm, waist radius of the Gaussian beam w_0

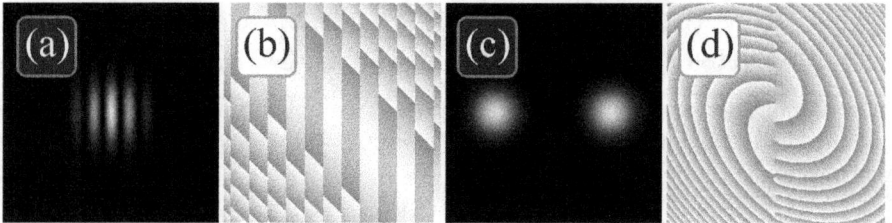

FIGURE 4.13 Intensity (a, c) and phase (b, d) distributions of the cosine Gaussian beam in the initial plane (a, b) and on the double focal distance of the cylindrical lens (c, d).

= 0.5 mm, the focal length of the cylindrical lens $f = z_0/2 \approx 73.8$ cm ($\gamma = 1$), period of the cosine function equals the waist radius (i.e. $\alpha_0 = w_0/(2\pi)$), computation domain in the input plane is $|x|, |y| \leq 1.5$ mm, computation domain in the output plane is $|\xi|$, $|\eta| \leq 3$ mm.

In conclusion, we investigated in this section OVs with the infinite TC [162]. Such OVs have a countable number of phase singularities (isolated intensity nulls), typically unitary-charge, located either equidistantly or not equidistantly on a straight line in the beam cross-section. Orbital angular momentum (OAM) of such OVs is finite, since only a finite number of screw dislocations are within the area of a notable intensity of the Gaussian beam, whereas the other phase singularities are in the periphery (and at infinity), where the intensity is almost zero. The OAM of such beams can be increased (theoretically, up to infinity) by simply increasing the Gaussian beam width. Note that for the beams with a finite number of optical vortices (i.e. with the finite TC), even the infinite increasing the Gaussian beam width does not lead to the infinite OAM.

4.4 OPTICAL VORTEX BEAMS WITH A SYMMETRIC AND ALMOST SYMMETRIC OAM-SPECTRUM

Recently, an increasing number of works have studied the OAM-spectrum of vortex beams [56]. It is known that, in addition to the topological charge (TC) [19] and the orbital angular momentum (OAM) [33], the vortex beams are characterized by the OAM-spectrum [65]. By OAM-spectrum, it is meant the distribution of vortex beam energy (power) by angular harmonics [65]. These harmonics are complex amplitudes of the form $\exp(in\varphi)$ with n being an integer number and (r, φ) being the polar coordinates in the beam cross-section. For the optical decomposition of a light field into the OAM-spectrum, several methods are known, which employ multi-order diffractive optical elements [65,205,206,207] or light modulators [208,209]. Light fields with a wide OAM-spectrum can be generated by using perturbed fork-gratings [200,37,46], circular diaphragms [210], complicated spiral phase plates [127], or separate holes in an opaque screen [211]. The OAM-spectrum of a light beam can be obtained from an interferogram [200] or simply from a measured beam intensity distribution [37,46]. Measuring the OAM by using the OAM-spectrum is investigated in many works [212,213,214,215,13,34]. The OAM of light beams can be measured by using an interferometer [212], by measuring an intensity distribution [213,34], by using cylindrical lenses [13], or by a single point photodetector [215]. Many methods of sorting the laser modes and optical vortices are based on the decomposition into the OAM-spectrum [216,217,218].

An optical vortex beam with rotational symmetry has in its transverse cross-section a single n-order intensity null in the center. Among such vortex beams, the most well-known are the Laguerre–Gaussian modes [219], Bessel–Gaussian beams [40], Hypergeeometric beams [41,220], Circular beams [42], and others. The TC of such beams is equal to their normalized-to-power OAM and equals n. The OAM-spectrum of such beams consists of a single angular harmonic. A simple shift of the screw dislocation center from the center of the host Gaussian beam allows the generation of a different OAM-spectrum [13].

An inverse problem is also of interest. Light fields are known with multiple separate screw dislocations distributed over the cross-section of the host Gaussian beam. Examples of these beams are form-invariant beams with their complex amplitude being described by a finite number of isolated intensity nulls [164]. Form-invariant (or structurally stable) beams are coaxial superpositions of the Laguerre–Gaussian optical vortices with the TCs of the same sign. As was proven in [164], such beams conserve their transverse structure on propagation, changing only in scale and rotation. Obtaining the OAM-spectrum of such beams with distributed screw dislocations is equivalent to constructing a polynomial from its roots. A beam with multiple screw dislocations can be obtained from a source rotationally symmetric vortex beam by interfering with a plane wave [109], by elliptical deformation [221], by a complex transverse shift of the complex amplitude function [43], or by an astigmatic transform [222].

In this section, we analyze theoretically a symmetrical OAM-spectrum of optical vortex beams. We show that the order of the central OAM-harmonic of a symmetrical OAM-spectrum is equal to the normalized orbital angular momentum of the light field. In addition, we study two form-invariant beams with an almost symmetrical OAM-spectrum and confirm that, indeed, the order (topological charge) of the average OAM-harmonic in the OAM-spectrum is approximately equal to the normalized-to-power OAM of the beam.

4.4.1 ORBITAL ANGULAR MOMENTUM OF A BEAM WITH A SYMMETRIC OAM-SPECTRUM

The complex amplitude of a paraxial light field in the initial plane can be expanded into an angular harmonic series:

$$E(r,\varphi) = \sum_{n=-\infty}^{\infty} E_n(r)\exp(in\varphi), \tag{4.61}$$

with (r, φ) being the polar coordinates in the plane $z = 0$. The complex amplitude of each angular harmonic is given by:

$$E_n(r) = \frac{1}{2\pi} \int_0^{2\pi} E(r,\varphi)\exp(-in\varphi)d\varphi. \tag{4.62}$$

Each angular harmonic has the following energy (power):

$$W_n = 2\pi \int_0^{\infty} |E_n(r)|^2 rdr, \tag{4.63}$$

while the total energy (power) of the beam is the sum of all the angular harmonic energies:

$$W = \sum_{n=-\infty}^{\infty} W_n. \tag{4.64}$$

The orbital angular momentum of the beam (4.61) can be obtained by the well-known expression [17,188]:

$$J_z = \text{Im} \int_{-\infty}^{\infty} \int_{-\infty}^{\infty} E^*(x,y) \left(x\frac{\partial}{\partial y} - y\frac{\partial}{\partial x} \right) E(x,y)dxdy$$

$$= \text{Im} \int_{0}^{\infty} \int_{0}^{2\pi} E^*(r,\varphi)\frac{\partial E(r,\varphi)}{\partial \varphi}rdrd\varphi, \tag{4.65}$$

where Im is the imaginary part of a complex number, E^* means the complex conjugation. Substituting Equation (4.61) into Equation (4.65) and using Equation (4.63), we obtain a well-known expression for the OAM of an arbitrary paraxial beam [223]:

$$J_z = \sum_{n=-\infty}^{\infty} nW_n. \tag{4.66}$$

In this section , we obtain some general properties of a symmetric OAM-spectrum and study several examples. We suppose that the powers of the angular harmonics (OAM-spectrum) are distributed symmetrically with respect to some average harmonic of the order n_0, i.e.:

$$W_{n_0+n} = W_{n_0-n}. \tag{4.67}$$

For such a symmetric OAM-spectrum, the order (TC) of the average angular harmonic equals the OAM of the whole beam, normalized to its power. Indeed:

$$J_z = \sum_{n=-\infty}^{\infty} nW_n = n_0W_{n_0} + \sum_{n=1}^{\infty} \left[(n_0+n)W_{n_0+n} + (n_0-n)W_{n_0-n} \right]$$

$$= n_0W_{n_0} + 2n_0\sum_{n=1}^{\infty} W_{n_0+n} = n_0W. \tag{4.68}$$

Equation (4.68) is derived by using Equations (4.64) and (4.67). This equation means that the normalized-to-power OAM of the beam with a symmetrical OAM-spectrum is equal to the TC of the average angular harmonic. According to Equation (4.68), all light fields with a symmetrical OAM-spectrum and with the same order of the central harmonic have the same normalized-to-power OAM, equal to the order of this central angular harmonic. Thus, Equation (4.68) leads to yet another method for changing the shape of a light field by changing the width and shape of its OAM-spectrum, but with the beam, the OAM remained unchanged. Other methods of varying the light field amplitude with the OAM remaining unchanged are also known [18].

4.4.2 Sample Beams with the Symmetric OAM-Spectrum

The complex amplitude of a light field with the symmetric OAM-spectrum can be obtained by shifting the OAM-spectrum, symmetric with respect to the 0th angular harmonic:

$$E_s(r,\varphi) = \sum_{n=-\infty}^{\infty} E_{|n|}(r)\exp(i(n+n_0)\varphi). \qquad (4.69)$$

The OAM of the beam (4.69) can be shown to equal $J_z = n_0 W$. Now we show that the field from Equation (4.69) has the OAM-spectrum, symmetric with respect to the central angular harmonic n_0. If $m = n_0 \pm n$, then the amplitude E_m of the mth OAM-harmonic of the field from Equation (4.69) is equal (the argument r is omitted for brevity):

$$E_m = \frac{1}{2\pi}\int_0^{2\pi} E_s(r,\varphi)\exp(-im\varphi)d\varphi$$

$$= \frac{1}{2\pi}\int_0^{2\pi}\sum_{p=-\infty}^{\infty} E_{|p|}(r)\exp\left[i(p+n_0-m)\varphi\right]d\varphi \qquad (4.70)$$

$$= E_{|m-n_0|} = E_{|n_0+n-n_0|} = E_{|n|} = E_{|n_0-n-n_0|} = E_{|-n|}.$$

For instance, we can consider a simple light field with the symmetric OAM-spectrum:

$$E_s(r,\varphi) = \exp\left(-\frac{r^2}{w^2}\right)\sum_{n=-\infty}^{\infty} C_n\left(\frac{\sqrt{2}r}{w}\right)^{|n|}\exp\left[i(n+n_0)\varphi\right], \qquad (4.71)$$

with w being the waist radius of the Gaussian beam and C_n being the superposition coefficients. For example, if $C_n = 1/|n|!$, each OAM-harmonic of the field from Equation (4.71) has the following energy in Equation (4.63):

$$W_n = \frac{\pi w^2}{2|n|!}. \qquad (4.72)$$

In this case, the total power from Equation (4.64) of the beam is $W = \pi w^2(e-1/2)$. The order n_0 of the central OAM-harmonic of the beam from Equation (4.71) is coinciding exactly with the normalized-to-power OAM $J_z/W = n_0$. Unfortunately, the beam (4.71) is not a mode and changes its transverse intensity shape on propagation, since its radial and azimuthal factors have different indices. In the Fresnel diffraction zone, each term in the sum from Equation (4.71) is proportional to the confluent hyper geometric function $_1F_1(a, b, \xi)$ [41,220].

4.4.3 FAMILY OF FORM-INVARIANT BEAMS WITH AN ALMOST SYMMETRIC OAM-SPECTRUM

As another example, we consider here a family of form-invariant light fields with the complex amplitude of the following form:

$$E_{ss}(r,\varphi) = \exp\left(-\frac{r^2}{w^2}\right)f(re^{i\varphi}) = \exp\left(-\frac{r^2}{w^2}\right)\sum_{n=0}^{\infty}\frac{C_n r^n}{n!}\exp(in\varphi). \qquad (4.73)$$

In Equation (4.73), $C_n = d^n f(z)/dz^n$ at $z = 0$. The function $f(re^{i\varphi})$ in Equation (4.73) is an entire function and, therefore, can be expanded into a converging Taylor series. As seen in Equation (4.73), the OAM-spectrum of the form-invariant field is positive, i.e. all angular harmonics have a nonnegative TC ($n \geq 0$). Such light fields preserve their transverse structure on propagation, only changing in scale and rotating around the optical axis. However, the OAM-spectrum of the beams from Equation (4.73) is not symmetric. Only for some fields, an almost symmetric spectrum can be obtained. For example, if we consider a function $f(re^{i\varphi}) = \cos(re^{i\varphi}/\alpha)$ (with α being some constant), then instead of Equation (4.73) we get:

$$E_c(r,\varphi) = \exp\left(-\frac{r^2}{w^2}\right)\cos\left(\frac{r}{\alpha}e^{i\varphi}\right)$$

$$= \exp\left(-\frac{r^2}{w^2}\right)\sum_{p=0}^{\infty}\frac{(-1)^p}{(2p)!}\left(\frac{r}{\alpha}\right)^{2p}\exp(2ip\varphi). \qquad (4.74)$$

It is seen from Equation (4.74) that the OAM-spectrum contains only even nonnegative angular harmonics. From Equation (4.74), the energy of each OAM-harmonic is given by:

$$W_{2p} = \frac{\pi w^2}{2(2p)!}\left(\frac{w^2}{2\alpha^2}\right)^{2p}. \qquad (4.75)$$

The total (summary) energy of the beam from Equation (4.74) is:

$$W = \sum_{p=0}^{\infty} W_{2p} = \frac{\pi w^2}{2} \cosh\left(\frac{w^2}{2\alpha^2}\right). \tag{4.76}$$

The orbital angular momentum of the beam from Equation (4.74) is equal to:

$$J_z = \sum_{n=-\infty}^{\infty} n W_n = \left(\frac{\pi w^2}{2}\right)\left(\frac{w^2}{2\alpha^2}\right) \sinh\left(\frac{w^2}{2\alpha^2}\right), \tag{4.77}$$

and the normalized-to-power OAM is thus given by:

$$\frac{J_z}{W} = \left(\frac{w^2}{2\alpha^2}\right) \tanh\left(\frac{w^2}{2\alpha^2}\right). \tag{4.78}$$

If the cosine period in the complex amplitude in Equation (4.74) is small, i.e. $w^2 \gg 2\alpha^2$, the hyperbolic tangent in Equation (4.78) is approximately equal to 1. Therefore, the OAM is nearly equal to $J_z/W \approx w^2/(2\alpha^2) \gg 1$. Now we show that if $w^2 \gg 2\alpha^2$, the order (TC) of the OAM-harmonic contributing the most to the OAM-spectrum of the beam from Equation (4.74), is equal to the normalized-to-power OAM of this beam. Indeed, if the cosine period is small (i.e. $w^2 \gg 2\alpha^2$), the energy in Equation (4.75) of each OAM-harmonic from Equation (4.75) is proportional to an expression $W_n \sim \xi^n/n!$ (with $\xi = w^2/(2\alpha^2)$), which achieves its maximal value, according to Stirling's formula ($n! \approx n^n$), at $n = \xi$.

The same considerations can also be done for another example, for comparison. If we consider a function $f(re^{i\varphi}) = \cos^2(re^{i\varphi}/\alpha)$, then, instead of Equation (4.74), we get:

$$E_c(r,\varphi) = \exp\left(-\frac{r^2}{w^2}\right)\cos^2\left(\frac{r}{\alpha}e^{i\varphi}\right)$$

$$= \exp\left(-\frac{r^2}{w^2}\right)\sum_{p=0}^{\infty}\frac{\theta_n(-1)^p}{2(2p)!}\left(\frac{2r}{\alpha}\right)^{2p}e^{2ip\varphi}, \tag{4.79}$$

with $\theta_n = \{1, n > 1; 2, n = 0\}$. Expression in Equation (4.79) differs from Equation (4.74) since the cosine function is squared. Light fields described by Equation (4.79) or by Equation (4,74) are examples of optical vortex beams with an almost symmetric OAM-spectrum. The OAM-spectrum is symmetric, but only approximately since only OAM-harmonics of nonnegative even orders are nonzero.

It is seen from Equation (4.79) that the OAM-spectrum has only even nonnegative angular harmonics. The energy of each OAM-harmonic in Equation (4.79) is given by:

$$W_{2p} = \frac{\pi w^2}{8(2p)!}\left(\frac{2w^2}{\alpha^2}\right)^{2p}, p > 0,$$

$$W_0 = \frac{\pi w^2}{2}, p = 0.$$

(4.80)

If $2w^2 > \alpha^2$, then with increasing p the energy of OAM-harmonics from Equation (4.80) at first grows, but then decays (because of the factorial in the denominator). Since, according to Stirling's formula, the factorial $n!$ at large n is approximately equal to n^n, the energy from Equation (4.80) is maximal when the numerator and denominator are nearly equal (i.e. $W_n \sim \eta^n/n! \approx (\eta/n)^n = 1$, where $\eta = 2w^2/\alpha^2$):

$$\frac{2w^2}{\alpha^2} \approx 2p.$$

(4.81)

Total energy of the beam from Equation (4.80) is equal to:

$$W = \sum_{n=-\infty}^{\infty} W_n = \frac{3\pi w^2}{8} + \left(\frac{\pi w^2}{8}\right)\cosh\left(\frac{2w^2}{\alpha^2}\right),$$

(4.82)

while the beam's OAM is:

$$J_z = \sum_{n=-\infty}^{\infty} nW_n = \left(\frac{\pi w^2}{8}\right)\left(\frac{2w^2}{\alpha^2}\right)\sinh\left(\frac{2w^2}{\alpha^2}\right).$$

(4.83)

If the cosine period is small, i.e. $2w^2 \gg \alpha^2$, the normalized-to-power OAM of the beam from Equation (4.79) equals:

$$\frac{J_z}{W} = \frac{\eta\sinh(\eta)}{3+\cosh(\eta)} \approx \eta = \frac{2w^2}{\alpha^2}.$$

(4.84)

The OAM-spectra in Equations (4.75) and (4.80) of the light fields from Equations (4.74) and (4.79) are symmetric only approximately, and the order of the central harmonic ($2w^2/\alpha^2 \approx 2p$), which contributes the most to the OAM-spectrum, coincides with the approximate expression for the normalized-to-power OAM: $J_z/W \approx 2w^2/\alpha^2 \gg 1$.

Thus, we have shown that the light fields from Equation (4.69) with the symmetric OAM-spectrum do not conserve their shape on space propagation, while the form-invariant fields from Equations (4.74) and (4.79) have the OAM-spectrum, only approximately symmetric. However, in all three given examples (Equations (4.71), (4.74), and (4.79)), the order (TC) of the average OAM-harmonic is coinciding (either exactly or approximately) with the normalized-to-power OAM of the vortex beam. We note that the beams from Equations (4.74) and (4.79) have a countable number of screw dislocations, each with TC = 1 (for beam from Equation (4.74)) or TC = 2 (for beam from Equation (4.79)). Therefore, both beams have an infinite TC. Thus, it seems that the beams cannot be compared by their TC. However, the beams from Equations (4.74) and (4.79) can be compared by their OAM. For instance, the normalized OAM of the beam from Equation (4.79) ($J_z/W \approx 2w^2/\alpha^2$) is four times greater than that of the beam from Equation (4.74) ($J_z/W \approx w^2/(2\alpha^2)$). Since the central OAM-harmonic of the OAM-spectrum of the beam from Equation (4.79) has the TC four times higher than the central OAM-harmonic of the beam from Equation (4.74), the TC of the central OAM-harmonic of the beam from Equation (4.79) is also four times greater than that of the beam from Equation (4.74).

4.4.4 NUMERICAL SIMULATION

Simulation of beams with a symmetric OAM-spectrum

In this section, we consider as an example a light beam from Equation (4.71) with normally distributed contributions of the OAM-harmonics, i.e. $C_n = \exp[-n^2/(2\sigma^2)]$. In the initial plane, the complex amplitude of such beam reads as:

$$E_s\left(r, \varphi\right) = \exp\left(-\frac{r^2}{w^2}\right) \sum_{n=-\infty}^{\infty} \exp\left(-\frac{n^2}{2\sigma^2}\right)\left(\frac{\sqrt{2}r}{w}\right)^{|n|} \exp\left[i(n+n_0)\varphi\right], \quad (4.85)$$

where σ defines the decay ratio of the OAM-harmonics contributions from the central n_0th harmonic to the periphery.

Figure 4.14 depicts the OAM-spectra, as well as the intensity and phase distributions of the beam from Equation (4.85) in the initial plane and at the Rayleigh distance, for several symmetrical OAM-spectra with the 10th-order central harmonic and with different decaying ratios of secondary (noncentral) harmonics contributions.

In the initial plane, maximal intensity of each harmonic in Equation (4.85) is on a ring with a radius $r_{\max,n} = w(|n|/2)^{1/2}$ and is equal to $I_{\max,n} = |n|^{|n|}\exp(-|n| - n^2/\sigma^2)$. Therefore, even at the slowest decaying of the angular harmonics contributions in Figure 4.14, i.e. at $\sigma = 1$, the maximal intensity of the $\pm$2nd harmonics is only 1% of the maximal intensity of the 0th harmonic, while the intensity of higher-order harmonics is even lower. Therefore, the intensity distribution in the initial plane consists mainly of the zero harmonic and is slightly distorted by $\pm$1st-order harmonics. As a result, the initial pattern for all four values σ has a Gaussian shape, which slowly

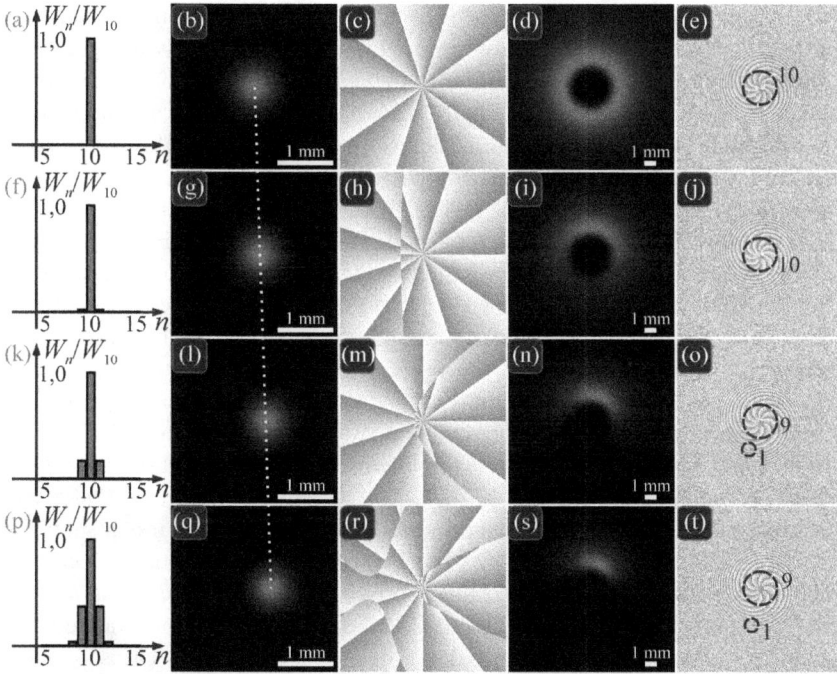

FIGURE 4.14 OAM-spectra (a, f, k, p), as well as distributions of the normalized-to-maximum intensity (b, d, g, i, l, n, q, s) and phase (c, e, h, j, m, o, r, t) in the initial plane (b, c, g, h, l, m, q, r) and at the Rayleigh distance (d, e, i, j, n, o, s, t) of the beam from Equation (4.85) for the following parameters: wavelength $\lambda = 532$ nm, Gaussian beam waist radius $w = 0.5$ mm, order of the central OAM-harmonic $n_0 = 10$, decaying ratio of the OAM-harmonics contributions $\sigma = 0.25$ (a–e), $\sigma = 0.5$ (f–j), $\sigma = 0.75$ (k–o), $\sigma = 1.00$ (p–t), computation domain |x|, |y| $\leq R$ with $R = 1.5$ mm (intensity in the initial plane), $R = 5$ mm (phase in the initial plane), and $R = 7.5$ mm (intensity and phase at the Rayleigh distance). The dashed circles on the phase distributions at the Rayleigh distance (e, j, o, t) show the area with optical vortices (their number is also written near the circles). The white dotted line (b, g, l, q) shows the shift of the Gaussian beam center.

moves to the right with increasing σ. For instance, at $\sigma = 0.25$, the superposition from Equation (4.85) is contributed mostly from the following three terms:

$$E_s(r,\varphi) = \exp\left(-\frac{r^2}{w^2}\right)\exp(i10\varphi) +$$

$$\exp(-8)\exp\left(-\frac{r^2}{w^2}\right)\left(\frac{\sqrt{2}r}{w}\right)\exp(i11\varphi) + \quad\quad (4.86)$$

$$\exp(-8)\exp\left(-\frac{r^2}{w^2}\right)\left(\frac{\sqrt{2}r}{w}\right)\exp(i9\varphi) + \ldots$$

It is seen from Equation (4.86) that the second and the third terms have much weaker amplitude than the first term. Therefore, in the initial plane, the vortex beam from Equation (4.85) has the shape of the Gaussian beam (Figure 4.14(b)). The beam from Equation (4.86) has a TC of 10 (Figure 4.14(c)), since for an arbitrary value of r the first term of the series (4.86) overwhelms all the rest terms. However, since the high-TC optical vortex is unstable, then, due to the other terms in Equation (4.86) in addition to the main first term, the central 10th-order vortex splits on propagation into 10 optical vortices with TC of +1, located close to each other (Figure 4.14(e)). We note that the separate vortices in the center of the phase distribution (Figure 4.14(e)) are not the additional vortices occurring due to the spin-orbit interaction [224]. We limit our study here by the paraxial scalar approximation and the beams do not have circular polarization as in [224].

The beams in Equation (4.85) are not modes and, on space propagation, other, noncentral, harmonics become more eminent. When σ is small, the OAM-spectrum is very narrow. Therefore, at $\sigma = 0.25$, the intensity distribution at the Rayleigh distance is rotationally symmetric (Figure 4.14(d)). When $\sigma = 0.5$, the OAM-spectrum is visually almost unchanged, but the intensity distribution acquires a shape of an asymmetric light ring (Figure 4.14(i)), which, with increasing contributions of secondary harmonics (i.e. with increasing σ), is broken up and becomes a crescent (Figure 4.14(n, s)). Despite the distorted intensity shape, the normalized OAM, according to Equation (4.68), should be equal to $n_0 = 10$ in all cases. Indeed, in all four phase distributions (Figure 4.14(e, j, o, t)), 10 optical vortices can be seen (marked by dashed circles in Figure 4.14), one of which moves away from the center with increasing σ. The normalized-to-power OAM, computed by Equation (4.65) and divided by the beam power, also has values close to 10: 9.35 (Figure 4.14 (b, c)), 9.46 (Figure 4.14 (d, e)), 9.37 (Figure 4.14 (g, h)), 9.43 (Figure 4.14 (i, j)), 9.48 (Figure 4.14 (l, m)), 9.45 (Figure 4.14 (n, o)), 9.58 (Figure 4.14 (q, r)), and 9.50 (Figure 4.14 (s, t)).

The shift of the Gaussian beam (Figure 4.14 (b, g, l, q)) in the initial plane to the right along the x-axis with increasing σ can be explained by Equation (4.86), which can be rewritten as:

$$E_s\left(r,\varphi\right) = \exp\left(-\frac{r^2}{w^2}\right)\exp(i10\varphi)\left(1+\frac{\alpha r}{w}\cos\varphi\right)+... \qquad (4.87)$$

with $\alpha = 2^{3/2}\exp[-1/(2\sigma^2)]$. According to Equation (4.87), the intensity maximum shifts along the axis $x = r\cos\varphi$ with increasing α :

$$I(r,\varphi) = \left|E_s\left(r,\varphi\right)\right|^2 \approx \exp\left(-\frac{2r^2}{w^2}+\frac{2\alpha x}{w}\right)+... \qquad (4.88)$$

Equation (4.87) also demonstrates that near $x \approx -w/\alpha$, the amplitude is zero, i.e. there is a phase jump by π (edge dislocation) along the vertical line $x = -w/\alpha$. This phase jump is clearly seen in Figure 4.14(h). With a further increase of α, the contribution of the other terms of the series from Equation (4.85) into Equation (4.87) is also increasing:

$$E_s\left(r,\varphi\right)=\exp\left(-\frac{r^2}{w^2}\right)\exp(i10\varphi)\left(1+\frac{\alpha r}{w}\cos\varphi+\frac{\beta r^2}{w^2}\cos 2\varphi\right)+... \qquad (4.89)$$

with $\beta=4\exp(-2/\sigma^2)$. According to Equation (4.89), the line of zero amplitude, where the phase has a jump by π (Figure 4.14 (m)), is now not a straight line as in Figure 4.14 (h), but a curve (Figure 4.14 (m)). The separation of one vortex with the TC +1 from the 10 vortices in the center (Figure 4.14 (o, t)) can be explained by the interaction between the edge dislocation with an optical vortex (screw dislocation) (Figure 4.14(m, r)) [225,226]. The difference from [225,226] is that, instead of the straight edge dislocation, it is bent in Figure 4.14(m, r). In order to give a detailed explanation of this phenomenon, the Fresnel transform of the function from Equation (4.89) should be obtained. This would lead to rather cumbersome expressions including the difference of two Bessel functions of half-integer orders [80], analysis of which would lead us far from the goal of this work.

In conclusion, in this section we give a general formula for the field from Equation (4.85), similar to Equation (4.89), which clearly illustrates that both the TC and the normalized-to-power OAM of the beam with a symmetric OAM-spectrum are equal to n_0:

$$E_s\left(r,\varphi\right)=\exp\left(-\frac{r^2}{w^2}\right)\exp\left(in_0\varphi\right)\left[1+\sum_{n=1}^{\infty}\frac{c_n r^n\cos n\varphi}{w^n}\right], \qquad (4.90)$$

where $c_n=2^{(|n|+2)/2}\exp\left(-\frac{n^2}{2\sigma^2}\right)$.

As seen from Equation (4.90), since the sum is real-valued, the TC of the field from Equation (4.90) is equal to n_0 and the OAM from Equation (4.65), normalized to the beam power in Equation (4.64), is also equal to $J_z/W=n_0$. In this section, we do not consider the intrinsic and extrinsic components of the OAM from Equation (4.65), and in all cases we calculate it relative to the origin. Of course, in the cases when the "center of mass" of the beam intensity distribution is shifted from the optical axis (i.e. the beam has an asymmetric intensity distribution), the OAM is a sum of two terms, only one of which is the intrinsic OAM.

Numerical simulation of form-invariant beams with an almost symmetric OAM-spectrum

In this subsection, we investigate numerically the form-invariant light beams from Equations (4.74) and (4.79). Figure 4.15 shows the OAM-spectra, as well as the intensity and phase distributions of these beams in the initial plane.

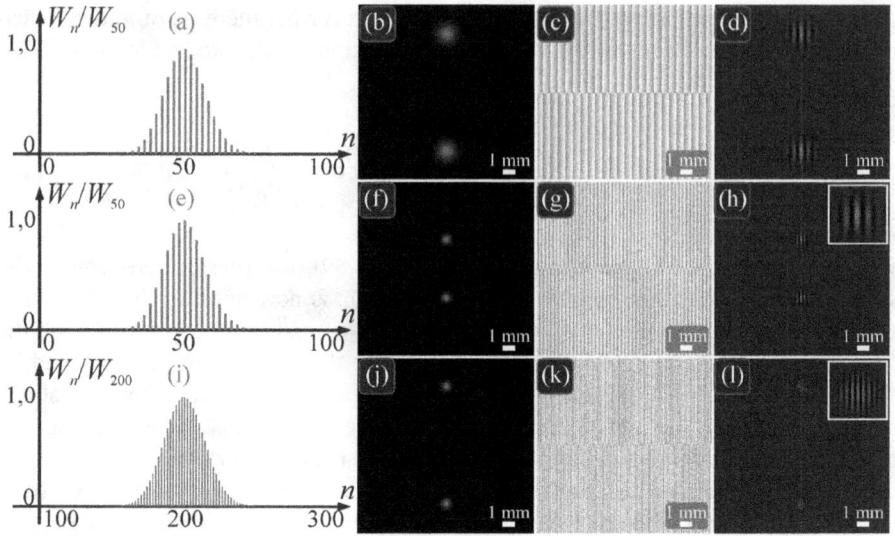

FIGURE 4.15 OAM-spectra (a, e, i), normalized-to-maximum intensity (b, f, j) and phase (c, g, k) distributions, as well as intensity distributions with an interfering plane wave (d, h, l) of the beams from Equation (4.74) (a–h) and (4.79) (i–l) in the initial plane for the following parameters: wavelength λ = 532 nm, Gaussian beam waist radius w = 1 mm (a–d) and w = 0.5 mm (e–l), scaling factor of the vortex cosine function α = w/10, computation domain $|x|$, $|y|$ $\leq R$ with R = 7.5 mm. The insets (h, l) show the 3× magnified fragments of the interference patterns

Figures 4.15(a–d) and (e–h) show the beams from Equation (4.74) with the different-waist radius w, but with the same ratio w/α = 10. Therefore, according to Equation (4.78), their OAM-spectra should be the same (up to a constant multiplier) and, indeed, the normalized-to-maximum OAM-spectra in Figure 4.15(a, e) look similar. The theory predicts that the normalized OAM equals $J_z/W \approx w^2/(2\alpha^2)$ = 50 in both cases. The OAM values, despite the different shapes of the beams from Equation (4.74) at w = 1 mm (Figure 4.15(a–d)) and w = 0.5 mm (Figure 4.15(e–h)), their effective TC is the same and the number of stripes in the interference patterns is also the same (in Figure 4.15(d, h), five stripes are clearly seen). The equal OAMs of the beams from Figure 4.15(a–d) and Figure 4.15(e–h) can be physically explained by an analogy with the mechanical torque. Compared to the beam with w = 0.5 mm (Figure 4.15(e–h)), the beam with w = 1 mm (Figure 4.15(a–d)) has two times greater leverage (the distance from the center of the pattern to the light spots), but two times lower transverse force (velocity of phase changes in the transverse plane). Equation (4.65) gives OAM 49.82 (Figure 4.15(a–d)) and 49.27 (Figure 4.15(e–h)).

According to the theory, the beam from Equation (4.79) (Figure 4.15(i–l)) should have the maximal-energy OAM-harmonic with the TC four times higher, and the normalized OAM should also increase four times. Numerical calculation gives the

OAM value 188.66, while the largest contribution is given by 200th OAM-harmonic (Figure 4.15 (i)). The interference pattern (Figure 4.15(l)) has two times more inter-ference stripes (clearly seen are nine stripes), but, at the same time, the light spots themselves in Figure 4.15(j) are also two times further from the center compared to Figure 4.15(f), i.e. by analogy with the mechanical torque, both the leverage and the force increase two times. The analogy between the transverse force, acting onto a particle in the light field, and the velocity of the phase change has the following explanation. The force is proportional to the transverse energy flow, which is propor-tional to the energy (intensity) density in the current point multiplied by the phase gradient. The greater the phase gradient, the greater the energy flow and the force. In [227], the energy flow, expressed via the phase gradient of a scalar field, is called an optical current. That is why the OAM increases four times. Similarly, it can be shown that the effective TC for the case from Figure 4.15(i–l) is four times greater than that for the case from Figure 4.15(e–h). Indeed, for a conventional (circular) optical vortex with the phase $n\varphi$, the TC equals the length of a circumference with the radius R divided by a period T of phase change by 2π : $TC = 2\pi R/T$. Since the period T of the phase changes by 2π in Figure 4.15(l) is two times smaller than that in Figure 4.15(h) (i.e. two times more interference stripes), and the distance R to the center of the Gaussian beam in Figure 4.15(l) is two times greater than in Figure 4.15(h), then the effective TC of the beam in Figure 4.15(l) is four times greater than TC of the beam from Figure 4.15(h).

Thus, we have shown theoretically and confirmed numerically that if a vortex beam has a symmetrical OAM-spectrum, then the TC of the central OAM-harmonic coincides with the beam's normalized-to-power OAM [228]. This general property of the vortex beams is also valid for beams with an approximately (almost) symmet-rical OAM-spectrum. For two examples of form-invariant beams with an infinite TC and with a finite OAM, we have demonstrated that they have an almost symmetrical OAM-spectrum and that the TC of the central angular harmonic is almost equal to the normalized-to-power OAM of these beams. Therefore, by varying the width of the symmetrical OAM-spectrum without changing the central OAM-harmonic (i.e. without changing its position), vortex beams can be obtained with different intensity distributions, but with the same OAM value. This is yet another method of changing the beam shape without changing its "effective" TC and orbital angular momentum. In practice, laser beams with the symmetric OAM-spectrum can be generated by using either an SLM or diffractive optical elements [229].

5 Transformation of an Edge Dislocation of a Wavefront into an Optical Vortex

5.1 CONVERTING AN ARRAY OF EDGE DISLOCATIONS INTO A MULTI-VORTEX BEAM

In recent years, optical vortices have been extensively studied [56]. Laser optical vortices can be generated in a number of different ways, which include, among others, digital holograms with a carrier frequency [230,231], amplitude diffraction gratings with a fork [232], mode converters [222], spiral phase plates [233,234], spiral axicons [235,236], q-plates [237], interferometers [238], light modulators [239], photonic components with metasurface [240], a microlaser [241], and a conventional beam splitter [191]. Below, we dwell on the simplest techniques. In their remarkable work, Abramochkin and Volostnikov [222] demonstrated how using an astigmatic converter, a non-vortex Hermite–Gaussian (HG) beam was converted into a vortex Laguerre–Gaussian (LG) beam. What is of interest to us is that after passing through a cylindrical lens and propagating in free space, straight lines of intensity nulls, which are present in the intensity distribution of the HG beam in the original plane, experience conversion, resulting in an isolated on-axis intensity null around which an optical LG vortex is generated. For instance, an optical beam studied in [222] had neither an isolated intensity null, nor optical vortex, nor topological charge (TC) in the original plane and at some on-axis distance. However, starting with a certain plane, an intensity null formed, around which an optical vortex with non-zero integer TC was generated. Thus, this example shows that, unlike OAM, TC is not conserved upon free-space propagation of a paraxial optical beam. In a number of publications, optical vortices were generated using nanoholes made in an opaque screen, which formed an array in the form of an Archimedes [242,243] or Fermat spiral [244,74]. However, a theoretical analysis of such structures is challenging. In [225,226,245,246], the interaction of an optical vortex with an edge dislocation (zero-intensity lines) embedded into a Gaussian beam was analyzed. The interaction was shown to lead to a split in the edge dislocation, resulting in the generation of additional optical vortices. These works are interesting and deserve further research but the use of an astigmatic transformation by means of a cylindrical lens was not discussed in [225,226,245,246].

DOI: 10.1201/9781003326304-5

We note that transforming the Hermite–Gaussian beams into the Laguerre–Gaussian beams using an astigmatic convertor made of two cylindrical lenses [222] was also studied in [247,248]. In [247], matrix formalism was developed to describe the transformation of an HG mode into an LG mode, whereas in [248], the authors showed that a cylindrical lens, which introduces the orbital angular momentum into the HG beam, is affected by a torque that tends to rotate the lens.

There are also recent works (2021 year) that investigate the asymmetric optical vortices [249], optical vortices generated by a superposition of vortex-free Ince–Gaussian beams [250], and an array of optical vortices generated by a metasurface [251].

In this section, using a simple example, we analyze in detail conditions at which the zero-intensity straight lines (edge dislocations) in the original plane can be converted into an isolated intensity null and an integer optical vortex using an astigmatic transformation implemented with a cylindrical lens tilted at 45 degrees to the Cartesian axes.

5.1.1 Complex Amplitude at the Double Focal Length

There may be two vertical zero-intensity lines (two edge dislocations) embedded into the waist of a Gaussian beam with an astigmatic phase. This is identical to an ideal thin cylindrical lens located in the Gaussian beam waist plane, with its generating line tilted at 45 degrees to the waist plane. The complex amplitude of such a light field in the initial plane takes the form:

$$E(x, y, z = 0) = \frac{\left(x^2 - a^2\right)}{w^2}$$

$$\times \exp\left(-\frac{x^2 + y^2}{w^2} - \frac{ik}{4f}(x + y)^2\right),$$

(5.1)

where k is the wavenumber of light, w is the waist of the Gaussian beam, (x, y) are the transverse Cartesian coordinates, z is the longitudinal on-axis coordinate, and f is the focal length of the cylindrical lens. Two vertical zero-intensity lines in Equation (5.1) are located at distances $+a$ and $-a$ from the vertical axis y. The second term in the argument of the exponential function in Equation (5.1) defines the phase profile of the parabolic cylindrical lens tilted at 45 degrees to the x- and y-axes.

We aim to determine conditions for generating an on-axis optical vortex with TC = 2. Indeed, based on results reported in [222], we may expect that under certain conditions beam from Equation (5.1) is converted into an LG beam (0,2) because at certain a, the first term in Equation (5.1) is given by a Hermite polynomial: $H_2(x) = x^2 - 1$, with the argument defined as $\sqrt{2}x / w$.

At distance $z = 2f$, the amplitude of light field from Equation (5.1) derived from a Fresnel transform:

$$E(x,y,z) = \frac{-ik}{2\pi z} \int\limits_{-\infty}^{\infty} \int\limits_{-\infty}^{\infty} E(x',y',z = 0)$$

$$\times \exp\left[\frac{ik}{2z}(x-x')^2 + \frac{ik}{2z}(y-y')^2 \right] dx'dy' \tag{5.2}$$

is:

$$E\left(\xi,\eta,z = 2f\right) = \left(\frac{-i\gamma}{\sqrt{1+\gamma^2}} \right) \left[\frac{1}{2(1+\gamma^2)} - \left(\frac{a}{w}\right)^2 - \frac{\gamma^2\left(\xi - i\gamma\eta\right)^2}{w^2\left(1+\gamma^2\right)^2} \right]$$

$$\times \exp\left(-\frac{\gamma^2\left(\xi^2+\eta^2\right)}{w^2\left(1+\gamma^2\right)} + \frac{i\gamma}{w^2}\left(\xi^2+\eta^2\right) + \frac{2i\gamma^3\xi\eta}{w^2\left(1+\gamma^2\right)} \right), \tag{5.3}$$

where $\gamma = z_0/(2f)$, $z_0 = kw^2/2$. In Equation (5.2), z_0 denotes the Rayleigh range. From Equation (5.3) it follows that at the double focal length, the waist of the Gaussian beam changes, becoming equal to $w(z = 2f) = (w/\gamma)(1 + \gamma^2)^{1/2}$, and an astigmatic phase emerges. The amplitude of the Gaussian beam from Equation (5.3) is multiplied by the function $F(\xi,\eta)$:

$$F(\xi,\eta) = A - B\xi^2 + C\eta^2 + iD\xi\eta, \tag{5.4}$$

where:

$$A = \frac{1}{2(1+\gamma^2)} - \left(\frac{a}{w}\right)^2, B = \frac{\gamma^2}{(1+\gamma^2)^2 w^2},$$

$$C = \frac{\gamma^4}{(1+\gamma^2)^2 w^2}, D = \frac{2\gamma^3}{(1+\gamma^2)^2 w^2}.$$

The beam in Equation (5.3) has two isolated intensity nulls located on the vertical axis η at $A < 0$, or on the horizontal axis ξ, at $A > 0$. It is only at $A = 0$ that beam from Equation (5.3) has a single isolated on-axis intensity null (at $\eta = \xi = 0$). The zero-intensity pairs for beam from Equation (5.3) can be derived from Equation (5.4):

$$1)\ \xi = 0,\ \ \eta = \pm\eta_0 = \pm\frac{\sqrt{\left(1+\gamma^2\right)\left(2a^2\left(1+\gamma^2\right)-w^2\right)}}{\sqrt{2}\gamma^2},$$

if $A < 0$;

$$2)\, \eta = 0, \quad \xi = \pm\xi_0 = \pm\frac{\sqrt{\left(1+\gamma^2\right)\left(w^2 - 2a^2\left(1+\gamma^2\right)\right)}}{\sqrt{2}\gamma};$$

if $A > 0$;   (5.5)

$3)\, \xi = \eta = 0$, if $A = 0$.

The condition $A > 0$ means that two vertical lines of intensity nulls in the original plane are located close to the η axis ($a < w/[2(1+\gamma^2)]^{1/2}$). At distance $z = 2f$, the two closely located zero-intensity lines produce two isolated intensity nulls on the ξ axis (second line in Equation (5.5)), with a pair of optical vortices with TC $=-1$ originating around these two intensity nulls. And vice versa, at $A < 0$, the two vertical zero-intensity lines in the original plane are found further from the η axis ($a > w/[2(1 + \gamma^2)]^{1/2}$). At the double focal length, these two zero-intensity lines also produce two isolated intensity nulls, but on the axis η (first line in Equation (5.5)). Centered at these two nulls, two optical vortices with TC $=-1$ are also generated. Finally, only at $A = 0$ and given the condition:

$$a = \frac{w}{\sqrt{2\left(1+\gamma^2\right)}},$$ (5.6)

the two zero-intensity lines in Equation (5.1) produce an isolated on-axis intensity null ($\xi = \eta = 0$), with an optical vortex with TC $=-2$ centered at it. From Equation (5.6) it is seen that for an on-axis optical vortex with a single intensity null to be produced, the zero-intensity lines in the original plane should not be located too far apart, because there is a maximum distance between them: $a < w/\sqrt{2}$ (at any γ).

From the above reasoning, the evolution of the intensity nulls in the plane at the double focal length from the origin as a function of the distance between the original-plane zero-intensity lines is as follows. May there be a single (2nd-order) zero-intensity line on the η axis, then at the output plane, there will be two intensity nulls on the horizontal axis at the maximum distance from the center, equal to $\xi_{max} = \pm w(1+\gamma^2)^{1/2}/(\sqrt{2}\gamma)$. Next, we assume that the zero-intensity line is split into two lines symmetrical to the vertical axis, with the between-line distance increasing. Then, two zeroes found in the output plane ($z = 2f$) at the maximum distance ξ_{max} begin moving to the center. When the half-distance between the two lines reaches a critical value of $x_k = a = w[2(1+\gamma^2)]^{-1/2}$ (condition in Equation (5.6)), the two output-plane zeroes will "merge" into a single 2nd-order intensity null at the center. If the between-zero distance in the original plane is further increased beyond the critical value, the 2nd-order intensity null will again split into two output-plane nulls that, this time, will be found on the vertical axis symmetrically to the origin. This process will be endless, because with the between-zero-line distance tending to infinity, the output-plane intensity nulls will also tend to infinity ($\pm\infty$) along the η axis.

If we look into the type of optical vortex produced at the double focus distance by the original beam from Equation (5.1) under the condition in Equation (5.6), Equation

(5.3) suggests that there occurs an on-axis elliptical vortex embedded into a Gaussian beam, with its astigmatic phase and an amplitude given by:

$$(\xi - i\gamma\eta)^2 = \rho^2 \exp(-i2\theta). \tag{5.7}$$

To attain a positive TC = +2, the cylindrical lens needs to be tilted at −45 degrees. In the polar elliptical coordinates, the right-hand side of Equation (5.7) takes the form:

$$\begin{cases} \xi = \rho\cos\theta, \\ \eta = \gamma^{-1}\rho\sin\theta. \end{cases} \tag{5.8}$$

The elliptical vortex changes into a canonical optical vortex at $\gamma = 1$ ($z_0 = 2f$, $a = w/2$):

$$(\xi - i\eta)^2 = \rho^2 \exp(-i2\theta), \tag{5.9}$$

with the light field becoming an LG beam (0,2) with an astigmatic phase:

$$E(\xi,\eta,z = z_0 = 2f) = i\frac{(\xi - i\eta)^2}{4\sqrt{2}w^2}$$
$$\times \exp\left(-\frac{\xi^2 + \eta^2}{2w^2} + \frac{ik}{4f}(\xi^2 + \eta^2 + \xi\eta)\right). \tag{5.10}$$

For beam from Equation (5.10) to remain an optical vortex with TC = −2 upon propagation, an extra cylindrical lens needs to be placed at the double focal length, tilted at 45 degrees to the axes, and with transmittance Exp[−ik(ξ + η)²/(8f)]. The extra lens serves to compensate for an astigmatic phase present in Equation (5.10) and has to have a focal length twice that of the lens in Equation (5.1).

Just behind the second cylindrical lens, instead of Equation (5.10), the field is given by:

$$E_2(\xi,\eta,z = z_0 = 2f) = i\frac{(\xi - i\eta)^2}{4\sqrt{2}w^2}$$
$$\times \exp\left(-\frac{\xi^2 + \eta^2}{2w^2} + \frac{ik}{8f}(\xi^2 + \eta^2)\right). \tag{5.11}$$

Within a constant, beam from Equation (5.11) is an LG mode (0, −2) at distance $z = z_0$ with its waist radius given by $w' = w[1 + (z/z_0)^2]^{1/2} = w\sqrt{2}$ and curvature radius $R = z[1 + (z_0/z)^2] = 2z = 4f$. We note; however, that beam from Equation (5.11) was deduced at $\gamma = 1$, at which the two original zero-intensity lines were found at a distance of half the waist radius from the optical axis ($a = \pm w/2$), as follows from from Equation (5.6). This distance differs from that of the origin of the roots of the Hermite polynomial $H_2\left(\sqrt{2}x / w\right) = 2w^{-2}\left(x^2 - w^2 / 2\right)$, which equals $a = \pm w / \sqrt{2}$.

Although elliptical optical Gaussian vortices with astigmatic phase were previously discussed [252,190] and the Hermite–Gaussian beam $(0, n)$ was converted with a tilted cylindrical lens [252], the effect of the distance between the zero-intensity lines on the generation of an optical vortex has never been studied. In [190], a mode beam made up of a canonical optical vortex with TC = n embedded into an elliptical astigmatic Gaussian beam was reported to conserve upon propagation without splitting into simple optical vortices.

To conclude this section, we note that instead of the initial light field given by Equation (5.1), another initial field can be considered, which has 2N 1st-order edge dislocations located symmetrically with respect to the optical axis (i.e. vertical zero-intensity lines intersecting with the horizontal axis at $x = \pm a_p$, $p = 1, 2, 3, \ldots, N$):

$$E(x, y, z = 0) = \prod_{p=1}^{N} \frac{\left(x^2 - a_p^2\right)}{w^2}$$

$$\times \exp\left(-\frac{x^2 + y^2}{w^2} - \frac{ik}{4f}(x + y)^2\right).$$

(5.12)

It can be shown that a cylindrical lens converts these 2N edge dislocations into 2N optical vortices. The expression for the complex amplitude of the light field at the distance $z = 2f$, similar to Equation (5.3), would be; however, cumbersome, and its detailed analysis would be difficult.

5.1.2 ORBITAL ANGULAR MOMENTUM

In the previous section, two straight zero-intensity lines in a Gaussian beam with an astigmatic phase in the original plane were shown to produce upon free-space propagation either a pair of 1st-order optical vortices or a single 2nd-order on-axis optical vortex. Hence, the TC of such a beam does not conserve upon propagation. In this section, we derive a relationship for the normalized OAM of beam from Equation (5.1). OAM of a paraxial beam is derived from the familiar relation [56]:

$$J_z = \text{Im} \int\limits_{-\infty}^{\infty} \int\limits_{-\infty}^{\infty} \bar{E}(x, y)$$

$$\times \left(x \frac{\partial E(x, y)}{\partial y} - y \frac{\partial E(x, y)}{\partial x}\right) dx dy,$$

(5.13)

$$W = \int\limits_{-\infty}^{\infty} \int\limits_{-\infty}^{\infty} \bar{E}(x, y) E(x, y) dx dy,$$

where Im is the imaginary part of the number, J_z is the OAM of the beam, W is the beam power, and $\overline{E}$ is the complex conjugate function to the function E. Because the OAM of the beam is conserved, it can be computed in the original plane. Substituting Equation (5.1) into Equation (5.13) yields the expression for the OAM normalized to the power:

$$\frac{J_z}{W} = -2\gamma w^2 \frac{3w^2 - 4a^2}{3w^4 - 8a^2 w^2 + 16a^4} \qquad . (5.14)$$

From Equation (5.14) it follows that at small a ($a < \sqrt{3}w/2$), OAM is negative: $J_z < 0$; at large a ($a > \sqrt{3}w/2$), OAM is positive: $J_z > 0$, and at $a = \sqrt{3}w/2$, OAM equals zero: $J_z = 0$.

This is a surprising result, as $J_z = 0$ means that two spatially separated zero-intensity vertical lines are able to "nullify" the OAM of an astigmatic Gaussian beam. It is also interesting that the "nullification" condition ($a = \sqrt{3}w/2$) for the OAM of vortex (5.3) is γ-independent. Among other things, this means that at any γ, OAM is "nullified" only when two intensity nulls (two −1st-order vortices) are on the vertical axis. With the OAM being contributed to by two components: a vortex one (here, negative) and an astigmatic one (here, negative and positive), at certain parameters the two components compensate for each other, making OAM equal to zero. The dynamics of the vortex OAM component are determined by the distance Δ between the two optical vortices with TC = −1 and can be described by the following expression [69] ($\gamma = 1$):

$$J_{\text{vor},z}/W = -4\left[2 + \left(\Delta/w\right)^2\right]^{-1}, \qquad (5.15)$$

whereas the dynamics of the astigmatic OAM component is determined by the ellipticity of the intensity distribution. For instance, the astigmatic OAM of an elliptic Gaussian beam with the waist radii w_x and w_y is given by [17]:

$$J_{\text{ast},z}/W = \left[k/\left(8f\right)\right]\left(w_y^2 - w_x^2\right). \qquad (5.16)$$

According to Equation (5.15), maximal (by modulus) vortex OAM is equal to −2 (at $\Delta = 0$) and, when Δ increases, it tends to zero, remaining negative. As seen from Equation (5.16), the astigmatic OAM is negative (or positive) when the beam intensity distribution is elongated along the axis x (or, respectively, y). The light field in Equation (5.3) has an axially symmetric Gaussian envelope. Therefore, at $\gamma = 1$, the ellipticity of the beam from Equation (5.3) is governed by the term A from Equation (5.4): if $A < 0$, the beam is elongated along the x-axis, if $A > 0$, the beam from Equation (5.3) is elongated along the y-axis, while at $A = 0$, the beam is circularly symmetric.

From Equation (5.14), it also follows that at $a = 0$ (the 2nd-order zero-intensity line lies on the η-axis), the OAM is $J_z/W = -2\gamma$ and at $\gamma = 1$, the OAM is $J_z/W = -2$. Hence, an optical vortex with TC=−2 can be produced using not only two zero-intensity lines combined with an astigmatic phase but also a single on-axis 2nd-order zero-intensity line ($a = 0$). This result is also surprising because it is not associated with the 2nd-order Hermite polynomial, which was utilized in [222] for producing a 2nd-order optical vortex. The difference is that at $a = 0$, the optical vortex produces two −1st order optical vortices lying on the horizontal axis ξ, as follows from Equation (5.4). At a tending to infinity, Equation (5.14) suggests that OAM equals zero. Thus, at $a = 0$, OAM is negative, at $a = \sqrt{3}w/2$, OAM is zero, and at infinitely large a, OAM is again zero. We can infer that at a certain a, OAM may be expected to have a positive maximum. Actually, the search of extremums of OAM from Equation (5.14) shows that the minimal negative and maximal positive values of OAM are, respectively, achieved at $a = (w/2)(3 - \sqrt{6})^{1/2} \approx 0.37w < w$ and $a = (w/2)(3 + \sqrt{6})^{1/2} \approx 1.17w > w$. The extreme OAM values equal $J_{z,min}/W = -(\gamma\sqrt{6}/2)(3 - \sqrt{6})^{-1} \approx -2.22\gamma$ and $J_{z,max}/W = (\gamma\sqrt{6}/2)(3 + \sqrt{6})^{-1} \approx 0.22\gamma$. The OAM is contributed to by the vortex component, which depends on the distance between the two vortices and varies from −2 to 0 [69], and by the astigmatic component, which is negative (positive) at $A < 0$ ($A > 0$), and equals zero at $A = 0$ [17].

If condition from Equation (5.6) is satisfied, a single on-axis 2nd-order intensity null is produced at the double focal length, and Equation (5.14) is rearranged to:

$$\frac{J_z}{W} = -2\gamma\frac{3\gamma^4 + 4\gamma^2 + 1}{3\gamma^4 + 2\gamma^2 + 3}. \tag{5.17}$$

Putting $\gamma = 1$, Equation (5.17) yields $J_z/W = -2$. Thus, given Equation (5.6) (i.e. $a = w/2$) and $\gamma = 1$, at the double focal length there will be a single 2nd-order intensity null with TC=−2, with the OAM of light field from Equation (5.3) also being equal to −2.

Remarkably, Equation (5.17) is similar to the relation for OAM:

$$\frac{J_z}{W} = 8\alpha\frac{\left(\alpha^2 + 1\right)}{\left(3\alpha^4 + 2\alpha^2 + 3\right)} \tag{5.18}$$

of an elliptical Gaussian vortex in the original plane:

$$E_2(x, y, z = 0) = \frac{\left(\alpha x + iy\right)^2}{w^2}\exp\left(-\frac{x^2 + y^2}{w^2}\right), \tag{5.19}$$

which was discussed in [221]. The difference is that the elliptical Gaussian vortex in Equation (5.19) has no astigmatic phase, which the beam in Equations (5.1) or (5.3) has. Hence, at any degree of ellipticity (any α), OAM from Equation (5.18) will not be larger than the TC of beam from Equation (5.19), which equals 2. On the contrary, at different γ, which defines an optical vortex ellipticity, the OAM of beam from Equation (5.3) can be both larger and smaller than −2.

5.1.3   Two Asymmetric Parallel Zero-Intensity Lines

In this section, we look into how the field pattern changes if the vertical zero-intensity lines in the original plane are asymmetric relative to the center. May the original light field be given by the amplitude:

$$E(x, y, z = 0) = \frac{(x-a)(x-b)}{w^2}$$

$$\times \exp\left(-\frac{x^2 + y^2}{w^2} - \frac{ik}{4f}(x+y)^2\right).$$

(5.20)

The only difference between field from Equation (5.20) and field from Equation (5.1) is in the location of zero-intensity lines. At the double focal length, field from Equation (5.20) takes the form:

$$E(\xi, \eta, z = 2f) = \frac{-i\gamma}{\sqrt{1+\gamma^2}} \exp\left\{-\frac{\gamma^2(\xi^2 + \eta^2)}{w^2(1+\gamma^2)}\right.$$

$$\left. + \frac{i\gamma}{w^2}(\xi^2 + \eta^2) + \frac{2i\gamma^3\xi\eta}{w^2(1+\gamma^2)}\right\}$$

(5.21)

$$\times \left[\frac{1}{2(1+\gamma^2)} + \frac{ab}{w^2} - \frac{\gamma^2(\xi - i\gamma\eta)^2}{w^2(1+\gamma^2)^2} + \frac{i\gamma(a+b)}{w^2}\frac{(\xi - i\gamma\eta)}{(1+\gamma^2)}\right].$$

At $a = -b$, the amplitudes in Equations (5.21) and (5.1) are the same. If $a \neq -b$, a Gaussian beam with an astigmatic phase in Equation (5.21) is produced, but it is multiplied by a function different from Equation (5.4):

$$F(\xi, \eta) = A - B\xi^2 + C\eta^2$$

$$+ iD\xi\eta + iE\xi + F\eta,$$

(5.22)

where:

$$A = \frac{1}{2(1+\gamma^2)} + \frac{ab}{w^2}, \quad B = \frac{\gamma^2}{w^2(1+\gamma^2)^2},$$

$$C = \frac{\gamma^4}{w^2(1+\gamma^2)^2}, \quad D = \frac{2\gamma^3}{w^2(1+\gamma^2)^2},$$

(5.23)

$$E = \frac{\gamma(a+b)}{w^2(1+\gamma^2)}, \quad F = \frac{\gamma^2(a+b)}{w^2(1+\gamma^2)}.$$

From Equation (5.23), it follows that for isolated intensity nulls of field from Equation (5.21) to be found in the plane $z = 2f$, the imaginary part of the function $F(\xi, \eta)$ should be equal to zero. This condition holds on two lines: at $\xi = 0$ and at $\eta = -(a+b)$ $(1 + \gamma^2)/(2\gamma^2)$. This means that at $A \neq 0$, at the center ($\xi = \eta = 0$) there cannot be a 2nd-order intensity null. At $A = 0$, when the zero-intensity lines in the original plane are on the opposite sides of the center and the product of the line-to-center distances is:

$$ab = \frac{-w^2}{2(1+\gamma^2)},\tag{5.24}$$

in the center of the plane at $z = 2f$ there is a -1st-order intensity null, instead of the -2nd order. And the second intensity null has the coordinates: $\xi = 0$, $\eta = -(a+b)(1+\gamma^2)/\gamma^2$.

The OAM of beam from Equation (5.20) is:

$$\frac{J_z}{W} = -2\gamma \frac{3w^4 + 2(a^2 + 4ab + b^2)w^2}{3w^4 + 4(a^2 + 4ab + b^2)w^2 + 16(ab)^2}.\tag{5.25}$$

At $a = -b$, the OAM in Equation (5.25) is equivalent to Equation (5.14). From Equation (5.25), it follows that at $ab > 0$, OAM is always negative, changes no sign, and never takes a zero value, as is OAM from Equation (5.14) for field in Equation (5.1).

5.1.4 An Astigmatic Cosine–Gauss Beam

Novel beams, known as Cosine–Gauss beams [253], have been proposed and first implemented as plasmonic beams propagating along the surface of the medium boundary. In this section, we analyze a different example of the Cosine–Gauss beam, which is generated from a Gaussian beam modulated with a cosine function by performing an astigmatic transformation with a cylindrical lens tilted at 45 degrees.

In the original plane, the complex amplitude of such a field is given by:

$$E(x,y,z=0) = \cos(\alpha x)$$
$$\times \exp\left(-\frac{x^2 + y^2}{w^2} - \frac{ik}{4f}(x+y)^2\right).\tag{5.26}$$

Remarkably, field from Equation (5.26) has an infinite number of zero-intensity lines parallel to the y-axis and separated by distance π/α. Let us look into the evolution of these zero-intensity lines ($\cos \alpha x = 0$) at the double length distance ($z = 2f$) from the original plane, where field from Equation (5.26) has the amplitude:

$$E(\xi,\eta,z=2f) = \frac{-i\gamma}{\sqrt{1+\gamma^2}} \cosh\left[\frac{\alpha\gamma}{1+\gamma^2}(\xi - i\gamma\eta)\right]$$
$$\times \exp\left[-\frac{\alpha^2 w^2}{4(1+\gamma^2)} - \frac{\gamma^2(\xi^2 + \eta^2)}{w^2(1+\gamma^2)} + \frac{ik}{4f}\left(\xi^2 + \eta^2 + \frac{2\gamma^2\xi\eta}{1+\gamma^2}\right)\right].\tag{5.27}$$

In Equation (5.27), the elliptical optical vortex $\xi - i\gamma\,\eta$ is under the sign of hyperbolic cosine function. At $\xi = 0$, the hyperbolic cosine changes to a conventional cosine, with the intensity nulls (optical vortices) being defined by the equation:

$$\cos\left(\frac{\alpha\gamma^2\eta}{1+\gamma^2}\right) = 0. \tag{5.28}$$

From Equation (5.28) it follows that the original-plane vertical zero-intensity lines have transformed into an infinite set of equidistant optical vortices with period $T = (\pi/\alpha)(1 + 1/\gamma^2)$ and TC $= -1$, found on the vertical axis η in the plane at the double focal distance $z = 2f$. Hence, field from Equation (5.27) has an infinite TC. When the zero-intensity lines are far apart and $\alpha \ll 1$, cosine in Equation (5.28) can be replaced by an approximate value: $\cos(\alpha x) \approx 1 - \alpha^2 x^2/2$. Thus, field from Equation (5.26) will behave in a similar manner to field from Equation (5.1) generating at the output plane on the vertical axis two intensity nulls far away from the center: $\eta = \pm\pi(1+\gamma^2)(2\alpha\gamma^2)^{-1}$.

Let us now derive the OAM of beam from Equation (5.27) by substituting Equation (5.26) into Equation (5.13):

$$J_z/W = \gamma p^2\left[1 + \exp\left(p^2\right)\right]^{-1}, \tag{5.29}$$

where $p = \alpha w/\sqrt{2}$. From Equation (5.29) it is seen that both at small ($p \ll 1$) and large ($p \gg 1$) values of the cosine period, the OAM of beam from Equations (5.26) and (5.27) tends to zero, with the OAM in Equation (5.29) taking a maximum value of about $\gamma/2$ on the interval $1 < p < 1.2$.

5.1.5   NUMERICAL MODELLING

Modeling the evolution of the optical vortices

Figure 5.1 depicts intensity and phase patterns in the original plane and at the double focus distance from the cylindrical lens for a Gaussian beam with two symmetrical zero-intensity lines (Equations (5.1) and (5.3)) for different between-the-line distances. The original-plane patterns were obtained using Equation (5.1) and the double focus ones – using a Fresnel transform in Equation (5.2) and Equation (5.3), with the patterns being visually indiscernible. The numerical simulation was conducted at the wavelength $\lambda = 532$ nm, the Gaussian beam waist radius $w = 0.5$ mm, the cylindrical lens focal length $f = 1$ m, the input plane computation domain $|x|, |y| \leq 1.5$ mm, the output plane computation domain $|\xi|, |\eta| \leq 3$ mm, with the between-zero-line distance taking values of $a = 0$ (single 2-order line, first row), $a = 0.8a_0$ (second row), $a = a_0$ (third row), $a = 1.2a_0$ (fourth row), where $a_0 = w/[2(1+\gamma^2)]^{1/2}$ is the distance of generating a 2nd-order on-axis vortex (Equation (5.6)).

Figure 5.1 corroborates the theoretical predictions. Actually, Figure 5.1 shows that at $a = 0$ (the central 2nd-order zero-intensity line), −1st-order optical vortices

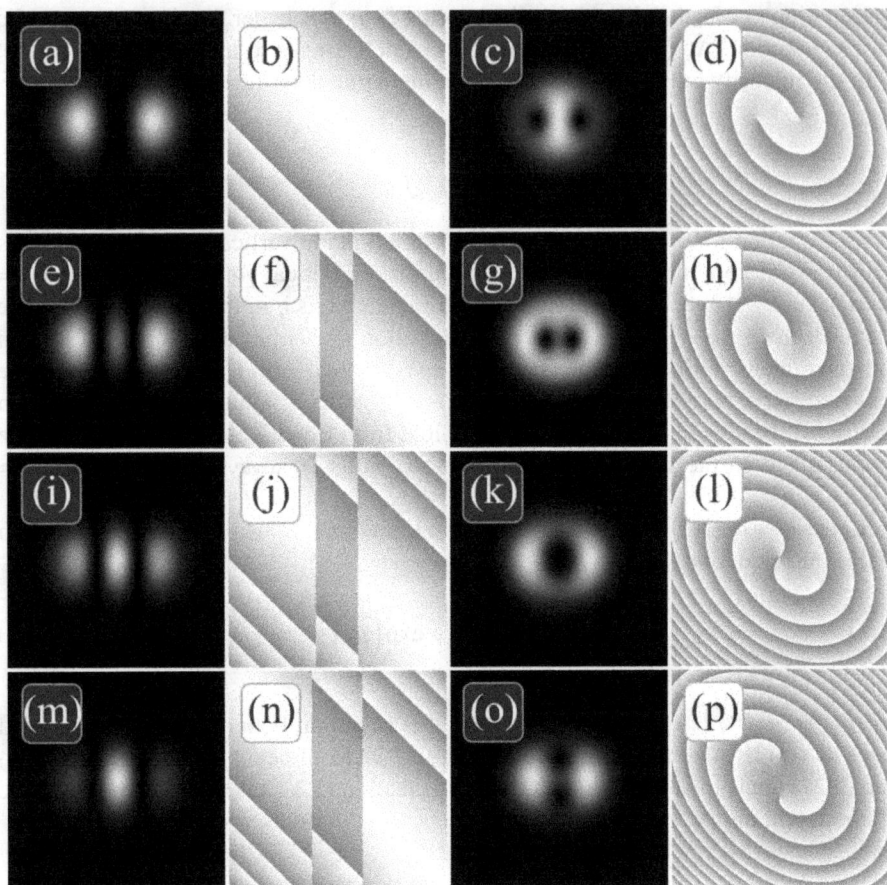

FIGURE 5.1 Patterns of intensity (a, c, e, g, i, k, m, o) and phase (b, d, f, h, j, l, n, p) in the original plane (a, b, e, f, i, j, m, n) and at the double focus distance of the cylindrical lens (c, d, g, h, k, l, o, p) for a Gaussian beam with two symmetrical zero-intensity lines (Equations (5.1) and (5.2)) with differences between-the-line distances.

in the output plane (isolated intensity nulls) are found on the horizontal axis, being maximally separated from each other (Figure 5.1, row 1). With increasing distance a between the zero-intensity lines in the original plane, the output-plane optical vortices get closer (Figure 5.1, row 2). Once the distance a becomes equal to that of Equation (5.6), both output vortices merge into an elliptical −2nd-order vortex (Figure 5.1, row 3), only to be split again into two −1st-order vortices, located on the vertical axis, as the distance a increases further (Figure 5.1, row 4).

Modeling the evolution of OAM

Figure 5.2 depicts intensity and phase patterns in the original plane and at the double focal distance from the cylindrical lens for a Gaussian beam with two symmetrical

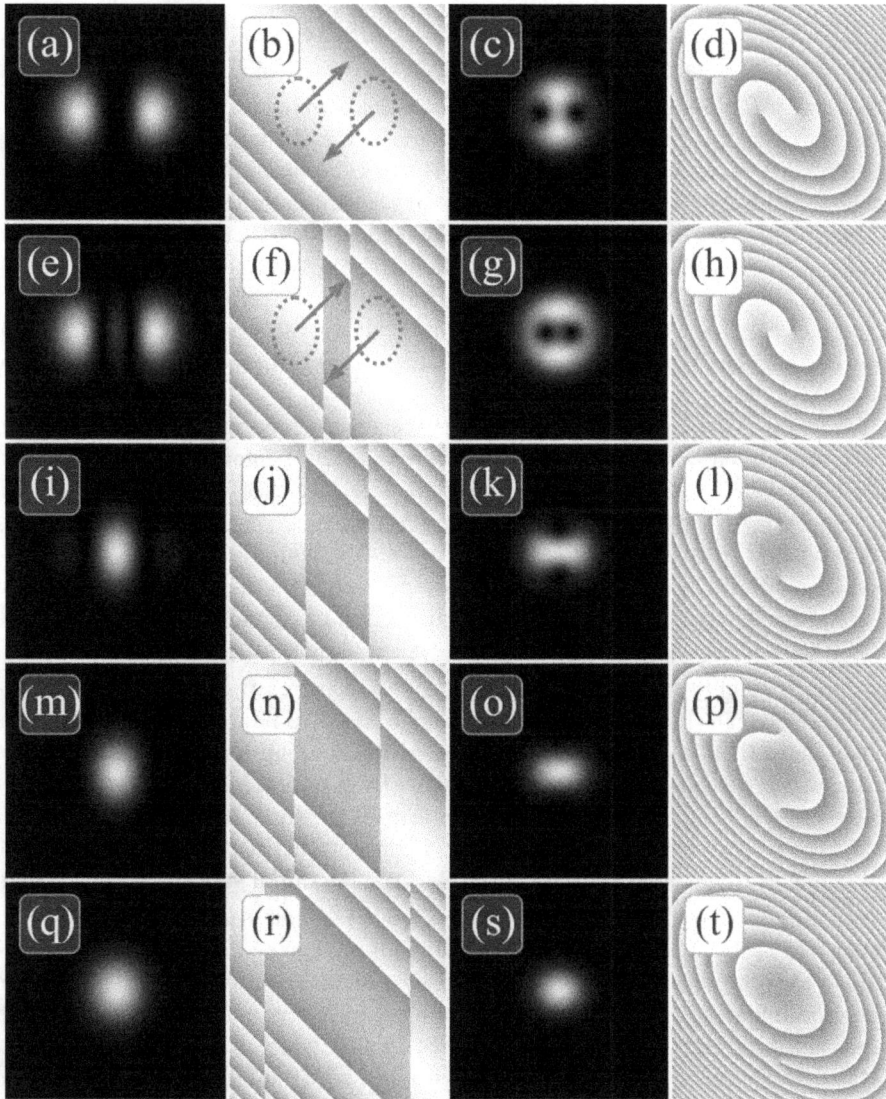

FIGURE 5.2 Patterns of intensity (a, c, e, g, i, k, m, o, q, s) and phase (b, d, f, h, j, l, n, p, r, t) in the original plane (a, b, e, f, i, j, m, n, q, r) and at the double focal distance of the cylindrical lens (c, d, g, h, k, l, o, p, s, t) for a Gaussian beam with two symmetrical zero-intensity lines (Equations (5.1) and (5.2)) with different values of the normalized OAM.

zero-intensity lines (Equations (5.1) and (5.3)) for different values of the normalized OAM. The modeling was conducted for the wavelength $\lambda = 532$ nm, the Gaussian beam waist radius $w = 0.5$ mm, the cylindrical lens focal length $f = z_0/2 \approx 73.8$ cm (i.e. at $\gamma = 1$), the input plane computation domain $|x|, |y| \leq 1.5$ mm, the output plane computation domain $|\xi|, |\eta| \leq 3$ mm, the distance between the zero-intensity lines

$a = 0$ (OAM = -2) (first row), $a = (w/2)(3-\sqrt{6})^{1/2} \approx 185$ μm (OAM is minimal and approximately equal to -2.22) (second row), $a = \sqrt{3}w/2 \approx 433$ μm (OAM = 0) (third row), $a = (w/2)(3+\sqrt{6})^{1/2} \approx 584$ μm (OAM is maximum and equal to 0.22) (fourth row), $a = 2w \approx 1$ mm (dropping from maximum, OAM approximately equals 0.11) (fifth row). Dashed ellipses and rows (b, f) mark the location of light spots in the original plane and the directions of the phase gradient in these spots.

In all patterns in Figure 5.2, normalized OAM values calculated using Equation (5.13) are the same as OAM values predicted theoretically using Equation (5.14). Actually, in row 1 of Figure 5.2, the theoretical and numerically simulated OAM values are, respectively, -2 and -1.998. For row 2, these values are -2.225 and -2.223, for row 3 they are 0 and 5×10^{-4}, for row 4 they are 0.2247 and 0.2248, and for row 5 they are 0.11454 and 0.11451.

Equation (5.1) suggests that there may be up to three intensity peaks in the original plane: a central one (if $a \neq 0$) at $x = y = 0$ (of intensity $(a/w)^4$) and two side-lobes at $x = \pm(a^2 + w^2)^{1/2}$ and $y = 0$ (of intensity $\exp[-2(1 + a^2/w^2)]$). Thus, we can infer that with increasing between-line distance a, a weak central intensity peak first appears, before starting to grow exponentially, meanwhile the side-lobe intensities exponentially drop. Figure 5.2 confirms this conclusion. Indeed, while at $a = 0$ (Figure 5.2, row 1) the original field comprises two intensity spots, at $a = (w/2)(3-\sqrt{6})^{1/2}$ (Figure 5.2, row 2), a weak third central intensity spot appears, gradually increasing its brightness with increasing a and then becoming brighter than the side-lobes, which finally cease to be seen (Figure 5.2, rows 3-5). These dynamics explain why the OAM is tending to a minimum (Figure 5.2, row 5) after passing its maximum value (Figure 5.2, row 4). The moment is defined as the product of a force and the arm of the force, but in the case under study, the largest proportion of intensity is concentrated at the center and, hence, the arm is near-zero. The drop of the OAM from the initial value of -2 (at $a = 0$) to the minimal value of -2.22 (at $a = (w/2)(3-\sqrt{6})^{1/2}$) can also be explained. With increasing a, the side-lobe intensities drop with increasing distance from the center. Up to a certain value of a, the increase of the arm outweighs the drop of the side-lobe intensities and so the OAM increases in magnitude. Considerably high values (circa -2) of OAM (Figure 5.2, rows 1 and 2) are due to the phase gradient in the original plane in the regions of high intensity. If microparticles were put in these regions, the phase gradient would tend to rotate them clockwise around the pattern center (Figure 5.2(b, f)).

Changing of the OAM in Figure 5.2 can also be explained by the formulae in Equations (5.15) and (5.16). The increase of the modulus of the negative OAM value in Figure 5.2(g), compared to that in Figure 5.2(c), is because the two intensity nulls become closer to each other, which is consistent with Equation (5.15). In Figure 5.2(i), the OAM is zero since the beam is elongated along the axis y and the positive astigmatic OAM component (Equation (5.16)) compensates for the negative vortex component (Equation (5.15)). Further, from Figure 5.2(i) to (q), the OAM decays since the beam becomes less elliptic and the distance between the intensity nulls increases.

Modeling a beam with two non-axisymmetric parallel zero-intensity lines

Shown in Figure 5.3 are intensity and phase patterns at the original plane and the double focal length of a cylindrical lens for a Gaussian beam with two non-axisymmetric zero-intensity lines (Equations (5.20) and (5.21)). The modeling was performed for the wavelength $\lambda = 532$ nm, the Gaussian beam waist radius $w = 0.5$ mm, the cylindrical beam focal length $f = z_0/2 \approx 73.8$ cm (i.e. at $\gamma = 1$), the computational domain at the input plane $|x|, |y| \leq 1.5$ mm, the computational domain at the output plane $|\xi|, |\eta| \leq 3$ mm, and the zero-intensity line coordinates $a = w$ and $b = -w/4$.

From Figure 5.3, the non-axisymmetric location of zero-intensity lines is seen to lead in the original plane to a pair of light spots of different sizes and intensity, thus producing a non-symmetric pattern at the output plane. For the above modeling parameters, the zero-intensity lines satisfy condition (5.24), resulting in a -1st-order intensity null at the center of the output plane and the second intensity null at point ($\xi = 0$, $\eta = -1.5w \approx -750\ \mu\mathrm{m}$), as corroborated by Figure 5.3(d). The normalized OAM in Figure 5.3, numerically simulated using Equation (5.13), is the same as the theoretical value derived from Equation (5.25) (-1.469 and -1.471, respectively).

Modeling an astigmatic Cosine–Gauss beam

Figure 5.4 depicts intensity and phase patterns at the original plane and the double focal length of a cylindrical lens for a Cosine–Gauss beam (Equations (5.26) and (5.27)). The modeling was conducted for the wavelength $\lambda = 532$ nm, the Gaussian beam waist $w = 0.5$ mm, the cylindrical lens focal length $f = z_0/2 \approx 73.8$ cm (i.e. at γ

FIGURE 5.3 Patterns of (a, c) intensity and (b, d) phase at the original plane (a, b) and the double focal length (c, d) for a Gaussian beam with two non-axisymmetric zero-intensity lines (Equations (5.20) and (5.21)).

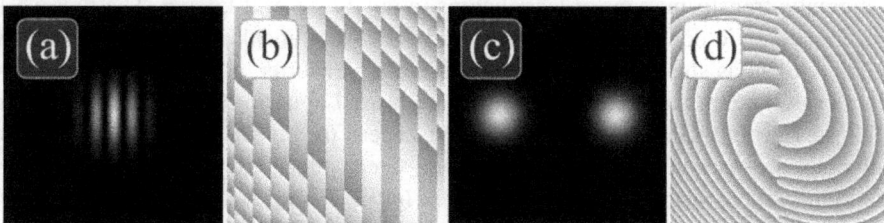

FIGURE 5.4 Patterns of (a, c) intensity and (b, d) phase at the original plane (a, b) and the double focal length (c, d) for the Cosine–Gauss beam of Equations (5.26) and (5.27).

= 1), the cosine function period put equal to the waist radius (i.e. $\alpha = 2\pi/w$), the computation domain in the input plane $|x|, |y| \leq 1.5$ mm, and the computation domain at the output plane $|\xi|, |\eta| \leq 3$ mm.

Figure 5.4 shows that on a vertical axis at the output plane, the Cosine–Gauss beam does produce a set of equidistant –1st-order optical vortices with period $T = (\pi /\alpha)(1 + 1/\gamma^2) = w = 0.5$ mm (with 12 vortices seen in Figure 5.4(d)). But theoretically speaking, the number of vortices may be infinite, leading to the infinitely large TC. At the modeling parameters assumed, the beam carries small OAM, with the numerically simulated and theoretical values being 0.004 and 5×10^{-8}).

Thus, we have theoretically and numerically shown that using a cylindrical lens, a Gaussian beam with a finite set of parallel zero-intensity lines (edge dislocations) can be converted into a vortex beam that carries OAM and has TC [254]. Using an example of two parallel center-symmetric zero-intensity lines, we analyzed the dynamics of the production of two intensity nulls at the double focal length, showing that with increasing distance between the vertical zero-intensity lines, two optical vortices are first generated on the horizontal axis, then converge at the center, before finally diverging again along the vertical axis. Remarkably, the resulting vortex has TC = –2 at any on-axis distance between the zero-intensity lines, except at the original plane. With a changing distance between the zero-intensity lines, the OAM of the beam also varies and can be negative and positive, or zero at a certain distance between the zero-intensity lines.

Using a particular example, we have demonstrated that the OAM and TC are different characteristics of the electromagnetic field. For example, an astigmatic Cosine–Gauss beam has infinite TC because behind the cylindrical lens the beam produces an infinite number of equidistant optical vortices with TC = –1, located on a straight line. In the meantime, all those vortices are in the region of near-zero intensity, which means that the OAM of such a beam is also near-zero.

We have shown that a vortex beam with two optical vortices, each with a topological charge –1, can be generated by a cylindrical lens and by an amplitude mask with the transmittance $T(x, y) = x^2$. Such a mask can be fabricated simply by a printer, or by using an amplitude spatial light modulator.

5.2 CONVERTING AN NTH-ORDER EDGE DISLOCATION TO A SET OF OPTICAL VORTICES

Astigmatic transformations of laser beams in optics are well-known. The first work on an astigmatic converter was published by E. Abramochkin and V. Volostnikov [222], which has shown that using an astigmatic converter, a vortex-free Hermite–Gauss (HG) beam can be converted into a vortex Laguerre–Gauss (LG) beam. In [222], a Hermite–Gauss beam $HG_{n,m}(x, y)$, where (n, m) are numbers of Hermite polynomial, was converted into a Laguerre–Gauss mode $LG_{n,m-n}(x, y)$, where n – the radial index, $m-n$ – the azimuthal index $(m > n)$. Using a single cylindrical lens. Numerous follow-up works studied the conversion of various laser beams using astigmatic transformations. For instance, the propagation of a Hermite–Gauss beam through a 4×4 optical system, including one with astigmatism, has been studied

[255,256]. The propagation of Hermite–Laguerre–Gauss beams through an astigmatic mode converter was analyzed [257] and focusing a high-order optical vortex
with an astigmatic lens was reported [258,259]. Transformations of an astigmatic
Sine–Gauss beam in a nonlinear medium were considered [260] and, an intracavity
astigmatic mode conversion was investigated [261,262]. Elliptical optical Gaussian
vortices with an astigmatic phase were considered earlier [252,190], with a Hermite–
Gaussian beam of order (0, n) reported to be transformed using a tilted cylindrical lens [252]. In [190], a canonical optical vortex with topological charge (TC) n
embedded in an elliptical astigmatic Gaussian beam was shown to conserve during
propagation, not splitting into simple optical vortices. In [221], the propagation of
elliptical optical vortices has been investigated. In [15] it was proposed that the TC
of an optical vortex should be measured using an astigmatic transformation and in
[70], an astigmatic vortex Hermite–Gaussian beam was analyzed.

On the other hand, works are also known that studied the behavior of a screw
[232] and edge dislocations in optical systems. For example, in [136], the evolution
of a beam containing a combination of edge (axial) and screw dislocations during
propagation was analyzed, where the combined optical vortex was shown to have a
fractional TC. In [225,226,245,246], the interaction of an optical vortex (screw dislocation) and an edge transverse dislocation (lines of zero-intensity), embedded in a
Gaussian beam was investigated. The interaction was shown to lead to the splitting
of the edge dislocation and the formation of additional optical vortices. In [263], the
astigmatic transformation of two edge dislocations was considered, and the simultaneous astigmatic transformation of an optical vortex and an edge dislocation of the
1st-order was investigated in [264].

As follows from the works reviewed, an astigmatic transformation of an nth-order
edge dislocation has not been considered. In this section, we show theoretically and
numerically that after passing through a cylindrical lens located in the waist and tilted
at 45 degrees to the axes, a vertical nth-order edge dislocation embedded in the waist
of a Gaussian beam, "disintegrates" into n isolated intensity nulls (screw dislocations).
Around these intensity nulls, n elliptical optical vortices with topological charge -1
are formed, lying on a straight line perpendicular to the edge dislocation, and located
at points with coordinates equal to the roots of an nth-order Hermite polynomial.

5.2.1 Complex Amplitude of Edge Dislocation at Double Focal Length

Consider a case of an edge dislocation embedded in the waist of a Gaussian beam
with an astigmatic phase. The dislocation is produced by placing an ideal thin cylindrical lens tilted at 45 degrees to the axes in the waist plane of the Gaussian beam
with an n-order edge dislocation passing through the center and coincident with the
vertical axis (see Figure 5.5).

The complex amplitude of the light field in the initial plane is expressed as:

$$E(x, y, z = 0) = \frac{x^n}{w^n} \exp\left(-\frac{x^2 + y^2}{w^2} - \frac{ik}{4f}(x + y)^2 \right), \tag{5.30}$$

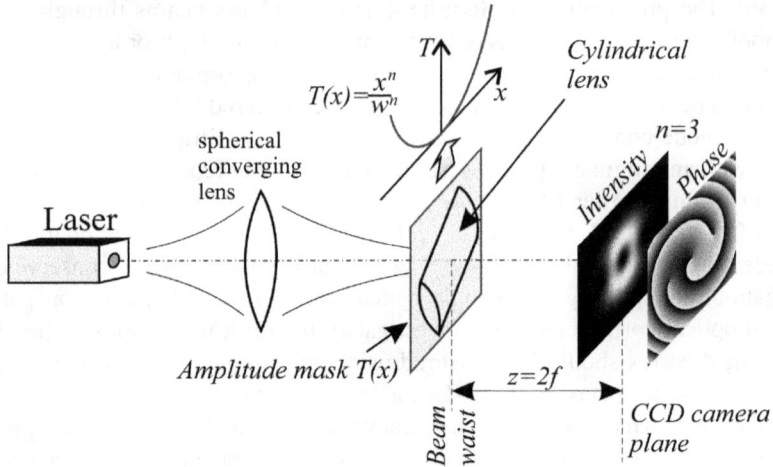

FIGURE 5.5 The scheme of the problem under consideration.

Where k is the wavenumber, w is the waist radius of the Gaussian beam, (x, y) are the Cartesian coordinates, z is the coordinate on the optical axis, and f is the focal distance of the cylindrical lens. The second term in the exponent in Equation (5.30) describes the phase distribution of light after passing a cylindrical lens with a parabolic profile, tilted at 45 degrees to the x- and y-axes. Our goal is to show that n optical vortices with TC $= -1$ are formed on the horizontal axis x at twice the focal distance from the cylindrical lens.

The amplitude of the light field (5.1) at distance $z = 2f$, obtained using a Fresnel transform, can be written as:

$$E\left(\xi,\eta,z=2f\right)=\left(\frac{(-i)^{n+1}\gamma}{2^{n}\left(1+\gamma^{2}\right)^{(n+1)/2}}\right)\exp\left(-\frac{\gamma^{2}\left(\xi^{2}+\eta^{2}\right)}{w^{2}\left(1+\gamma^{2}\right)}\right.$$

$$\left.+\frac{i\gamma}{w^{2}}\left(\xi^{2}+\eta^{2}\right)+\frac{2i\gamma^{3}\xi\eta}{w^{2}\left(1+\gamma^{2}\right)}\right)H_{n}\left[-\frac{\gamma\left(\xi-i\gamma\eta\right)}{w\sqrt{1+\gamma^{2}}}\right],$$

(5.31)

where $\gamma = z_0/(2f)$, $z_0 = kw^2/2$. In Equation (5.31) z_0 is the Rayleigh distance, $H_n(x)$ is a Hermite polynomial. It follows from (5.31), that at the double focal distance, the waist radius of the Gaussian beam has changed: $w(z = 2f) = \overline{w} = w\gamma^{-1}\sqrt{1+\gamma^2}$ and an astigmatic phase appeared. It also follows from Equation (5.31) that for $\eta = 0$ the argument of the Hermite polynomial takes a real value. Equating this argument to the values of the roots of the polynomial $H_n(\sigma_n) = 0$, we obtain coordinates of the centers of elliptical optical vortices (screw dislocations) located on the horizontal axis ξ:

$$\xi_n = -w\sqrt{1+\gamma^2}\left(\gamma\right)^{-1}\sigma_n$$

(5.32)

Some special cases follow on from the formula from Equation (5.32). Let there be a 2nd-order edge dislocation coincident with the η axis. Then there will be just two intensity nulls on the horizontal axis in the output plane at a distance of $\xi_{1,2} = \pm w\left(1+\gamma^2\right)^{1/2}\left(\sqrt{2}\gamma\right)^{-1}$ from the center. This follows from the fact that the 2nd-order Hermite polynomial has only two roots: $\sigma_{1,2} = \pm 1/\sqrt{2}$. A third-order edge dislocation will produce three optical vortices at the output: one on the optical axis, and two others at points with the coordinates: $\xi_{1,2} = \pm w\left(1+\gamma^2\right)^{1/2}\left(\gamma\sqrt{2/3}\right)^{-1}$, since the third-order Hermite polynomial has three roots: $\sigma_0 = 0$, $\sigma_{1,2} = \pm\sqrt{3/2}$, etc.

5.2.2 STRUCTURALLY STABLE VORTEX BEAMS

Let us now look at what optical vortices are formed at the double focal length for the initial beam of Equation (5.30). It follows from Equation (5.31) that around the roots of the Hermite polynomial on the horizontal axis, elliptical vortices with TC = −1 and elliptical amplitude near the intensity nulls are formed:

$$\xi - i\gamma\eta = \rho\exp(-i\theta). \tag{5.33}$$

The right-handed side in Equation (5.33) is obtained in the polar elliptic coordinates:

$$\begin{cases} \xi = \rho\cos\theta, \\ \eta = \gamma^{-1}\rho\sin\theta. \end{cases} \tag{5.34}$$

An elliptical vortex becomes a canonical one when $\gamma = 1$ ($z_0 = 2f$, $a = w/2$):

$$\xi - i\eta = \rho\exp(-i\theta), \tag{5.35}$$

and the amplitude of the light field from Equation (5.31) will be given by ($\gamma = 1$)

$$E\left(\xi,\eta,z = z_0 = 2f\right) = \left(\frac{i^{n-1}}{2^{(3n+1)/2}}\right)\exp\left(-\frac{\left(\xi^2+\eta^2\right)}{2w^2}\right.$$

$$\left. +\frac{ik}{4f}\left(\xi^2+\eta^2+\xi\eta\right)\right)H_n\left[\frac{\left(\xi-i\eta\right)}{w\sqrt{2}}\right]. \tag{5.36}$$

In order for the beam from Equation (5.36) to retain its structure (becoming a structurally stable beam) during further propagation, an extra cylindrical lens needs to be placed at the double focal length, tilted at 45 degrees to the axes, with its transmission given by $\exp\left[-ik\left(\xi+\eta\right)^2/(8f)\right]$. This lens serves to compensate for the astigmatic phase present in Equation (5.36), and its focal length needs to be twice that of the cylindrical lens in Equation (5.30). Immediately behind the second cylindrical lens, instead of Equation (5.36) we derive:

$$E\left(\xi,\eta,z=z_0=2f\right)=\left(\frac{i^{n-1}}{2^{(3n+1)/2}}\right)$$

$$\times\exp\left(-\frac{\left(\xi^2+\eta^2\right)}{2w^2}+\frac{ik}{8f}\left(\xi^2+\eta^2\right)\right)H_n\left[\frac{\left(\xi-i\eta\right)}{w\sqrt{2}}\right]. \tag{5.37}$$

The beam from Equation (5.37) is an example of a family of structurally stable beams [257], which rotate upon propagation, changing only in scale. Within designations, the beam from Equation (5.37) is identical to the Hermite vortex beam in [70].

It follows from the theory of structurally stable Gaussian laser beams [257] that a beam with the initial amplitude:

$$E\left(\xi,\eta,z=0\right)=H_n\left[\alpha\left(\xi\pm i\eta\right)\right]\exp\left(-\beta\left(\xi^2+\eta^2\right)\right), \tag{5.38}$$

will preserve its structure upon propagation, rotating and changing only in scale. Here α and β are complex constants and $\mathrm{Re}\beta>0$. This is general proof that the beam from Equation (5.37) also preserves its structure during propagation.

5.2.3 ORBITAL ANGULAR MOMENTUM

In this section, we obtain an expression for the normalized orbital angular momentum of the beam from Equation (5.30). The orbital angular momentum (OAM) of the paraxial beam can be found from the well-known expressions [252]:

$$J_z=\mathrm{Im}\int_{-\infty}^{\infty}\int_{-\infty}^{\infty}\bar{E}(x,y)\left(x\frac{\partial E(x,y)}{\partial y}-y\frac{\partial E(x,y)}{\partial x}\right)dxdy,$$

$$W=\int_{-\infty}^{\infty}\int_{-\infty}^{\infty}\bar{E}(x,y)E(x,y)dxdy, \tag{5.39}$$

where Im is the imaginary part of the number, J_z is OAM of the beam, W is the power, $\bar{E}$ is the complex conjugated function to the function E. Since the OAM of the beam is preserved upon propagation, let us calculate it in the initial plane. We substitute function from Equation (5.30) into Equation (5.39) and obtain a simple expression for the OAM normalized to the power:

$$\frac{J_z}{W}=-\gamma n. \tag{5.40}$$

When $\gamma = 1$, OAM of the field from Equation (5.30) coincides with its topological charge (TC) and equals $J_z / W = -n$. Thus, a vortex beam with TC and OAM equal to $-n$ can be formed using an nth-order edge dislocation and two cylindrical lenses.

5.2.4 NUMERICAL SIMULATION

Shown in Figure 5.6 are intensity and phase patterns in the initial plane and at the double focal length of the cylindrical lens, the incident light is a Gaussian beam with an nth-order zero-intensity line (Equations (5.30) and (5.31)).

Distributions at the initial plane have been obtained using Equation (5.30), and at the double focal lens – using Equation (5.31). Simulation parameters are as follows: wavelength is $\lambda = 532$ nm, the Gaussian beam waist radius is $w = 0.5$ mm, the focal distance of the cylindrical lens is $f = 1$ m, the computation domain at the initial plane is $|x|, |y| \leq 1.5$ mm, the computation domain at the output plane is $|\xi|, |\eta| \leq 3$ mm for $n = 3$ and 7, and $|\xi|, |\eta| \leq 4$ mm for $n = 10$. It is seen from Figure 5.6 that a vertically located edge dislocation of orders $n = 3, 7$, and 10, on which the phase is "disrupted", is transformed at the double focal length into elliptical optical vortices (isolated intensity nulls) of the order -1 and the distance between them on the horizontal axis is expressed by Equation (5.32). The number of the vortices are, respectively, $n = 3, 7$, and 10. At $n = 3$, coordinates of the dislocations can also be expressed as: $\xi_{1,2} = \pm w \left(1 + \gamma^2\right)^{1/2} \left(\gamma \sqrt{2/3}\right)^{-1}$. The result of this formula coincides with the location of the dislocations in Figure 5.6(b): $\xi_{1,2} = \pm 1031$ μm.

We have shown theoretically that the astigmatic transformation of an edge dislocation (straight line of zero-intensity) of the nth-order forms n elliptical optical vortices (screw dislocations) with TC = -1 at twice the focal distance from the cylindrical

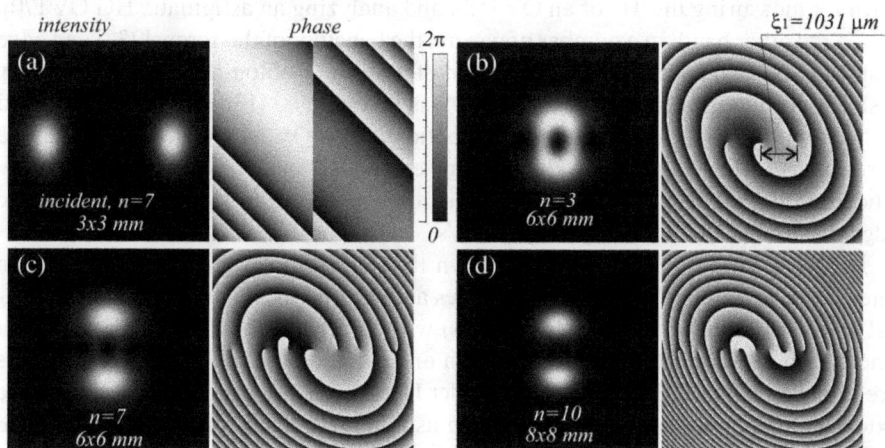

FIGURE 5.6 Intensity (first column) and phase (second column) distributions in the initial plane of a Gaussian beam with an edge dislocation with $n = 7$ (a) and at twice the focal length of a cylindrical lens with order $n = 3$ (b), $n = 7$ (c), $n = 10$ (d)

lens [265]. The vortices are located on a straight line perpendicular to the edge dislocation points, its coordinates are the roots of an nth-order Hermite polynomial. The orbital angular momentum of an edge dislocation with an astigmatic phase has been found to be proportional to n.

5.3   ASTIGMATIC TRANSFORMATION OF A FRACTIONAL ORDER EDGE DISLOCATION

Astigmatic transforms have been fairly well-known in optics. The first ever work to deal with an astigmatic converter was published by E. Abramochkin and V. Volostnikov [83] and discussed their use to convert a non-vortex Hermite–Gaussian (HG) into a vortex Laguerre–Gaussian (LG) beam. More specifically, Abramochkin et al. described the conversion of an HG beam (n,m) into an LG mode $(n,\ m\text{-}n,$ at $m>n)$ with the aid of two cylindrical lenses. Later on, numerous authors studied the use of astigmatic transforms for the conversion of various types of laser beams. For instance, the propagation of an HG beam in a 4×4 optical system, including an astigmatic one, was investigated in [255,256], whereas [257] was concerned with a Hermite–Laguerre–Gaussian beam traveling in an astigmatic mode converter. Also, focusing of a high-order optical vortex (OV) with an astigmatic lens [258,259], conversion of an astigmatic Sin–Gaussian beam in a nonlinear medium [260], and astigmatic mode conversion in a laser cavity [261,262] have been reported. Elliptic optical Gaussian beams with astigmatic phase were previously considered in [252,190], with the former dealing with the conversion of an HG beam $(0, n)$ by means of a tilted cylindrical lens and the latter studying a mode beam in which a conventional OV with the charge n embedded into an elliptic astigmatic Gaussian beam remained unchanged upon propagation, not splitting into simple OVs. Other topic-related papers discussed the propagation of elliptic OVs [221], the use of an astigmatic transform for measuring the TC of an OV [15], and analyzing an astigmatic HG OV [70].

On the other hand, in a number of papers, the behavior of the screw [232] and edge dislocations were analyzed. The evolution of a mixture of (on-axis) edge and screw dislocations upon free-space propagation was investigated [136], with the combined OV shown to have a fractional TC. Interactions of an OV (screw dislocation) and a zero-intensity line (transverse dislocation) embedded into a Gaussian beam were studied in [225,226,245,246]. Such interaction was shown to result in a split of the edge dislocation and the birth of extra OVs.

In a recent work [265], we have shown that an astigmatic transformation of an integer nth-order edge dislocation produces at twice the focal length from a cylindrical lens n elliptic OVs (screw dislocations) with the charge 1, arranged on a straight line perpendicular to the edge dislocation and located at points whose coordinates are proportional to the roots of an nth-order Hermite polynomial. In this section, we extend the reported finding, analyzing an astigmatic transformation of a fractional ν th-order edge dislocation (ν is real). We remind that Michael Berry was the first to discuss the generation of an OV with initial fractional TC [19], with the OV being described as an infinite series decomposed in terms of OAM-modes or integer-TC angular harmonics. In [19], it was also demonstrated that in addition to the main

OVs, a countable number of alternating-sign OVs with charge 1 (Hilbert hotel [135]) were born, thus keeping the whole TC finite.

Compared to Berry's paper [19], this work offers essentially novel findings, as we for the first time introduce Tricomi functions to describe in a closed form the whole infinite set of extra OVs, showing that the beam produced by the fractional order edge dislocation has an infinite TC. We also show theoretically and numerically that a vertical ν th-order edge dislocation ($\nu = n + \alpha$, n is the integer part and $0 \leq \alpha \leq 1$ is the fractional part of the number) embedded into the Gaussian beam waist located in a cylindrical lens tilted at 45 degrees to the coordinate axes is disintegrated into n isolated intensity nulls (screw dislocations). Remarkably, these nulls serve as centers of n newly born elliptic OVs with TC = -1, which are arranged along a straight line perpendicular to the edge dislocation and located at points described by the roots of the Tricomi function. Moreover, alongside these "main" edge dislocations, there appears an "extra" OV located on the same line but somewhat aloof. Considering that with α tending to 1, it tends to join the "main" OVs, we have termed this extra OV a "companion". Interestingly, the "companion" is followed by an infinite number of same-sign "escort" OVs with a charge 1, which are not located on the straight line of the main and extra "companion" OVs. Instead, they are arranged on diverging hyperbolic curves located above and under the straight line of the main OVs. We have called these OVs an "escort" because of their behavior with varying $\nu = n + \alpha \in [n, n+1]$: with α growing from 0 till α_0 (the "parting" moment), they approach the main OVs together with the "companion" OVs, leaving the "companion" and main OVs behind as α continues to grow from α_0 till 1. At $\alpha = 0$, both the "companion" and the "escort" are at infinity. At $\alpha = 1$, it is only the "escort" OVs that are at infinity, whereas the "companion" keeps close to the main OVs. At the next stage, namely at $\nu \in [n+1, n+2]$, the former "companion" joins the "flock" of the main OVs, bringing their number to $n+1$, before the "escorting procession" is re-enacted for the next "companion".

5.3.1 COMPLEX AMPLITUDE OF A FIELD WITH A FRACTIONAL ORDER EDGE DISLOCATION

It would be of interest to analyze an edge dislocation embedded into a Gaussian beam with an astigmatic phase. Such a configuration can be implemented if at the waist plane center of a Gaussian beam with ν th-order vertical dislocation is placed an ideal cylindrical lens tilted at 45 degrees relative to the coordinate axes. In the source plane, the complex amplitude of this field is given by:

$$E(x, y, z = 0) = \left(\frac{x}{w}\right)^{\nu} \exp\left(-\frac{x^2 + y^2}{w^2} - \frac{ik}{4f}(x+y)^2\right), \qquad (5.41)$$

where k is the wave number of light, w is the Gaussian beam waist radius, (x, y) are the transverse Cartesian coordinates, z is the longitudinal coordinate along the optical axis, and f is the focal length of the cylindrical lens. In Equation (5.41), the second term in the power exponent describes the phase distribution of a parabolic-profile cylindrical lens tilted at 45 degrees to the x- and y-axes.

Using a Fresnel transform, we can derive the complex amplitude of the initial field from Equation (5.41) at a distance of z:

$$E(x,y,z) = \frac{1}{\sqrt{\sigma(\sigma - 2\gamma Z)}} \left(\frac{Z(\sigma - \gamma Z)}{i\sigma(\sigma - 2\gamma Z)} \right)^{v/2}$$

$$\times \exp\left(-\frac{X^2 + Y^2}{\sigma} - \frac{i\gamma(X+Y)^2}{\sigma(\sigma - 2\gamma Z)} \right) \qquad (5.42)$$

$$\times \Psi\left[\frac{-v}{2}, \frac{1}{2}, \frac{i(\sigma - \gamma Z)}{\sigma Z(\sigma - 2\gamma Z)} \left(X + \frac{\gamma ZY}{(\sigma - \gamma Z)} \right)^2 \right],$$

where $X = x/w$, $X = y/w$, $Z = z/z_0$ are dimensionless variables, $z_0 = kw^2/2$ is the Rayleigh range, $\sigma = 1 + iZ$, $\gamma = z_0/(2f)$ are auxiliary parameters, and $\Psi\,(a,b,s)$ is the newly introduced Tricomi function [81]:

$$\Psi(a,b,s) = \frac{\Gamma(1-b)}{\Gamma(1+a-b)} \cdot {}_1F_1(a,b;s) + \frac{\Gamma(b-1)}{\Gamma(a)} s^{1-b} \cdot {}_1F_1(1+a-b, 2-b; s). \quad (5.43)$$

Here, $\Gamma\,(a)$ is the gamma-function and ${}_1F_1(a,b,s)$ is a Kummer's (confluent hypergeometric) function [81]:

$$_1F_1(a,b,s) = M(a,b,s) = \frac{\Gamma(b)}{\Gamma(a)} \sum_{n=0}^{\infty} \frac{\Gamma(a+n)}{\Gamma(b+n)} \cdot \frac{s^n}{n!}. \qquad (5.44)$$

Thus, in our case the Tricomi function takes the form:

$$\Psi\left(\frac{-v}{2}, \frac{1}{2}, s \right) = \frac{\sqrt{\pi}}{\Gamma\left(\frac{1}{2}(1-v)\right)} \cdot {}_1F_1\left(\frac{-v}{2}, \frac{1}{2}, s \right) - \frac{2\sqrt{\pi s}}{\Gamma\left(-\frac{1}{2}v\right)} \cdot {}_1F_1\left(\frac{1-v}{2}, \frac{3}{2}, s \right). (5.45)$$

In particular, if $v = n$ is an integer nonnegative number, the Tricomi function is reduced to a Hermite polynomial [81]:

$$\Psi\left(\frac{-n}{2}, \frac{1}{2}, s \right) = 2^{-n} H_n(\sqrt{s}). \qquad (5.46)$$

5.3.2 Complex Amplitude of the Field at Twice the Focal Length

At twice the focal length from the cylindrical lens ($z = 2f$), $\gamma Z = 1$, and Equation (5.42) is simplified to:

$$E(x,y,2f) = \gamma \left(\frac{i}{\sqrt{1+\gamma^2}}\right)^{\nu+1} \exp\left(\left[\frac{-\gamma^2}{1+\gamma^2} + i\gamma\right]\frac{x^2+y^2}{w^2} + \frac{i\gamma^3}{1+\gamma^2} \cdot \frac{2xy}{w^2}\right)$$

$$\times \Psi\left(\frac{-\nu}{2},\frac{1}{2},\frac{\gamma^2}{1+\gamma^2} \cdot \frac{(x-i\gamma y)^2}{w^2}\right).$$

(5.47)

In a particular case of $\nu=n$, where n is an integer nonnegative number, Equation (5.47) becomes similar to the formula earlier proposed in [265]:

$$E(\xi,\eta,z=2f) = \left(\frac{(i)^{n+1}\gamma}{2^n\left(1+\gamma^2\right)^{(n+1)/2}}\right)\exp\left(-\frac{\gamma^2\left(\xi^2+\eta^2\right)}{w^2\left(1+\gamma^2\right)}\right.$$

(5.48)

$$\left.+\frac{i\gamma}{w^2}\left(\xi^2+\eta^2\right)+\frac{2i\gamma^3\xi\eta}{w^2\left(1+\gamma^2\right)}\right)H_n\left[\frac{\gamma\left(\xi-i\eta\right)}{w\sqrt{1+\gamma^2}}\right].$$

In Equation (5.48), a complex-argument Hermite polynomial is found in place of the Tricomi function because Equation (5.46) is valid. It is possible to ascertain that at $\nu = n$, Equation (5.48) becomes identical to Equation (5.47) from [265]. Hence, we may infer that when the edge dislocation order becomes equal to $\nu = n$ (n is an integer positive number), at twice the focal length from the cylindrical lens, n elliptic screw dislocations (OVs) with the charge -1 are born in the beam cross-section on the horizontal axis. For the TC of the newly born OVs to be positive (TC = +1), the cylindrical lens in Equation (5.41) needs to be rotated by 90 degrees, which gives a transmittance of $\exp\left(-(ik/4f)(x-y)^2\right)$.

5.3.3 KUMMER'S AND TRICOMI FUNCTION ZEROS

An asymptotic relation for the Tricomi function in Equation (5.45), which appears as a factor in the amplitude relation from Equation (5.48), takes the form [81]:

$$\Psi\left(-\frac{\nu}{2},\frac{1}{2},\zeta^2\right) \approx \zeta^\nu\left(1-\frac{\nu(\nu-1)}{4\zeta^2}+\frac{\nu(\nu-1)(\nu-2)(\nu-3)}{16\zeta^4}-\ldots\right), \quad \zeta \gg 1, \text{ (5.49)}$$

where $\zeta^2 = \frac{\gamma^2}{w^2(1+\gamma^2)}(x-i\gamma y)^2$. From Equation (5.49), it is seen that, first, the Tricomi function is diverging as degree ν of a complex argument ζ and, second, at large ζ, the Tricomi function would seemingly have at the origin a ν-times degenerate zero. If the first two terms in Equations (5.39) are equated to zero, we reveal that on the horizontal axis there is an extra $(n+1)th$ zero of the Tricomi function, whose coordinate can approximately be described by:

$$\xi_\nu \approx \frac{w\sqrt{1+\gamma^2}}{2\gamma}\sqrt{\nu(\nu-1)}, \quad \nu \gg 1. \tag{5.50}$$

From Equation (5.50), it follows that the approximate coordinate of the extra intensity null on the horizontal axis is similar to the relation for the maximum-number positive root of the Hermite polynomial from Equation (5.48):

$$\xi_n \approx \frac{w\sqrt{1+\gamma^2}}{2\gamma}\sqrt{n(n-1)}. \tag{5.51}$$

Roots of the Hermite polynomial have symmetric pairs relative to the origin, with the Hermite polynomial roots H_n alternating with the roots H_{n+1}. Hence, considering that $n < \nu < (n+1)$, the roots from Equation (5.50) need to be located between the maximum-number zeros of the Hermite polynomial, H_n and H_{n+1}.

At large-order fractional edge dislocation, $\nu \gg 1$, the Tricomi function in Equation (5.45) is described by a different asymptotic relation on the horizontal axis [81]:

$$\Psi\left(-\frac{\nu}{2},\frac{1}{2},\bar{x}^2\right) \approx \pi^{-1/2}\Gamma\left(\frac{1-\nu}{2}\right)\frac{e^{\bar{x}^2/2}}{\bar{x}}\cos\left(\bar{x}\sqrt{1+2\nu}-\frac{\nu\pi}{2}\right), \quad \nu \gg 1. \tag{5.52}$$

where $\bar{x} = \dfrac{\gamma x}{w\sqrt{1+\gamma^2}}$. From Equation (5.52) follows an approximate relation that defines the Tricomi function roots as zeros of a cosine function:

$$\bar{x}_p = \frac{\pi w\sqrt{1+\gamma^2}}{2\gamma\sqrt{2\nu+1}}(\nu+p+1) \tag{5.53}$$

Equation (5.53) for the positive coordinates of zeros is similar to the Kummer's function zeros, which are defined through zeros of the Bessel function with the number $(-1/2)$ and, thus, assuming that the Kummer's function is given by $M\left(-\dfrac{\nu}{2},\dfrac{1}{2},\bar{x}^2\right)$ we find that [81]:

$$\bar{x}_p = \frac{\pi w\sqrt{1+\gamma^2}}{2\gamma\sqrt{2\nu+1}}(2p-1). \tag{5.54}$$

From the fact that the roots of Kummer's function in Equation (5.54) and Tricomi function in Equation (5.53) have near-same coordinates, it is not evident that these functions have a finite number of zeros on the horizontal axis nor that there may be zeros lying off the horizontal axis. Actually, for Kummer's function zeros lying off the horizontal axis to be found, complex roots need to be derived. Let us equate the complex argument of Kummer's function in Equations (5.44) to the complex value of the root. Asymptotic relations for the complex roots of Kummer's function were derived in [266] and utilized later in [68], which was concerned with an asymmetric Kummer's beam:

$$M(a,c,\gamma_p)=0\,,\gamma_p\approx 2\pi ip+\left(1+\frac{c-2a}{2\pi ip}\right)$$

$$\times\left[(c-2a)\log 2\pi\,|p|+\log\frac{\Gamma(a)}{\Gamma(c-a)}\pm i\frac{\pi}{2}(c-2)\right] \qquad (5.55)$$

$$+\frac{2a(a-c)-c}{2\pi ip}+O\left(\frac{\log|p|}{p^2}\right),\,p\to\pm\infty.$$

Putting large $p \gg 1$ in Equation (5.55), we can derive real and imaginary coordinates of roots of two Kummer's functions by entering the formula for the Tricomi function in Equation (5.55):

$$\begin{cases} a_p\approx\left(\frac{1}{2}+v\right)\ln\left(2\pi\,|p|\right)+\ln\left(\frac{\Gamma\left(-v/2\right)}{\Gamma(1/2+v/2)}\right)\mp\frac{3(1+2v)}{16p}, \\[2mm] b_p\approx 2\pi\,p-\frac{v(v+1)-1}{4\pi\,p}\mp\frac{3\pi}{4}-\frac{(2v+1)}{4\pi\,p}\left[\left(\frac{1}{2}+v\right)\ln\left(2\pi\,|p|\right)\right. \\[2mm] \left.+\ln\left(\frac{\Gamma\left(-v/2\right)}{\Gamma(1/2+v/2)}\right)\right], \end{cases} \qquad (5.56)$$

for the Kummer's function given by $M\left(-\frac{v}{2},\frac{1}{2},\zeta^2\right)$ and:

$$M(a,c,\gamma_p)=0\,,\gamma_p\approx 2\pi ip+\left(1+\frac{c-2a}{2\pi ip}\right)$$

$$\times\left[(c-2a)\log 2\pi\,|p|+\log\frac{\Gamma(a)}{\Gamma(c-a)}\pm i\frac{\pi}{2}(c-2)\right] \qquad (5.57)$$

$$+\frac{2a(a-c)-c}{2\pi ip}+O\left(\frac{\log|p|}{p^2}\right),\,p\to\pm\infty.$$

$$\begin{cases} a_p\approx\left(\frac{1}{2}+v\right)\ln\left(2\pi\,|p|\right)+\ln\left(\frac{\Gamma\left(1/2-v/2\right)}{\Gamma(1+v/2)}\right), \\[2mm] b_p\approx 2\pi\,p-\frac{(v+2)(v-1)-3}{4\pi\,p}\mp\frac{\pi}{4}-\frac{(2v+1)}{4\pi\,p}\left[\left(\frac{1}{2}+v\right)\ln\left(2\pi p\right)+\ln\left(\frac{\Gamma\left(1/2-v/2\right)}{\Gamma(1+v/2)}\right)\right], \end{cases}$$

$$(5.58)$$

for the second Kummer's function given by $M\left(\dfrac{1-v}{2},\dfrac{3}{2},\zeta^2\right)$, see Equation (5.45). At large $p \gg 1$, i.e. for zeros lying off the horizontal axis and on the periphery of the beam cross-section, from Equations (5.56) and (5.57) both Kummer's functions in Equation (5.45) are seen to be described by the same asymptotic relations, meaning that zeros of the Tricomi function can approximately be given by:

$$\begin{cases} a_p \approx \left(\dfrac{1}{2}+v\right)\ln\left(2\pi p\right), \\[2mm] b_p \approx 2\pi p. \end{cases} \qquad (5.58)$$

The relation from Equation (5.58) holds at fractional v values as for even integer $v = 2k$, the γ-function in the logarithm argument in Equation (5.56) tends to infinity: $\Gamma(-k)\rightarrow\infty$, with all the off-horizontal-axis roots "departing" to infinity. At odd integer $v = 2k+1$, the γ-function in the logarithm argument in Equation (5.57) tends to infinity: $\Gamma(-k))\rightarrow\infty$, so that all the off-horizontal-axis roots "depart" to infinity. We note that with the argument of the Tricomi function in Equation (5.58) given by x-$i\gamma y$, elliptic phase singularities with the charge -1 occur around all the zeros.

From Equation (5.58), the Cartesian coordinates of peripheral roots of the Tricomi function can be expressed through the parameters (w, v, γ) and the root number p as:

$$\begin{cases} x_p^2 - y_p^2 = \dfrac{w\,(1+\gamma\,)}{\gamma^2}\left(\dfrac{1}{2}+v\right)\ln\left(2\pi|p|\right), \\[3mm] x_p y_p = -\dfrac{\pi w\,(1+\gamma\,)}{\gamma^2}\,p. \end{cases} \qquad (5.59)$$

Equation (5.59) suggests that, first, with increasing v , peripheral off-horizontal-axis zeros depart from the origin of coordinates as x_p increases. Second, because the right-hand side of Equation (5.59) is positive, we find that $x_p > y_p$ and with increasing p, the peripheral intensity nulls at the beam cross-section move away from the abscissa while keeping below the diagonal of Quadrants I and IV ($x_p > 0$). In Equation (5.59), the first equation contains an absolute value of p, because the logarithm argument needs to be positive, whereas in the second equation there is no absolute value of p. From the first equation in Equation (5.59), the peripheral nulls are seen to be arranged in symmetric pairs relative to the horizontal axis for both positive and negative x values. Meanwhile, the second equation in Equation (5.59) suggests that for the positive abscissa values ($x_p > 0$), the intensity nulls with positive numbers ($p > 0$) have negative ordinate values ($y_p < 0$), and the intensity nulls with negative numbers ($p < 0$) have positive ordinates ($y_p > 0$). At negative abscissa values ($x_p < 0$), the numbers swap places: the positive p is at the top and the negative p is at the bottom.

5.3.4 RESULTS OF THE NUMERICAL SIMULATION

To verify the calculation results, phase patterns of the light field in Equation (5.48) with integer $n = 5$ were numerically simulated in the source plane using Equation (5.41) (Figure 5.7(a)) and at twice the focal length from a cylindrical lens ($z = 2f$) with the help of a Rayleigh–Sommerfeld integral (Figure 5.7(b)), a Fresnel transform (Figure 5.7(c)), and Equation (5.48) (Figure 5.7(d)). The numerical simulation was conducted at the wavelength of light $\lambda = 532$ nm, the focal length of the cylindrical lens $f = 1$ m, the calculation domain 6 mm× 6 mm (300 × 300 pixels), and the Gaussian beam waist radius $w = 500$ μm.

From Figure 5.7, all three phase patterns (b–d) are seen to be in agreement and have five screw dislocations (OVs) whose centers are on the horizontal axis. Shown in Figure 5.8 are phase patterns numerically simulated using Equation (5.47) with the aid of the Tricomi function at (a) integer $\nu = 5$ (Figure 5.8(a)) and (b) fractional $\nu = 5.1$. The simulation parameters are the same as in Figure 5.7, except that the calculation domain twice as large is taken. From Figure 5.7(b)–(d) and Figure 5.8(a), Equations (5.48) and (5.47) are seen to give similar results, with five screw dislocations ($\nu = 5$) found on the horizontal axis. It is seen in Figure 5.8(b) that at small values of the fractional part of the initial edge dislocation ($\nu = 5.1$), on the right and left from the already existing five dislocations with TC = −1, there occur a multitude of

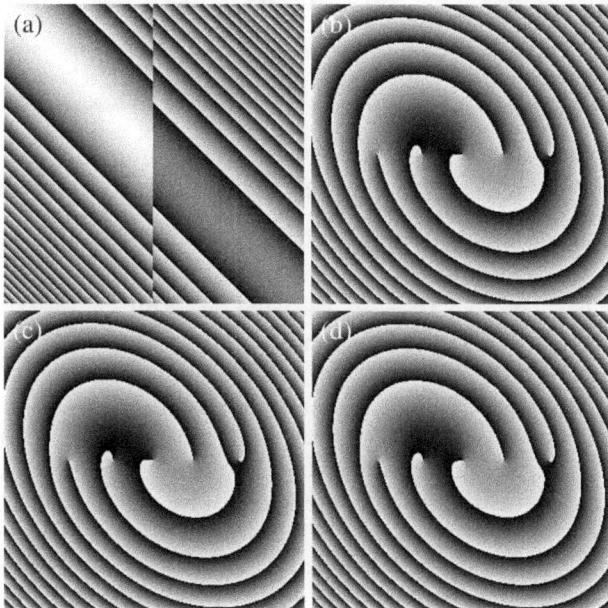

FIGURE 5.7 Phase patterns of field from Equation (5.48) (a) in the source plane and at twice the focal length from a cylindrical lens numerically simulated with the help of (b) a Rayleigh–Sommerfeld integral, (c) an integral Fresnel transform, and (d) an analytical formula in Equation (5.48).

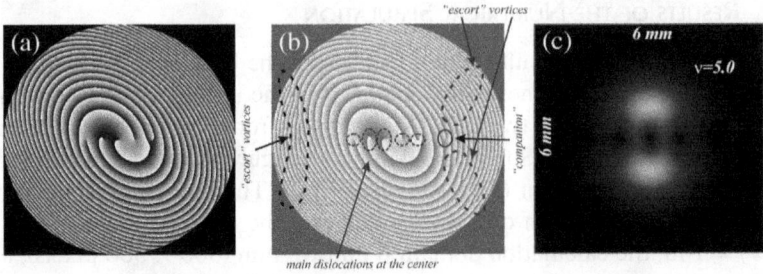

FIGURE 5.8 Phase patterns of field from Equation (5.41) at twice the focal length from a cylindrical lens ($z = 2f = 2m$) deduced using Equation (5.47) for (a) $\nu = 5$ and (b) $\nu = 5.1$, and (c) intensity pattern calculated from the beam with phase in Figure 5.7(a). Computation domain – 12 × 12 mm (600 × 600 pixels), wavelength – $\lambda = 532$ nm, Gaussian beam waist radius – $w = 500$ μm.

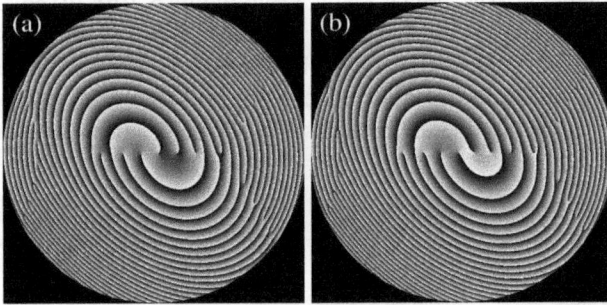

FIGURE 5.9 Phase patterns of field from Equation (5.41) at twice the focal length from the cylindrical lens ($z = 2f = 2m$, derived using Equation (5.49) for (a) $\nu = 6.5$ and (b) $\nu = 7.5$. Computation domain 12 × 12 mm (600 × 600 pixels), a wavelength of light $\lambda = 532$ nm, and Gaussian beam radius $w = 500$ μm.

extra screw dislocations with TC = −1. The extra dislocation closest to the center on the right is found on the horizontal axis and is "being prepared" to become the sixth dislocation as soon as the fractional part gets close to 1. Figure 5.8(c) depicts the field amplitude in Equation (5.47), with its phase presented in Figure 5.8(a). Here, just three central intensity nulls can be seen on the horizontal axis and although the rest nulls are not seen, but their position can be extracted from the phase distribution in Figure 5.8(a). Shown in Figure 5.9 are similar phase patterns for the field from Equation (5.47) at (a) $\nu = 6.5$ and (b) $\nu = 7.5$, the rest parameters being the same as those in Figure 5.8. It can be seen that the first of the extra vortices has approached the initial six OVs (Figure 5.9(a)) and the initial seven OVs (Figure 5.9(b)).

Shown in Figure 5.9 are phase patterns of field from Equation (5.41) at twice the focal length from the cylindrical lens ($z = 2f = 2m$, derived using Equation (5.49) for (a) $\nu = 6.5$ and (b) $\nu = 7.5$. The computation domain was 12 × 12 mm (600 × 600 pixels), the wavelength of light was $\lambda = 532$ nm, and the Gaussian beam radius was $w = 500$ μm.

From Figure 5.10, with increasing fractional part of ν , the sixth dislocation ("a new companion"), which appears in Figure 5.8(b), is seen to approach the five main dislocations (observed in Figure 5.8(a)).

Figure 5.11 depicts the evolution of phase patterns as the increasing fractional number ν tends to the next integer 6 (i.e. α tends to 1), with the dislocations on the

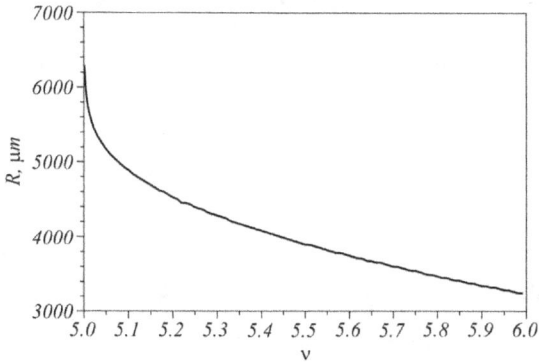

FIGURE 5.10   Distance of the sixth dislocation center to the beam center (μm) against the value of the fractional part of ν on the interval $5 < \nu < 6$.

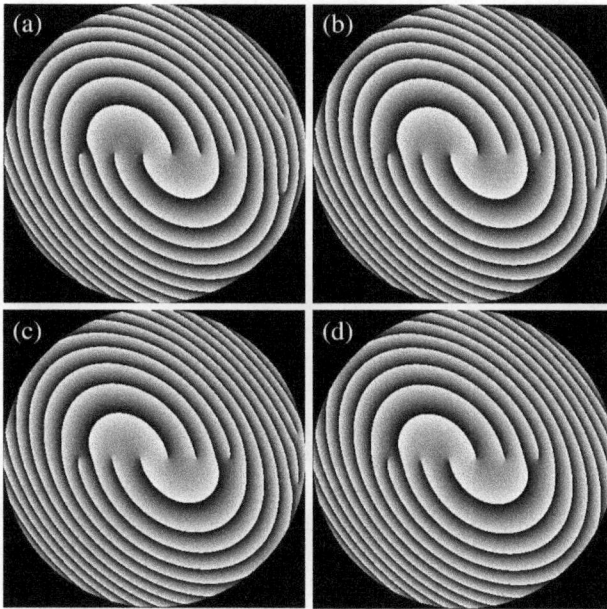

FIGURE 5.11   Phase patterns, similar to those presented in Figures 5.7 and 5.8 but calculated using a Fresnel transform for fractional ν values: (a) 5.9, (b) 5.99, (c) 5.999, and (d) 6. Computation domains 8.7 × 8.7 mm (435 × 435 pixels).

right-hand side seen to start departing to the periphery. We see that at $\nu = 5.9$, four OVs can be located on the right of the center. At $\nu = 5.99$, just two dislocations remain to be observed on the right, getting closer to the right-hand side at $\nu = 5.999$, before disappearing out of sight at $\nu = 6$.

Summing up, at small α values, extra dislocations appear, including an extra "companion" OV, which is closest to the center, and "escort" OVs, which are found above and under the horizontal axis. With α increasing toward 0.5, all extra dislocations (both the "companion" and the "escort") draw closer to the center. With a further increase in α, the "companion" (first horizontal dislocation) keeps close to the five main dislocations, whereas the "escort" OVs gradually depart to the periphery, disappearing out of sight at $\nu = 6$.

5.3.5 Discussion of Results

A significant conclusion that can be made from the numerical simulation results is as follows: considering that the fractional number ν enters Equation (5.47) only as a term of the constant before the exponential function and the first parameter of the Tricomi function, it is the latter which is responsible for the birth of extra dislocations and their evolution with a varying fractional part of ν. From Equation (5.47), the OVs are seen to be born at Tricomi function zeros lying on the horizontal axis because the Tricomi function argument becomes real at $y = 0$. We have shown that at an integer-order edge dislocation, $\nu = n$, the Tricomi function zeros coincide with real zeros of a Hermite polynomial $H_n(x)$, lying on the horizontal axis at the cross-section of beam (5.48). In a previous section, using asymptotic relations for the Tricomi function in Equation (5.49), we derived approximate coordinates of an extra, $(n+1)$-th intensity null on the horizontal axis, Equation (5.50). Hence we infer that considering that (i) $n < \nu < (n+1)$ and (ii) the relation to describe its coordinate is similar to that of the extreme Hermite polynomial zero, the extra intensity null in Equation (5.50) may be expected to lie between the extreme zeros of the Hermite polynomials H_n and H_{n+1}. Then, using asymptotic relations for complex zeros of two Kummer's functions in Equation (5.55), which enter the Tricomi function as constituent terms, we have deduced asymptotic relations for peripheral Tricomi function zeros in Equation (5.58). From Equation (5.58) follows asymptotic formulae (5.59) to define the coordinates (x, y) of a countable number of OVs not lying on the horizontal axis (where belong main n OVs and an extra $(n+1)$th OV, termed a "companion"), instead being located in symmetric pairs above and under the horizontal axis and departing from the axis with the growing zero number. All the OVs, including the main ones, the "companion", and the "escort" have the same charge –1. From Equations (5.56) and (5.57), it follows that with ν tending to an integer, all the "escort" OVs depart to infinity.

We have theoretically and numerically described stages of the generation and evolution of a pattern of intensity nulls and associated OVs (screw dislocations) produced as a result of an astigmatic transformation of a νth-order edge dislocation (a vertical straight line of intensity nulls), where $\nu = n+\alpha$ is a real positive number, n is an integer, and $0 < \alpha < 1$ is the fractional part of ν. The astigmatic transformation was implemented with a cylindrical lens tilted to the axis at an angle of 45 degrees. We have demonstrated that at twice the focal length from the cylindrical lens, there occur n elliptic OVs with charge -1, arranged on a straight line perpendicular to the edge dislocation at points with coordinates defined by zeros of a Tricomi function. At $\alpha = 0$ or $\alpha = 1$, the Tricomi function zeros coincide with the zeros of nth- and $(n+1)$th-order Hermite polynomials. Some distance away from the main OVs, an extra OV with charge -1 is born, which departs to the periphery if α tends to 0, or draws close to the main n OVs if α tends to 1, and is termed a "companion" OVs. Using an asymptotic representation of the Tricomi function zeros, we have demonstrated the birth of a countable number of intensity nulls (OVs with charge -1) arranged along four diverging hyperbolic curves equidistant from the straight line of n main intensity nulls. These "escort" OVs come closer to the beam center, accompanying the extra "companion" OV if $0 < \alpha < 0.5$, or depart to the periphery leaving the "companion" behind, near the main OVs, if $0.5 < \alpha < 1$. At $\alpha = 0$ and $\alpha = 1$, the "escort" OVs reside at infinity. The theoretical conclusions have been shown to agree with the numerical simulation. The modeling was conducted using three independent techniques: a Fresnel transform, Rayleigh–Sommerfeld integrals, and a Tricomi-function-based analytical formula for the complex amplitude of a light field. All three techniques have produced near-same results.

6 Fourier-Invariant and Structurally Stable Optical Vortex Beams

6.1 FRACTIONAL-ORDER-BESSEL FOURIER-INVARIANT OPTICAL VORTICES

In optics, propagation-invariant light beams are of special interest. These beams either propagate without diffraction or preserve their transverse shape (up to scale and rotation) on propagation in free space or in optical systems.

Among diffraction-free beams, the most well-known are Bessel beams [94], Mathieu beams [3], and parabolic beams [2]. All these beams are exact solutions of the Helmholtz equation (i.e. nonparaxial), have infinite energy, and thus can be generated only approximately. For example, Bessel beams are typically generated by using an axicon [267,268], while its far-field image (perfect vortex) was shown to be more effectively generated by another phase-only element [269]. In 1987, when Bessel beams were discovered, finite-energy Bessel–Gaussian beams [40] had also been proposed. Both Bessel and Bessel–Gaussian beams are not Fourier-invariant since the Fourier-transform of the Bessel beams yields the annular Dirac delta-function, whereas the Fourier-transform of the Bessel–Gaussian beams is described by the modified Bessel function. In later years, various-kind families of vortex beams appeared, with their amplitude being proportional to the Bessel function. These families include the Bessel beams with quadratic radial dependence [270], generalized Bessel–Gaussian beams [271], Hankel–Bessel beams [272], asymmetric Bessel [43], and Bessel–Gaussian [67] beams, or Bessel beams of half-integer order [228]. However, all the above mentioned types of Bessel beams [94,40,270,271,272,43,67,228] are not Fourier-invariant. In [273], shape-invariant Bessel-like beams (including higher-order ones) were described, which were experimentally generated from a Gaussian beam by using two axicons.

Paraxial propagation in free space and in optical systems is typically described by some integral transforms, such as the Fresnel transform, ABCD-transform, or the Fourier transform. The latter describes propagation from the front focal plane of a spherical lens to the back focal plane, or in a cavity with concave mirrors. For non-vortex fields, eigenfunctions of the mentioned linear canonical transforms are studied in detail in [274]. Among the optical vortices, the most well-known Fourier-invariant field is the Laguerre–Gaussian (LG) beam [4]. Conventional LG beams have a wide dark area in the center. This is undesirable for interferometry tasks, and beams with narrower dark spots would be useful. Such beams could also be adopted

DOI: 10.1201/9781003326304-6

for effective rotation of metallic (absorbing) particles since they are trapped in dark areas. To rotate a particle, it should be large enough if a narrow vortex is used. If the vortex ring is wide and its intensity grows faster with the distance from the optical axis, small particles also can be rotated.

In this section, we found, analytically, a light field, which is invariant to the Fourier transform, contains an optical vortex, and has low sidelobes. This field obeys the paraxial wave equation, but we found its complex amplitude only in the Fraunhofer diffraction zone. This beam is physically realizable, since it is of finite energy, although it does not contain the Gaussian envelope. This is the first time when a finite-energy beam is found with its complex amplitude containing only the Bessel function. Radial distribution of the complex amplitude is proportional to the Bessel function of fractional order. Compared to the half-integer Bessel-like beams [228], our fractional-order Bessel (FOB) beams are described by the Bessel function having the order of an odd integer number divided by six. The main geometrical difference between FOB modes and LG modes with zero radial number is that the dark spot of the former is smaller. The main difference from the shape-invariant Bessel-like beams [273] is low sidelobes.

6.1.1 FRACTIONAL-ORDER BESSEL FOURIER-MODES

Let the initial field have the following complex amplitude in the initial plane (front focal plane of a spherical lens):

$$E_1(r,\varphi) = \left(\sqrt{\alpha}\,r\right)^{-(m+1)/3} J_{(2m-1)/6}\left(\alpha r^2\right)\exp\left(i\alpha r^2 + im\varphi\right), \qquad (6.1)$$

where (r, φ) are the polar coordinates, m is the topological charge of the optical vortex, and α is a scaling factor.

According to [158] (expression (5-19) in [158]), in the far field (back focal plane of the spherical lens), the complex amplitude reads as:

$$E_2(\rho,\theta) = \frac{-ik}{2\pi f} \int\limits_0^\infty \int\limits_0^{2\pi} E_1(r,\varphi)\exp\left\{-i\frac{k}{f}\rho r\cos(\varphi-\theta)\right\} r\,dr\,d\varphi, \qquad (6.2)$$

where (ρ, θ) are the polar coordinates in the far field, $k = 2\pi/\lambda$ is the wavenumber of light with the wavelength of λ, f is the focal length of the lens.

Substituting Equation (6.1) into Equation (6.2) yields:

$$E_2(\rho,\theta) = \frac{-ik}{2\pi f} \int\limits_0^\infty \int\limits_0^{2\pi} \left(\sqrt{\alpha}\,r\right)^{-(m+1)/3} J_{(2m-1)/6}\left(\alpha r^2\right)$$

$$\times \exp\left\{i\alpha r^2 + im\varphi - i\frac{k}{f}\rho r\cos(\varphi-\theta)\right\} r\,dr\,d\varphi. \qquad (6.3)$$

Evaluating the integral over φ, we get:

$$E_2(\rho,\theta) = (-i)^{m+1} \frac{k}{f\alpha^{(m+1)/6}} \exp(im\theta)$$

$$\times \int_0^\infty r^{(2-m)/3} J_{(2m-1)/6}(\alpha r^2) \exp(i\alpha r^2) J_m\left(\frac{k}{f}\rho r\right) dr. \tag{6.4}$$

In [275], there is a reference integral (expression 2.12.40.15):

$$\int_0^\infty x^{(2-v)/3} \begin{Bmatrix} \sin bx^2 \\ \cos bx^2 \end{Bmatrix} J_{(2v-1)/6}(bx^2) J_v(cx) dx$$

$$= (4b)^{(v-2)/3} c^{-(v+1)/3} \begin{Bmatrix} \sin\varphi \\ \cos\varphi \end{Bmatrix} J_{(2v-1)/6}\left(\frac{c^2}{16b}\right), \tag{6.5}$$

where b, $c > 0$, Re $v > -(7 \pm 3)/4$ and:

$$\varphi = (v+1)\frac{\pi}{6} - \frac{c^2}{16b}. \tag{6.6}$$

Using this integral, we get from Equation (6.4):

$$E_2(\rho,\theta) = i^{-2(m+1)/3} \left(\frac{k}{4f}\right)^{(2-m)/3} \alpha^{(m-5)/6} \rho^{-(m+1)/3}$$

$$\times \exp\left(-\frac{ik^2\rho^2}{16\alpha f^2} + im\theta\right) J_{(2m-1)/6}\left(\frac{k^2\rho^2}{16\alpha f^2}\right). \tag{6.7}$$

According to Equations (6.1) and (6.7), both the input and output fields have the same rotationally symmetric intensity structure: $I(r,\varphi) = |E(r,\varphi)|^2 \sim \xi^{-(m+1)/3} J_{(2m-1)/6}^2(\xi)$, where $\xi = \beta r^2$ and β is some real constant.

It is also seen in Equations (6.1) and (6.7) that if $\alpha = k/(4f)$ then the input and output beams have the same radius.

In some specific cases, when $m = 2, 5, 8, 11, \ldots$, the Bessel function acquires half-integer order and the complex amplitude can be expressed via elementary functions. For example, at $m = 2$:

$$E_1(r,\varphi) = \sqrt{2/\pi}\,\xi_1^{-1} \sin\xi_1 \exp(i\xi_1 + 2i\varphi), \tag{6.8}$$

$$E_2(\rho,\theta) = -\frac{k}{\sqrt{8\pi}\,f\alpha} \xi_2^{-1} \sin\xi_2 \exp(-i\xi_2 + 2i\theta), \tag{6.9}$$

where:

$$\xi_1 = \alpha r^2, \tag{6.10}$$

$$\xi_2 = \frac{k^2 \rho^2}{16\alpha f^2}. \tag{6.11}$$

We note; however, that the Fourier-invariant field from Equations (6.1) and (6.7) has a shortcoming. The complex amplitude of a light field should be continuous. To avoid the singularity at $r = 0$, the topological charge should meet the condition $m \geq 2$. However, it is seen in Equations (6.8) and (6.9) that even at $m = 2$ there is a second-order vortex without the intensity null, since the complex amplitudes from Equation (6.8), (6.9) near the optical axis read as:

$$E_1(r \to 0, \varphi) = \sqrt{2/\pi} \exp(i\alpha r^2 + 2i\varphi), \tag{6.12}$$

$$E_2(\rho \to 0, \theta) = -\frac{k}{\sqrt{8\pi}\,f\alpha} \exp\left(-\frac{ik^2\rho^2}{16\alpha f^2} + 2i\theta\right). \tag{6.13}$$

In addition, not only the field itself, but its derivative should be also continuous (since the magnetic field should be continuous). The expression for $\partial E/\partial r$ is rather cumbersome and we don't write it here, but it shows that for avoiding the singularity (without intensity null) we should suppose that $m \geq 5$. At $m = 5$, complex amplitude reads as:

$$E_1(r, \varphi) = \sqrt{2/\pi}\,\xi_1^{-5/2} \left(\sin\xi_1 - \xi_1 \cos\xi_1\right)\exp(i\xi_1 + 5i\varphi), \tag{6.14}$$

$$E_2(\rho, \theta) = \frac{k}{\sqrt{8\pi}\,f\alpha}\,\xi_2^{-5/2} \left(\sin\xi_2 - \xi_2 \cos\xi_2\right)\exp(-i\xi_2 + 5i\theta), \tag{6.15}$$

with ξ_1 and ξ_2 defined in Equations (6.10) and (6.11).

6.1.2 ENERGY OF THE FRACTIONAL-ORDER BESSEL FOURIER-MODES

Further, we derive the power of FOB Fourier-modes. Equation (6.7) can be rewritten as:

$$E_2(\rho, \theta) = i^{-2(m+1)/3} \frac{k}{4\alpha f}\,\xi^{-(m+1)/6} \exp(-i\xi + im\theta) J_{(2m-1)/6}(\xi) \tag{6.16}$$

with

$$\xi = \frac{k^2\rho^2}{16\alpha f^2}. \tag{6.17}$$

The beam power thus is given by

$$W = \int_0^\infty \int_0^{2\pi} \left| E_2(\rho,\theta) \right|^2 \rho d\rho d\theta = \frac{\pi}{\alpha} \int_0^\infty \xi^{-(m+1)/3} J_{(2m-1)/6}^2 (\xi) d\xi. \tag{6.18}$$

Using the reference integral (expression 2.12.31.2 in [275]), we get the beam power:

$$W = \frac{\pi^{3/2}}{3^{m/2}} \frac{\Gamma\left(\dfrac{m+1}{6}\right)}{\Gamma\left(\dfrac{m+3}{6}\right)\Gamma\left(\dfrac{m+4}{6}\right)\Gamma\left(\dfrac{m+5}{6}\right)} \frac{1}{\alpha}. \tag{6.19}$$

Thus, in contrast to the conventional Bessel modes, power of FOB Fourier-modes is finite and therefore the FOB Fourier-modes are physically realizable.

6.1.3 NUMERICAL SIMULATION

In this section, we measure three parameters characterizing the main ring on the intensity pattern (Figure 6.1): r_0 is the radius of the light ring measured as the radius of the circumference of maximal intensity, b is the FWHM (full width at half-max-imum level) of the light ring, and d is the diameter of the dark circle inside the ring measured also at the half-maximum level.

Beam width

Here, we compare the FOB Fourier-mode with the well-known LG mode, which is conserved on propagation in space, and in the far field particularly. In the initial plane, the complex amplitude of the LG mode (in the polar coordinates) reads as:

$$E_1(r,\varphi) = \left(\frac{\sqrt{2}r}{w_1}\right)^m L_p^m \left(\frac{2r^2}{w_1^2}\right) \exp\left(-\frac{r^2}{w_1^2} + im\varphi\right), \tag{6.20}$$

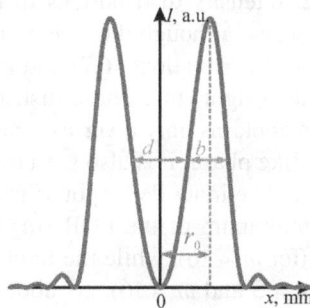

FIGURE 6.1 Parameters of the intensity distribution measured in simulation.

where w_1 is the waist radius of the Gaussian beam, m and p are respectively azimuthal (topological charge) and radial mode indices. Below we suppose that the radial index p is zero to avoid the peripheral light rings (since the FOB Fourier-modes, as shown below, also have one overwhelming intensity ring with almost no sidelobes). When such a beam is in the front focal plane of a spherical lens, in the back focal plane there is the following Fourier transformed the distribution of the complex amplitude:

$$E_2\left(\rho,\theta\right) = \left(-i\right)^{m+1}\left(\frac{z_R}{f}\right)\left(\frac{\sqrt{2}\rho}{w_2}\right)^m \exp\left(-\frac{\rho^2}{w_2^2} + im\theta\right), \tag{6.21}$$

where f is the focal length of the lens, and:

$$z_R = \frac{kw_1^2}{2},$$

$$w_2 = w_1 \frac{f}{z_R}. \tag{6.22}$$

Figure 6.2 shows the intensity and phase distributions of the FOB and LG Fourier-modes in the initial plane (front focal plane of the lens) and in the far field (back focal plane of the lens). We used the following parameters: wavelength $\lambda = 532$ nm, focal length of the lens $f = 1$ m, topological charge of the optical vortex $m = 5$, and $m = 16$. The scaling factor of the FOB Fourier-mode and the waist radius of the LG mode are respectively $\alpha = k(4f) \approx 2.95$ mm^{-2} and $w_1 = (2f/k)^{1/2} \approx 411$ µm. At these values, the beam in the output plane has just the same size as in the initial plane. The computation domain is $-R \le x, y \le R$ ($R = 2$ mm). Intensity and phase distributions in the initial plane were computed by using Equations (6.1) and (6.20), whereas the intensity distributions in the output plane were obtained by using the Fresnel transform implemented as a convolution with utilizing the fast Fourier transform.

We obtain the output intensity distributions numerically by the Fresnel transform instead of using the analytical Equations (6.7) and (6.21) in order to validate the above theory. Theoretically, the output distributions should coincide with the input ones. As seen in Figure 6.2, intensity distributions in the output plane look very similar to those in the input plane, although they were obtained numerically by two Fresnel transforms rather than by Equations (6.7) and (6.21). Both beams have the transverse intensity shaped as a light ring. Phase distributions are; however, quite different. Whereas LG mode contains only a vortex phase, FOB Fourier-mode has, in addition, a parabolic lens–like phase. It is also seen that the light ring of the FOB Fourier-mode is much thicker. The inner dark spot (d from Figure 6.1) is, therefore, smaller. Full width at half-maximum of the FOB ring (b from Figure 6.1) is 551 µm (for $m = 5$) and 446 µm (for $m = 16$), while the light ring of the LG mode has a width of 344 µm (for both $m = 5$ and $m = 16$), i.e. about 1.6 times narrower (for $m = 5$). Besides, LG mode is without any sidelobes, while the FOB Fourier-mode has sidelobes 4.6% from the maximal intensity.

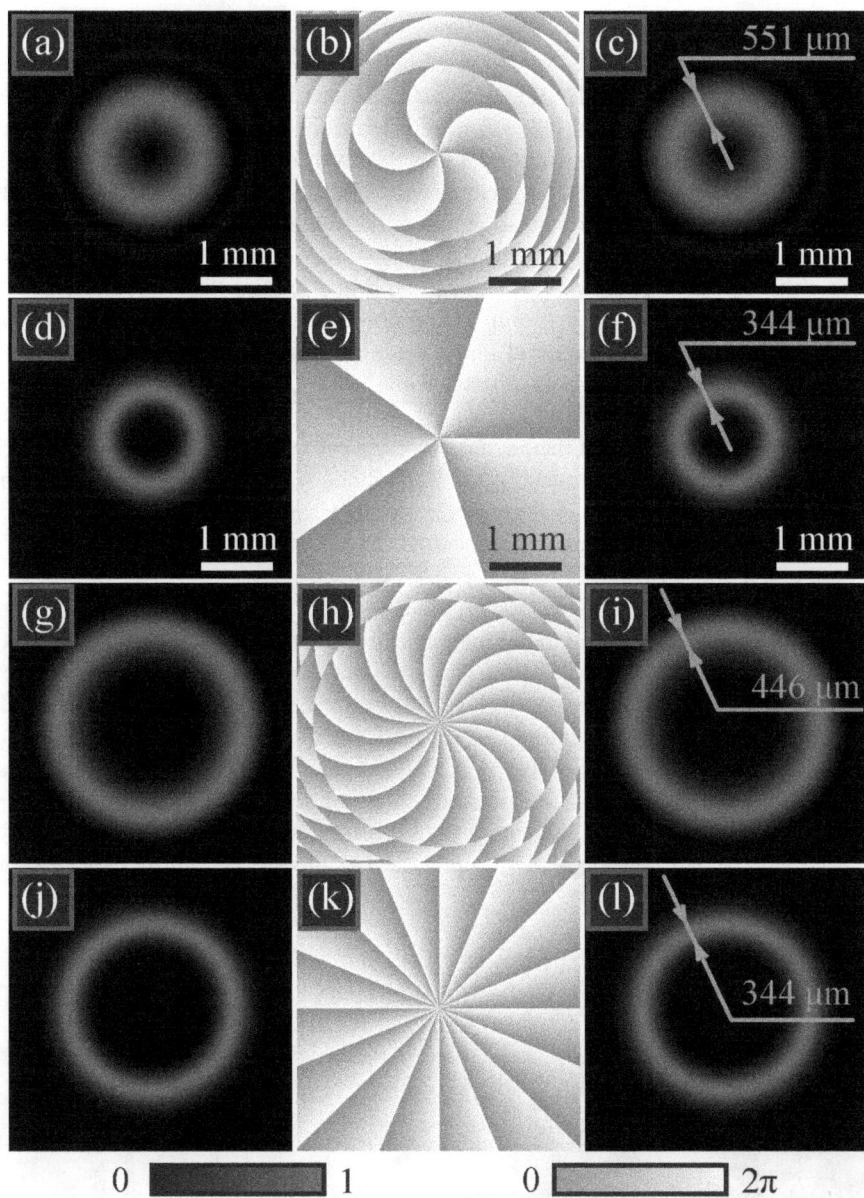

FIGURE 6.2 Intensity (a, c, d, f, g, i, j, l) and phase (b, e, h, k) distributions of the FOB (a–c, g–i) and LG (d–f, j–l) Fourier-modes in the initial plane (front focal plane of the lens) (a, b, d, e, g, h, j, k) and in the far field (back focal plane of the lens) (c, f, i, l) for $m = 5$ and $m = 16$.

Beam divergence

In addition, we studied the divergence of both kinds of modes. We have chosen parameters so that both FOB and LG Fourier-modes had the same ring radius (r_0 from Figure 6.1) in the initial plane: 1 mm. Figure 6.3 shows the intensity distributions of the FOB and LG Fourier-modes in the front and back focal planes of the lens. We used the following parameters: wavelength $\lambda = 532$ nm, focal length of the lens $f = 1$ m, and the topological charge of the optical vortex $m = 5$. The scaling factor of the FOB Fourier-mode and the waist radius of the LG mode are respectively $\alpha \approx 1.526$ mm^{-1} and $w_1 \approx 634$ μm (such α and w_1 give the ring radius of $r_0 = 1$ mm in the initial plane). The computation domain is $-R \leq x, y \leq R$ ($R = 3$ mm). Intensity distributions in the initial plane were computed by using Equations (6.1) and (6.20), whereas the intensity distributions in the output plane we obtained by using Equations (6.7) and (6.19).

As computation shows, in the far field (back focal plane), the radius of the maximal intensity ring is 23% greater for the FOB mode ($r_0 \approx 519$ μm) than that of the LG mode ($r_0 \approx 422$ μm). As in Figure 6.2, the light ring of the FOB mode is much thicker than that of the LG mode (full widths at half-maximum are respectively $b = 404$ μm and $b = 220$ μm).

In addition, in the input plane (Figure 6.3(a, e, c, g)), where the ring radius of both beams is the same, we measured the diameter of the inner dark spot (at the intensity half-maximum). It equals $d = 1.14$ mm (Figure 6.3(a)) and $d = 1.5$ mm (Figure 6.3(e)) for the FOB and LG modes respectively. Thus, the dark spot area is almost two times less (58%) for the FOB mode. Even in the back focal plane (Figure 6.3(b, f, d, h)), despite the larger ring radius of the FOB mode, the diameter of the dark spot is less than that of the LG mode ($d = 589$ μm and $d = 633$ μm respectively).

The larger radius of the light ring of the FOB modes compared to that of the LG modes can be explained theoretically. For the LG modes, according to Equations

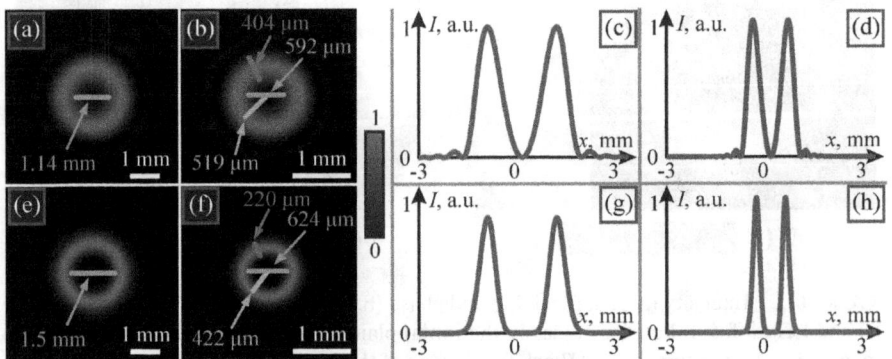

FIGURE 6.3 Intensity distributions (a, b, e, f) and their horizontal cross-sections (c, d, g, h) of the FOB (a–d) and LG (e–h) Fourier-modes in the initial plane (front focal plane of the lens) (a, c, e, g) and in the far field (back focal plane of the lens) (b, d, f, h). Whit lines show radii of the maximal intensity.

(6.20) and (6.21), radial dependence of the amplitude near the center is $|E| \sim r^m$. Equations (6.1) and (6.7) show that, on the contrary, for the FOB modes $|E| \sim r^{(m-2)/3}$. This means that at $r \ll 1$ the intensity near the center grows much faster for the FOB modes than for the LG modes.

Beam propagation in free space

In this subsection, using the Fresnel transform, we numerically compute the intensity patterns in several different transverse planes in order to see the free-space propagation dynamics of the FOB modes. Figure 6.4 illustrates these patterns for the following parameters: wavelength $\lambda = 532$ nm, topological charge of the optical vortex $m = 5$, scaling factor of the FOB Fourier-mode $\alpha \approx 1.526$ mm^{-1} (so that the light ring radius in the initial plane is $r_0 = 1$ mm), propagation distances are 0, 0.5 m, 1 m, 3 m, 5 m, 10 m, 30 m, and 50 m. The computation domain is $-R \leq x, y \leq R$ with $R = 3$ mm ($z = 0$ and $z = 0.5$ m), $R = 4$ mm ($z = 1$ m), $R = 5$ mm ($z = 3$ m), $R = 10$ mm ($z = 5$ m), $R = 20$ mm ($z = 10$ m), $R = 50$ mm ($z = 30$ m), $R = 75$ mm ($z = 50$ m).

As seen in Figure 6.4, the FOB Fourier-mode does not look like a propagation-invariant beam. Its intensity pattern consists of several rings in some transverse planes, and even in the plane with one ring (Figure 6.4(d)), the relative ring width (width of the ring b from Figure 6.1 divided by its radius r_0) is much less than in the initial plane (Figure 6.4(a)). However, on further propagation to the far field (Figure 6.4(h)), the intensity pattern acquires the shape exactly the same (but scaled) as in the initial plane.

In this section, we found analytically a finite-energy light field, which contains an optical vortex, has low sidelobes, and is invariant to the Fourier transform, i.e. its intensity distribution in the back focal plane of a spherical lens (i.e. in the far field) is similar to that in the front focal plane [276]. The radial distribution of the

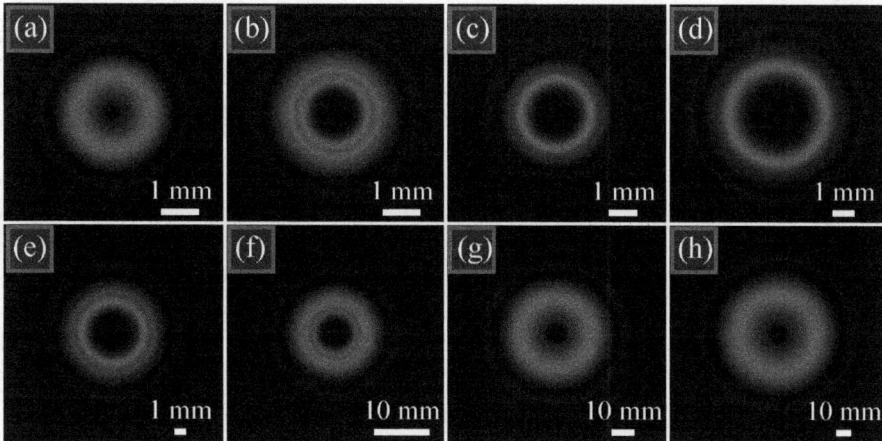

FIGURE 6.4 Transverse intensity distributions of the FOB Fourier-mode in several transverse planes: $z = 0$ (initial plane) (a), 0.5 m (b), 1 m (c), 3 m (d), 5 m (e), 10 m (f), 30 m (g), and 50 m (h).

complex amplitude is proportional to the Bessel function of a fractional order (odd integer number divided by six). Similar to previously known Bessel-modulated beams, e.g. the Bessel–Gaussian beams [40] or the Bessel–Gaussian beams with quadratic radial dependence [270], the complex amplitude of our beams is derived based on a reference integral (from [275]). To this reference expression, we added an analytically obtained energy of such beams, along with some of their physical properties, like geometrical parameters of the light ring on the intensity pattern. These beams have finite energy, but they don't have the Gaussian envelope. Compared to the LG modes with zero radial number, such FOB modes have a smaller inner dark spot. Potential applications are in optical communications, interferometry, and trapping.

6.2 NEW TYPE OF ELEGANT LASER BEAMS: SINUSOIDAL GAUSSIAN OPTICAL VORTEX

Hypergeometric laser beams, known in optics since 2007 [41,220], still attract great attention. They are described by a function that is the exact solution of the paraxial Helmholtz equation (Schrödinger-type equations). These beams represent a wide class of vortex beams that have several internal parameters and contain, in special cases, modernized elegant Laguerre–Gauss beams, Bessel beams, and others [277,278]. These beams can be generated using a light modulator or diffractive optical elements [279,280]. The nonparaxial version of hypergeometric beams [281] and their sharp focusing [282] were investigated. The passage of such beams through a medium with a hyperbolic refractive index [283] and with a parabolic refractive index [286,287], through an optical system performing a fractional Fourier transform [284], and an optical system with a high numerical aperture [289], through an ABCD optical system [288] and a uniaxial crystal [285] was considered. Hypergeometric beams proved to be stable while propagating in an atmosphere with non-Kolmogorov turbulence [290] and while passing through oceanic turbulence [291,292]. The passage of such beams through a nonlocal nonlinear medium was investigated [293,294]. In [295,296], hypergeometric beams with sharp autofocusing were studied. In [297], the diffraction of a hypergeometric beam on a grating with a fork was investigated. The forces that act on a Rayleigh particle captured at the focus of a hypergeometric beam were considered in [298]. In [68], an asymmetric hypergeometric beam was proposed.

In this section, an axial superposition of two Gaussian hypergeometric beams with different initial wavefront curvatures is considered. It is shown that if each beam has a singularity in the center of the initial plane and its energy is not limited, then the superposition of these beams has no singularity and its energy is limited. Such a superposition with a unit topological charge, which we named a sinusoidal Gaussian beam, is investigated in detail. It was shown that the radius of the main ring of such a beam is determined by the radius of the wavefront curvature and is almost independent of the Gaussian beam waist radius.

6.2.1 HYPERGEOMETRIC BEAMS WITH A PARABOLIC WAVEFRONT

There are well-known hypergeometric beams [41], in which complex amplitude in the initial plane has the form:

$$E(r,\varphi,z=0) = w\exp\left(-r^2/w^2 + in\varphi\right)/r, \qquad (6.23)$$

where (r, φ, z) is the cylindrical coordinates, w is the Gaussian beam waist radius, and n is the integer topological charge of the beam from Equation (6.23). The beam from Equation (6.23) has a singularity at zero ($r = 0$), and despite the presence of a Gaussian beam, the energy of the beam from Equation (6.23) is infinite. However, the singularity disappears while propagating these beams in space. Let us consider a similar beam with a quadratic phase in the initial plane instead of the light field from Equation (6.23). It is equal to placing a spiral phase plate (its transmission function is $\exp(in\varphi)$) not in the waist of a Gaussian beam, i.e. in the place where the beam converges or diverges. Then the initial field instead of Equation (6.23) will look like:

$$E_{\pm}(r,\varphi,z=0) = \left(-iw/2r\right)\exp\left(\pm ikr^2/2f - r^2/w^2 + in\varphi\right), \qquad (6.24)$$

where $k = 2\pi/\lambda$ is the wavenumber of light with a wavelength of λ, f is the focal length of a spherical (parabolic) lens. The propagation of the light field from Equation (6.24) is described by the Fresnel transform:

$$E(\rho,\theta,z) = \left(-ik/2\pi z\right)\int_0^{\infty}\int_0^{2\pi} E\left(r,\varphi,z=0\right)\exp\left[ik\left(r^2 + \rho^2\right)/2z\right.$$
$$\left. - ikr\rho\cos\left(\varphi-\theta\right)/z\right]rdrd\varphi, \qquad (6.25)$$

where (ρ, θ) are polar coordinates in the observation plane. Substituting the field from Equation (6.24) into Equation (6.25) and using the reference integral [277]:

$$\int_0^{\infty}\exp\left(-pr^2\right)J_n(cr)dr = 0.5\sqrt{\pi/p}\exp\left(-c^2/8p\right)I_{\frac{n}{2}}\left(c^2/8p\right), \qquad (6.26)$$

where $J_n(x)$ and $I_{n/2}(x)$ are the integer Bessel function and modified half-integer Bessel function, respectively, we obtain:

$$E_{\pm}(\rho,\theta,z) = \left(-i\right)^{n+2}\left(\sqrt{\pi}z_0/2z\right)\exp\left(ik\rho^2/2z + in\theta\right)$$
$$\times q_{\pm}^{-1/2}(z)\exp\left(-x_{\pm}\right)I_{n/2}(x_{\pm}), \qquad (6.27)$$

where $z_0 = kw^2/2$:

$$q_\pm(z) = 1 - iz_0 \left(z^{-1} \pm f^{-1} \right),$$

$$x_\pm = \left(z_0/z \right)^2 \left(\rho/w \right)^2 \big/ 2q_\pm(z).$$

(6.28)

It can be seen from Equation (6.27) that the field has no singularities for any $z > 0$.

Next, we show that the linear combination of two initial complex amplitudes in Equation (6.24) will not have a singularity at the origin and the initial field will have finite energy. Indeed, the difference between two fields from Equation (6.24) with plus and minus signs is equal to:

$$E = E_+ - E_- = \left(w/r \right) \sin\left(br^2 \right) \exp\left(-r^2/w^2 + in\varphi \right),$$

(6.29)

where $b = k/(2f)$. It can be seen from Equation (6.29) that at $r = 0$ the amplitude is equal to zero, i.e. the field from Equation (6.29) does not have a singularity as the field in Equation (6.24) at the origin. We find the energy (power) of the beam from Equation (6.29) by the formula:

$$W = 2\pi \int_0^\infty |E|^2 \, r dr = \pi w^2 \int_0^\infty \left[\sin^2(bx) \exp\left(-px \right)/x \right] dx$$

$$= \left(\pi w^2/4 \right) \ln\left(1 + 4b^2/p^2 \right) = \pi w^2 \ln\left(1 + z_0^2/f^2 \right)/4.$$

(6.30)

The complex field amplitude from Equation (6.29) at any distance z can be found using Equation (6.27):

$$E(\rho,\theta,z) = \left(-i \right)^{n+2} \left(\sqrt{\pi} z_0/2z \right) \exp\left(ik\rho^2/2z + in\theta \right)$$

$$\times \left[q_+^{-1/2}(z) e^{-x_+} I_{n/2}(x_+) - q_-^{-1/2}(z) e^{-x_-} I_{n/2}(x_-) \right].$$

(6.31)

Although the solution to the paraxial Helmholtz equation, Equation (6.31), is closed, it is difficult to analyze it, since the modified Bessel functions have an arbitrary order and complex arguments of different magnitudes. However, the Bessel functions in Equation (6.31) are expressed in terms of the elementary sine function for $n = 1$. Therefore, the analysis of the field from Equation (6.29) in the Fresnel diffraction zone for $n = 1$ can be carried out in more detail.

6.2.2 Special Case: An Elegant Sinusoidal Gaussian Vortex

For convenience, we rewrite the initial field from Equation (6.29) for $n = 1$ and replace the common factor w by $b^{-1/2}$ (this is a dimensional constant and does not affect the final results):

$$E_1(r,\varphi,z=0) = \sin(br^2)\exp(-r^2/w^2 + i\varphi)/r\sqrt{b}. \qquad (6.32)$$

In the Fresnel diffraction zone, the complex amplitude is found by substituting Equation (6.32) into Equation (6.25) and is equal to:

$$E_1(\rho,\theta,z) = -\sin(b(z)\rho^2)\exp(-\rho^2/w^2(z) + ik\rho^2/2R(z) + i\theta)/\rho\sqrt{b}, \quad (6.33)$$

where:

$$w^2(z) = w^2\left[4z_0^2 z^2 + (\gamma z^2 - z_0^2)^2\right]/\left[z_0^2(z_0^2 + \gamma z^2)\right],$$

$$R(z) = \left[4z_0^2 z^2 + (\gamma z^2 - z_0^2)^2\right]/\left[\gamma z(z_0^2 + \gamma z^2)\right],$$

$$b(z) = bz_0^2\left[(\gamma z^2 - z_0^2) + 2iz_0 z\right]/\left[4z_0^2 z^2 + (\gamma z^2 - z_0^2)^2\right],$$

$$\gamma = 1 + b^2 w^4 = 1 + z_0^2/f^2. \qquad\qquad\qquad\qquad (6.34)$$

A comparison of Equations (6.32) and (6.33) shows that the field retains its structure in the Fresnel diffraction zone, but since the sine argument in Equation (6.33) is complex, beam from Equation (6.33) cannot be called the mode. This beam is an elegant mode, that is, a mode that retains the form of the analytical expression in Equation (6.33) by which it is described. However, its intensity distribution changes during propagation. A redistribution of energy occurs between the rings of the diffraction pattern, although the pattern itself remains axisymmetric. The elegant Laguerre–Gauss modes, in which amplitude is described by Laguerre polynomials but with a complex argument are known [278]. Therefore, the intensity of such beams changes during propagation, although not significantly. We should note that the dependences of the beam radius $w(z)$ and the curvature radius $R(z)$ in Equation (6.34) differ from the waist radius and the curvature radius for a Gaussian beam since the dependence of these quantities on z in Equation (6.34) is only partial. The other part depending on z is present in the argument of the sin function and is "hidden" in the function $b(z)$. An interesting feature of the beam from Equations (6.32) and (6.33) is that the radius of the first (main) intensity ring is almost independent of the Gaussian beam waist radius w. The radius of the first ring in the initial plane can be estimated as half the radius at which the sine function in Equation (6.32) has zero ($br^2 = \pi$):

$$r_{max} \simeq \sqrt{\lambda f/2}. \qquad (6.35)$$

A more precise expression for the radius of the first intensity ring can be found from the condition of the equality to zero of the derivative of the field intensity from Equation (6.32) with respect to the argument:

$$\frac{\partial}{\partial x} I(x) = \frac{\partial}{\partial x}\left[\sin^2(x)\exp(-2x/x_0)/x\right] = 0, \qquad (6.36)$$

where $x = br^2$, $x_0 = bw^2$.

Simplifying Equation (6.36), we obtain the transcendental equation for finding the radius of the first intensity ring in the initial plane:

$$\tan x = 2xx_0 / (2x + x_0). \tag{6.37}$$

For comparison, we should recall that the radius of the Laguerre–Gauss mode (0, 1) ring:

$$E_{LG1}(r, \varphi, z = 0) = r \exp(-r^2/w^2 + i\varphi)/w \tag{6.38}$$

depends on the waist radius and is equal to the expression:

$$r_{\max}^{LG} = w/\sqrt{2}. \tag{6.39}$$

Another feature of the beam from Equation (6.33) is that it consists of two almost Gaussian beams with different divergences. Indeed, amplitude from Equation (6.33) can be represented as:

$$E_1(\rho, \theta, z) = \left(i/2\rho\sqrt{b} \right) \exp\left(-\rho^2/w^2(z) + ik\rho^2/2R(z) + i\theta \right)$$
$$\times \left[\exp\left(ib_1(z)\rho^2 - b_2(z)\rho^2 \right) - \exp\left(-ib_1(z)\rho^2 + b_2(z)\rho^2 \right) \right], \tag{6.40}$$

where $b_1(z) = \text{Re } b(z)$, $b_2(z) = \text{Im } b(z)$. From Equation (6.40), we obtain an equation for the Gaussian beam radius depending on the distance z for both terms:

$$\overline{w}_{\pm}^2(z) = w^2 \left[4z_0^2 z^2 + (\gamma z^2 - z_0^2)^2 \right] / \left[z_0^2(z_0^2 + \gamma z^2 \pm 2bw^2 z_0 z) \right]. \tag{6.41}$$

The plus sign in Equation (6.41) refers to the first term in Equation (6.40), and the minus sign refers to the second. It can be shown that the presence of a minus in the denominator of the Equation (6.41) does not lead to zero and negative values of the denominator. Although the radii of Gaussian beams in Equation (6.41) are different, both radii increase in the far zone at $z \gg z_0$ equally and linearly with z:

$$\overline{w}_{\pm}(z \gg z_0) = w_G\sqrt{\gamma} = w_G\sqrt{1 + z_0^2/f^2}, \tag{6.42}$$

where w_G is far-field Gaussian beam radius:

$$w_G(z \gg z_0) = \lambda z/\pi w. \tag{6.43}$$

It can be seen from Equation (6.42) that the divergence of the beam (6.33) is in a square root of "gamma" times greater than the divergence of an ordinary Gaussian beam. This was to be expected since the sine argument in Equation (6.32) is the phase of a spherical lens with a focal length f. And this lens changes (always increases, since it is located in the waist) the divergence of the initial Gaussian beam.

However, if we compare the beam from Equation (6.33) not with a Gaussian beam, but with the Laguerre–Gauss mode (0, 1), then the beam divergence in Equation (6.37) will be equal to:

$$w_{LG1}(z \gg z_0) = \sqrt{2}\,w_G(z \gg z_0) = \sqrt{2}\,\lambda z/\pi w. \qquad (6.44)$$

This follows from the general formula for the divergence of the Laguerre–Gauss mode (p, l) [299]:

$$w_{LG}(z) = w\sqrt{2p+l+1}\left(1 + z^2/z_0^2\right)^{1/2}. \qquad (6.45)$$

Thus, if $z_0 < f$, then, as follows from a comparison of Equations (6.42) and (6.44), the divergence of the optical vortex from Equation (6.33) is less than the divergence of the Laguerre–Gauss beam from Equation (6.38).

6.2.3 NUMERICAL SIMULATION

Figure 6.5 shows the intensity and the phase of the beam from Equation (6.32) with a waist radius $w_0 = 500$ μm and $w_0 = 5000$ μm in the initial plane $z = 0$ m and at a distance $z = 2$ m. The rest calculation parameters are as follows: wavelength $\lambda = 532$ nm, focal length $f = 1$ m.

It can be obvious from Figure 6.5 that the intensity ring radius is almost independent of the Gaussian beam waist radius: it was 310 μm and 442 μm in the initial plane and 711 μm and 764 μm at a distance of $z = 2$ m for $w_0 = 500$ μm and $w_0 = 5000$ μm, respectively. That is, a change in the waist radius by a factor of 10 led to an increase in the intensity ring radius in the initial plane by a factor of 1.43 and by only 7% at a distance of $z = 2$ m.

In this study, hypergeometric beams with a parabolic wavefront in the initial plane are considered. Although hypergeometric beams have a singularity in the center of the initial plane and infinite energy the superposition of two such beams does not have a singularity and has finite energy. A particular case of such a superposition as a sinusoidal Gaussian beam with a unit topological charge is considered in detail. This beam belongs to the type of elegant laser beams, since it is described by the same functions with a complex argument both in the initial plane and in the Fresnel diffraction zone. The diameter of the first light ring in a sinusoidal Gaussian beam is almost independent of the Gaussian beam waist radius. The obtained beams could find their applications in optical micromanipulation [300,301,302], free-space optical communications [303,304].

6.3 PROPAGATION-INVARIANT OFF-AXIS ELLIPTIC GAUSSIAN BEAMS WITH THE ORBITAL ANGULAR MOMENTUM

Among different-kind light fields, form-invariants are fields of special interest (or, propagation-invariants). The transverse intensity shape of such beams remains

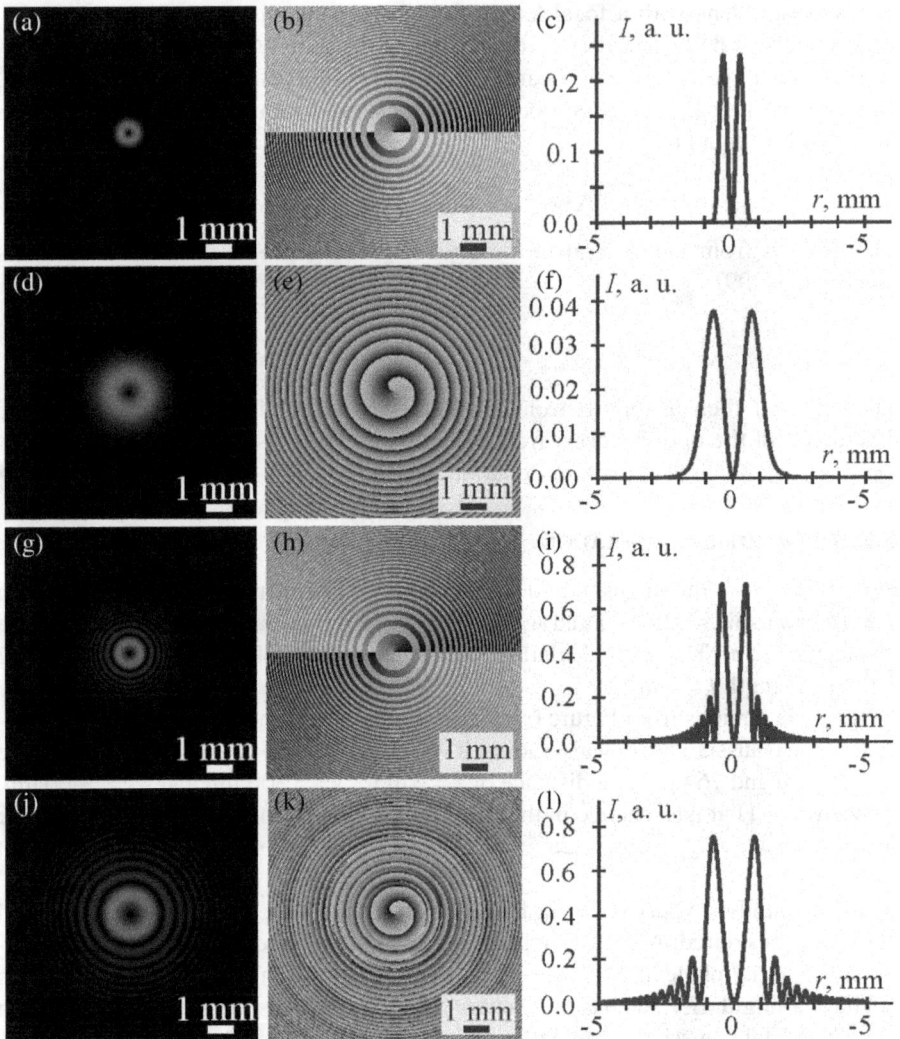

FIGURE 6.5 The intensity (left column, negative), the phase (middle column), and the intensity cross-sections (right column) of a beam from Equation (6.32) with a waist radius w_0 = 500 μm (a–f) and w_0 = 5000 μm (g–l) in the initial plane $z = 0$ m (a–c, g–i), and at a distance $z = 2$ m (d–f, j–l). Other calculation parameters: wavelength $\lambda = 532$ nm, focal length $f = 1$ m.

unchanged on free-space propagation (up to scale and rotation). Nonparaxial examples of such fields are plane waves, diffraction-free Bessel beams [1], Mathieu beams [3], and parabolic beams [2]. Well-known paraxial examples are the Hermite–Gaussian and Laguerre–Gaussian beams [4], Gaussian beams with arbitrary located optical vortices [49], and some superpositions of such beams [305,306]. In [189], a general procedure is described for calculating the paraxial propagation-invariant beams, whose transverse intensity distribution has a shape of an arbitrary curve. The procedure is based on using an elementary spiral beam (off-axis Gaussian beam with a wavefront tilt) as a building block. In [307], other building blocks, highly localized wavepackets, were suggested to construct nondiffracting pulsed beams.

In some tasks; however, using light beams with an elliptic intensity cross-section can be advantageous. For example, as shown in [308], adopting elliptic vortex beams for optical data transmission in a turbulent atmosphere reduces the scintillation index [309]. In [310], elliptic vortex beams with a fractional topological charge (number of 2π phase jumps along a contour encompassing either a single optical vortex, i.e. local topological charge, or the whole beam [19]) are used in free-space information transfer to increase the throughput and security. In [311], partially coherent four-petal elliptic Gaussian beams are investigated, also in the turbulent atmosphere. It is shown that the number of petals can change on propagation. In optical trapping, an elongated intensity distribution is convenient for preventing a particle from moving along one coordinate [312]. In [313,314,315], elliptic Gaussian beams are used to increase the recording throughput of thermochemical laser-induced periodic surface structures.

The interest in the elliptic beams motivates their analytical studies. Generally, the paraxial propagation of a light beam is described by the integral Fresnel transform [158], whose kernel is a quadratic exponential. Thus, an analytical description of a Gaussian beam with an arbitrary ellipticity, tilt angle, center position, and parabolic wavefront curvature is possible, in principle. Such a description is given in [316]. However, such beams are not propagation-invariant and their transverse cross-section changes on space propagation. In [190], propagation-invariant elliptic Gaussian beams were considered with an optical vortex in the center, and their orbital angular momentum was derived. On propagation in space, such beams rotate around their center.

In this section, based on the theory of paraxial structurally stable light beams developed in [189], we study paraxial propagation-invariant elliptic Gaussian beams, similar to [190], but with an arbitrary position of the ellipse center in the transverse plane and with an arbitrary ellipse tilt. For this purpose, such as in [189,307], we also use simple beams as building blocks and construct a continuous superposition of elementary spiral light beams [189] on a plane. Then, solving a nonlinear system of five equations to determine the weight coefficients of this superposition, we obtain an analytical expression for the complex amplitude distribution. In addition, we obtain a formula for the orbital angular momentum of such beams. Similarly to the Steiner theorem in mechanics, it consists of two terms. The first one coincides with Equation (4) in [317] and describes the intrinsic OAM with respect to the beam center. This term grows with the beam ellipticity. The second term depends parabolically on the distance between the beam's "center of mass" and the optical axis. It turns out that the ellipse orientation in the transverse plane (tilt angle of the ellipse's axes to the Cartesian coordinate axes) does not affect the normalized orbital angular momentum.

6.3.1 Propagation-Invariant Off-Axis Gaussian Beams

As has been shown in [189] (formula (6.1)), any function of the following form:

$$E_{\pm}\left(x,y,z\right)=\frac{1}{q}\exp\left(-\frac{x^2+y^2}{qw_0^2}\right)f\left(\frac{x\pm iy}{qw_0}\right),\qquad(6.46)$$

where (x, y, z) are the Cartesian coordinates, w_0 is the Gaussian beam waist radius, $q = 1 + iz/z_0$, $z_0 = kw_0^2 / 2$ is the Rayleigh range, $k = 2\pi /\lambda$ is the wavenumber of light with the wavelength λ , $f(x \pm iy)$ is an arbitrary entire analytical function, is a solution of the paraxial wave equation:

$$2ik\frac{\partial E}{\partial z}+\frac{\partial^2 E}{\partial x^2}+\frac{\partial^2 E}{\partial y^2}=0\qquad(6.47)$$

and describes a propagation-invariant light field (traveling along the optical axis z), i.e. a field with its transverse intensity shape conserving on free-space propagation, changing only in scale and rotating. Of course, it is meant that the propagation distance is much greater than the beam width, i.e. $z >> w_0|q|$, so that the paraxial approximation is valid.

The freedom in choosing the functions $f(.)$ allows a description of optical fields with very different physical properties. For example, if we choose the cosine function:

$$E\left(x,y,z\right)=\frac{1}{q}\cos\left(\frac{x+iy}{\alpha_0 q}\right)\exp\left(-\frac{x^2+y^2}{qw_0^2}\right),\qquad(6.48)$$

with α_0 being a complex-valued parameter, then we obtain a cosine vortex beam [162] with its intensity distribution consisting of two light spots (Figure 6.6(a, c)) and with an infinite topological charge, since the beam contains an infinite number of optical vortices (Figure 6.6(b, d)).

If, instead of the cosine, we choose the exponential, i.e. if we consider a light beam:

$$E\left(x,y,z\right)=\frac{1}{q}\exp\left(\frac{x+iy}{\alpha_0 q}\right)\exp\left(-\frac{x^2+y^2}{qw_0^2}\right),\qquad(6.49)$$

then, it is expectable that the beam should contain one light spot, rather than two. Such a beam is described in [189] and is called an elementary spiral beam. In contrast to the cosine vortex beam from Figure 6.6(a–d), the intensity distribution of the beam from Equation (6.49) contains a single off-axis spot (Figure 6.6(e, g)) and the topological charge of such a beam is zero, since the beam does not contain optical vortices (Figure 6.6(f, h)). Thus, despite the cosine being just a sum of two exponentials, using the functions in Equation (6.46) leads to light beams with drastically different phase distributions.

If we denote $1/\alpha_0 = (2q/w^2)(x_c - iy_c)$ with x_c and y_c being real numbers (z-dependent), then, separating in Equatiom (6.49) the real and the imaginary parts in the

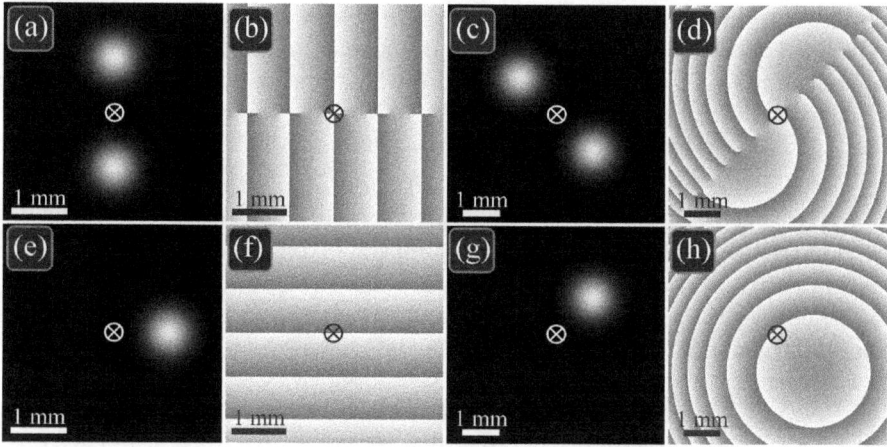

FIGURE 6.6 Intensity (a, c, e, g) and phase (b, d, f, h) distributions of the vortex cosine beam from Equation (6.48) and of the elementary spiral beam from Equation (6.49) in the initial plane (a, b, e, f) and at the Rayleigh distance (c, d, g, h) for the following parameters: wavelength $\lambda = 532$ nm, Gaussian beam waist radius $w_0 = 0.5$ mm, scaling factor $\alpha_0 = w_0/4$, computation domain $|x|, |y| \leq R$ with $R = 2$ mm (in the initial plane), and $R = 3$ mm (at the Rayleigh distance). The cross in the center shows the optical axis, around which the diffraction pattern rotates on propagation (at the Rayleigh distance, the rotation angle is $\pi/4$).

exponentials, and dividing by the constant multiplier $\exp[(x_c^2 + y_c^2)/w^2] = \exp[|w_0/(2\alpha_0)|^2]$, we can rewrite Equation (6.49) as follows:

$$E(x,y,z) = \frac{w_0}{w} \exp\left[-\frac{(x-x_c)^2 + (y-y_c)^2}{w^2} + i\frac{2}{w^2}(x_c y - y_c x) + \frac{ik}{2R}(x^2 + y^2) - i\zeta \right],$$

(6.50)

with $w(z)$, $R(z)$, and $\zeta(z)$ being respectively the beam width, the wavefront curvature radius, and the Gouy phase at the distance z [4]:

$$w(z) = w_0\sqrt{1 + z^2/z_0^2},$$

$$R(z) = z(1 + z_0^2/z^2),$$

(6.51)

$$\zeta(z) = \arctan(z/z_0),$$

As seen from Equation (6.50), x_c and y_c are the Cartesian coordinates of the intensity maximum (center of the Gaussian beam) at the distance z, which are related to the Cartesian coordinates of the beam center (x_{c0}, y_{c0}) in the initial plane as follows:

$$x_c = (w/w_0)(x_{c0}\cos\zeta - y_{c0}\sin\zeta),$$

$$y_c = (w/w_0)(y_{c0}\cos\zeta + x_{c0}\sin\zeta).$$

(6.52)

It is also seen in Equation (6.50) that the light field is an off-axis Gaussian beam and an inclined plane wave is added to the wavefront with the inclination angle matching the shift of the light spot and inclination direction orthogonal to the direction from the optical axis to this spot (Figure 6.6(e, f)). It is this matching that makes the beam propagation-invariant, since, on space propagation, the widening of the Gaussian beam is proportional to its moving away from the optical axis. For example, at the Rayleigh distance, $z = z_0$, the beam width grows $\sqrt{2}$ times, and its distance from the optical axis also grows $\sqrt{2}$ times, and it is rotated by the angle $\pi/4$. Thus, all such beams would together make a pattern that does not change on propagation.

The beam given by Equations (6.49) and (6.50) is a basic beam for constructing other propagation-invariant light beams. In [189], a general procedure is described for constructing the propagation-invariant beams with a shape of an arbitrary curve. Below we investigate analytically a superposition of the beams from Equation (6.50), but in the whole transverse plane, rather than along a curve (such as in [189]). We show that it leads to the analytical description of an off-axis elliptic Gaussian beam, which is invariant to space propagation and rotating around the optical axis.

6.3.2 PROPAGATION-INVARIANT ELLIPTIC GAUSSIAN BEAMS

Using Equation (6.52), Equation (6.50) can be rearranged and written via the coordinates of the beam center in the initial plane:

$$
E(x,y,z) = \frac{w_0}{w} \exp\left[-\frac{x^2 + y^2}{w^2} + \frac{ik}{2R}(x^2 + y^2) - i\zeta - \frac{x_{c0}^2 + y_{c0}^2}{w_0^2} \right.
$$

$$
\left. + \frac{2}{qw_0^2}(x + iy)(x_{c0} - iy_{c0}) \right].
$$

(6.53)

Using the beams from Equation (6.53) as a basis, we can construct a continuous superposition with the weight coefficients in a Gaussian-like (quadratic exponential) dependence on the beams' central positions:

$$
E(x,y,z) = \frac{w_0}{w} \exp\left[-\frac{x^2 + y^2}{w^2} + \frac{ik}{2R}(x^2 + y^2) - i\zeta \right]
$$

$$
\times \int_{-\infty}^{+\infty}\int_{-\infty}^{+\infty} \exp\left[-p_{xx}\frac{x_{c0}^2}{w_0^2} - p_{yy}\frac{y_{c0}^2}{w_0^2} - 2p_{xy}\frac{x_{c0}y_{c0}}{w_0^2} - 2p_x\frac{x_{c0}}{w_0} - 2p_y\frac{y_{c0}}{w_0} \right]
$$

$$
\times \exp\left[-\frac{x_{c0}^2 + y_{c0}^2}{w_0^2} + \frac{2}{qw_0^2}(x + iy)(x_{c0} - iy_{c0}) \right] dx_{c0}dy_{c0}.
$$

(6.54)

where p_{xx}, p_{yy}, p_{xy}, p_x, and p_y are some (generally, complex) numbers that define the most contributing constituent beam from Equation (6.53) in the superposition in Equation (6.54) and how fast the contributions decay for other constituent beams.

Rearrangement of the exponentials in Equation (6.54) yields

$$E(x,y,z) = \frac{w_0}{w} \exp\left[-\frac{x^2+y^2}{w^2} + \frac{ik}{2R}\left(x^2+y^2\right) - i\zeta \right]$$

$$\times \int_{-\infty}^{+\infty}\int_{-\infty}^{+\infty} \exp\left[-\left(1+p_{xx}\right)\frac{x_{c0}^2}{w_0^2} - \left(1+p_{yy}\right)\frac{y_{c0}^2}{w_0^2} - 2p_{xy}\frac{x_{c0}y_{c0}}{w_0^2} \right] \qquad (6.55)$$

$$\times \exp\left[-\frac{2}{w_0}\left(p_x - \frac{x+iy}{qw_0} \right)x_{c0} - \frac{2}{w_0}\left(p_y + i\frac{x+iy}{qw_0} \right)y_{c0} \right] dx_{c0}dy_{c0}.$$

The integral of the quadratic exponential is well-known and, therefore:

$$E(x,y,z) = \frac{w_0^3}{w}\frac{\pi}{\sqrt{G}} \exp\left[-\frac{x^2+y^2}{w^2} + \frac{ik}{2R}\left(x^2+y^2\right) - i\zeta \right]$$

$$\times \exp\left\{ \frac{1+p_{xx}}{G}\left(p_y + i\frac{x+iy}{qw_0} \right)^2 + \frac{1+p_{yy}}{G}\left(p_x - \frac{x+iy}{qw_0} \right)^2 \right\} \qquad (6.56)$$

$$\times \exp\left\{ -\frac{2p_{xy}}{G}\left(p_x - \frac{x+iy}{qw_0} \right)\left(p_y + i\frac{x+iy}{qw_0} \right) \right\},$$

with $G = (1+p_{xx})(1+p_{yy}) - (p_{xy})^2$.

In the dimensionless Cartesian coordinates, rotated by an angle equal to the Gouy phase and normalized by the beam width w:

$$\begin{pmatrix} u \\ v \end{pmatrix} = \frac{1}{w}\begin{pmatrix} \cos\zeta & \sin\zeta \\ -\sin\zeta & \cos\zeta \end{pmatrix}\begin{pmatrix} x \\ y \end{pmatrix}, \qquad (6.57)$$

the complex amplitude from Equation (6.56) reads as:

$$E(x,y,z) = \frac{w_0^3}{w}\frac{\pi}{\sqrt{G}} \exp\left[-u^2 - v^2 + iz_0|q|^2 R^{-1}\left(u^2+v^2\right) - i\zeta \right]$$

$$\times \exp\left\{ \frac{1+p_{xx}}{G}\left[p_y + i(u+iv) \right]^2 + \frac{1+p_{yy}}{G}\left[p_x - (u+iv) \right]^2 \right\} \qquad (6.58)$$

$$\times \exp\left\{ -\frac{2p_{xy}}{G}\left[p_x - (u+iv) \right]\left[p_y + i(u+iv) \right] \right\},$$

or:

$$E(x,y,z) = C_0 \exp\left(-A_{xx}u^2 - A_{yy}v^2 - A_{xy}uv - A_x u - A_y v \right), \qquad (6.59)$$

with:

$$C_0 = \frac{w_0^3}{w}\frac{\pi}{\sqrt{G}}\exp\left(-i\zeta + \frac{1+p_{xx}}{G}p_y^2 + \frac{1+p_{yy}}{G}p_x^2 - \frac{2p_{xy}}{G}p_xp_y\right),$$

$$A_{xx} = 1 - iz_0|q|^2 R^{-1} + \frac{p_{xx}-p_{yy}-2ip_{xy}}{G},$$

$$A_{yy} = 1 - iz_0|q|^2 R^{-1} - \frac{p_{xx}-p_{yy}-2ip_{xy}}{G},$$

$$A_{xy} = 2i\frac{p_{xx}-p_{yy}-2ip_{xy}}{G},$$

$$A_x = \frac{-2}{G}\left[ip_y\left(1+p_{xx}\right)-p_x\left(1+p_{yy}\right)-ip_{xy}\left(p_x+ip_y\right)\right],$$

$$A_y = i\frac{-2}{G}\left[ip_y\left(1+p_{xx}\right)-p_x\left(1+p_{yy}\right)-ip_{xy}\left(p_x+ip_y\right)\right].$$

(6.60)

Real parts of these coefficients define the position, sizes, and orientation of the elliptic light spot. On the other hand, if an elliptic Gaussian light spot in the initial plane is located in the point (x_0, y_0) (spot center), is tilted by the angle α from the x-axis, and has the sizes σ_x and σ_y (Figure 6.7), its complex amplitude is given by:

$$E(x,y,0) = \exp\left(-\frac{x''^2}{\sigma_x^2}-\frac{y''^2}{\sigma_y^2}\right)$$

$$= \exp\left\{-\frac{\left[(x-x_0)\cos\alpha+(y-y_0)\sin\alpha\right]^2}{\sigma_x^2}-\frac{\left[(y-y_0)\cos\alpha-(x-x_0)\sin\alpha\right]^2}{\sigma_y^2}\right\}.$$

(6.61)

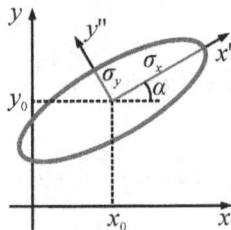

FIGURE 6.7 Geometric parameters of an elliptic light spot.

Thus, for the beam from Equation (6.59) to describe such an elliptic spot, the following conditions should be fulfilled (we suppose that all the superposition coefficients, p_{xx}, p_{yy}, p_{xy}, p_x, and p_y are real-values):

$$
\begin{cases}
\dfrac{1}{w_0^2}\left(1+\dfrac{p_{xx}-p_{yy}}{G}\right)=\dfrac{\cos^2\alpha}{\sigma_x^2}+\dfrac{\sin^2\alpha}{\sigma_y^2}, \\[2ex]
\dfrac{1}{w_0^2}\left(1-\dfrac{p_{xx}-p_{yy}}{G}\right)=\dfrac{\cos^2\alpha}{\sigma_y^2}+\dfrac{\sin^2\alpha}{\sigma_x^2}, \\[2ex]
\dfrac{1}{w_0^2}\dfrac{4p_{xy}}{G}=\left(\dfrac{1}{\sigma_x^2}-\dfrac{1}{\sigma_y^2}\right)\sin 2\alpha, \\[2ex]
\dfrac{1}{w_0}\dfrac{2}{G}\left[\left(1+p_{yy}\right)p_x-p_{xy}p_y\right]=-2\left(\dfrac{\cos^2\alpha}{\sigma_x^2}+\dfrac{\sin^2\alpha}{\sigma_y^2}\right)x_0-\left(\dfrac{1}{\sigma_x^2}-\dfrac{1}{\sigma_y^2}\right)y_0\sin 2\alpha, \\[2ex]
\dfrac{1}{w_0}\dfrac{2}{G}\left[\left(1+p_{xx}\right)p_y-p_{xy}p_x\right]=-2\left(\dfrac{\sin^2\alpha}{\sigma_x^2}+\dfrac{\cos^2\alpha}{\sigma_y^2}\right)y_0-\left(\dfrac{1}{\sigma_x^2}-\dfrac{1}{\sigma_y^2}\right)x_0\sin 2\alpha.
\end{cases}
\tag{6.62}
$$

The system in Equation (6.62) consists of 5 nonlinear equations. However, it turned out that it can be solved analytically. Summing the first two equations, we get the following condition:

$$
\frac{1}{\sigma_x^2}+\frac{1}{\sigma_y^2}=\frac{2}{w_0^2}.
\tag{6.63}
$$

This condition means that the beams from Equation (6.50) do not allow constructing an elliptic beam with an arbitrary size. The transverse sizes of the ellipse should be related by Equation (6.63), according to which, the –2th power mean [318] of these sizes (i.e. $[(\sigma_x^{-2}+\sigma_y^{-2})/2]^{-1/2}$) equals the waist radius w_0 of the elementary spiral beams from Equations (6.49), (6.50), used in constructing the superposition. We call this radius w_0 an effective waist radius, since the elliptic Gaussian beam propagates in space with the same phase velocity (Gouy phase) as does a circular Gaussian beam with the waist radius w_0.

As turns out, the rest parameters of the ellipse (x_0, y_0, α) can be arbitrary, and, for any of their values, the system in Equation (6.62) has a solution. Substituting the derived values p_{xx}, p_{yy}, p_{xy}, p_x, and p_y into Equation (6.58), we get:

$$
E(x,y,z)=\frac{w_0}{w}\exp\left[-\left(1-iz_0\left|q\right|^2 R^{-1}-\mu\right)u^2-\left(1-iz_0\left|q\right|^2 R^{-1}+\mu\right)v^2+2i\mu uv\right]
$$
$$
\times\exp\left\{-2\left[\frac{x_0}{w_0}(\mu-1)+i\frac{y_0}{w_0}(\mu+1)\right](u+iv)-i\zeta\right\},
\tag{6.64}
$$

with:

$$\mu = \gamma e^{-2i\alpha}, \tag{6.65}$$

$$\gamma = 1 - \frac{w_0^2}{\sigma_x^2} = \frac{w_0^2}{\sigma_y^2} - 1. \tag{6.66}$$

Returning from the coordinates from Equation (6.57) to the original coordinates, we get the final expression for the complex amplitude:

$$E(x,y,z) = \frac{w_0}{w} \exp\left[-\left(\frac{1}{w^2} - \frac{ik}{2R} - \frac{\mu}{w^2} e^{-2i\zeta} \right) x^2 - \left(\frac{1}{w^2} - \frac{ik}{2R} + \frac{\mu}{w^2} e^{-2i\zeta} \right) y^2 \right]$$

$$\times \exp\left\{ 2i\frac{\mu}{w^2} e^{-2i\zeta} xy - 2\left(\frac{1}{w^2} - \frac{ik}{2R} \right)\left[x_0(\mu-1) + iy_0(\mu+1) \right](x+iy) - i\zeta \right\}. \tag{6.67}$$

Figure 6.8 illustrates the intensity and phase distributions of two elliptic beams from Equation (6.67) in the initial plane and at the Rayleigh distance.

Figure 6.8 confirms that, after free-space propagation, the intensity distributions change only in scale and are rotated with respect to the origin (optical axis), i.e. the solution (6.67) of the paraxial wave Equation (6.47) is stable.

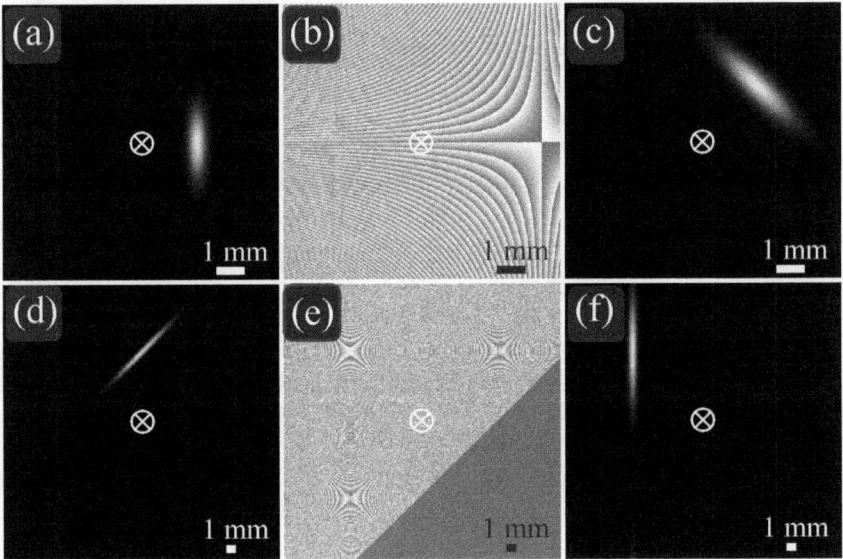

FIGURE 6.8 Intensity (a, c, d, f) and phase (b, e) distributions of the propagation-invariant off-axis elliptic Gaussian beam in the initial plane (a, b, d, e) and at the Rayleigh distance $z = z_0$ (c, f) for the following parameters: wavelength $\lambda = 532$ nm, Gaussian beam waist radii $\sigma_x = 1425$ μm and $\sigma_y = 365$ μm ($w_0 = 500$ μm) (a–c) and $\sigma_x = 4583$ μm, and $\sigma_y = 355$ μm ($w_0 = 500$ μm) (d–f), coordinates of the beam center $(x_0, y_0) = (2, 0)$ mm (a–c) and $(x_0, y_0) = (0, 7.5)$ mm (d–f), tilt angle of the major ellipse axis to the x-axis $\alpha = \pi/2$ (a–c) and $\alpha = \pi/4$ (d–f), computation domain $|x|, |y| \leq R$ with $R = 5$ mm (a–c), and $R = 15$ mm (d–f).

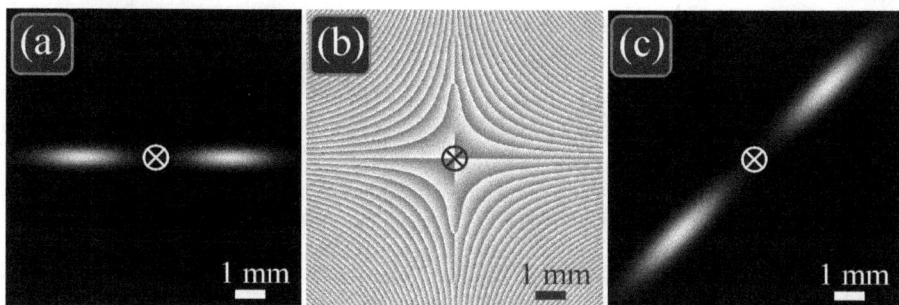

FIGURE 6.9 Distributions of intensity (a, c) and phase (b) of the propagation-invariant two-petal beam in the initial plane (a, b) and at the Rayleigh distance $z = z_0$ (c) for the following parameters: wavelength $\lambda = 532$ nm, waist radii of the two opposite elliptic Gaussian beams $\sigma_x = 1617$ μm and $\sigma_y = 362$ μm (i.e. $w_0 = 500$ μm), positions of the beam centers $(x_0, y_0) = (\pm 2.5, 0)$ mm, tilt angle of the major ellipse axis to the x-axis $\alpha = 0$, computation domain $|x|, |y| \leq R$ ($R = 5$ mm). The cross in the center shows the optical axis, around which the diffraction pattern rotates on propagation.

On propagation, both these beams acquire the same Gouy phase as a circular Gaussian beam with the waist radius $w_0 = 500$ μm. Therefore, the superposition of such beams also does not change its shape on propagation.

For instance, such beams can be used to construct a propagation-invariant two-petal superposition.

Such fields are used in single-molecule microscopy and for improving the longitudinal resolution in imaging systems [192]. In addition, controllable rotation makes it possible to use these beams for distance measurement. Figure 6.9 depicts the intensity and phase distributions of the two-petal beam, composed of two opposite elliptic Gaussian beams from Equation (6.67), in the initial plane and at the Rayleigh distance (Figure 6.9).

In the same way, we can construct, for instance, triangular or square propagation-invariant beams. Figure 6.10 depicts the intensity and phase distributions of the triangular beam, composed of three elliptic Gaussian beams from Equation (6.67), in the initial plane, near field ($z = z_0/2$), at the Rayleigh distance ($z = z_0$), and in the far field ($z = 5z_0$).

Figure 6.11 depicts the intensity and phase distributions of the off-axis square beam, composed of four elliptic Gaussian beams from Equation (6.67), in the same planes.

As seen in Figures 6.10 and 6.11, both beams rotate on propagation (by $\pi/4$ at the Rayleigh distance and almost by $\pi/2$ in the far field) around the optical axis, but their transverse shape remains the same as it was in the initial plane.

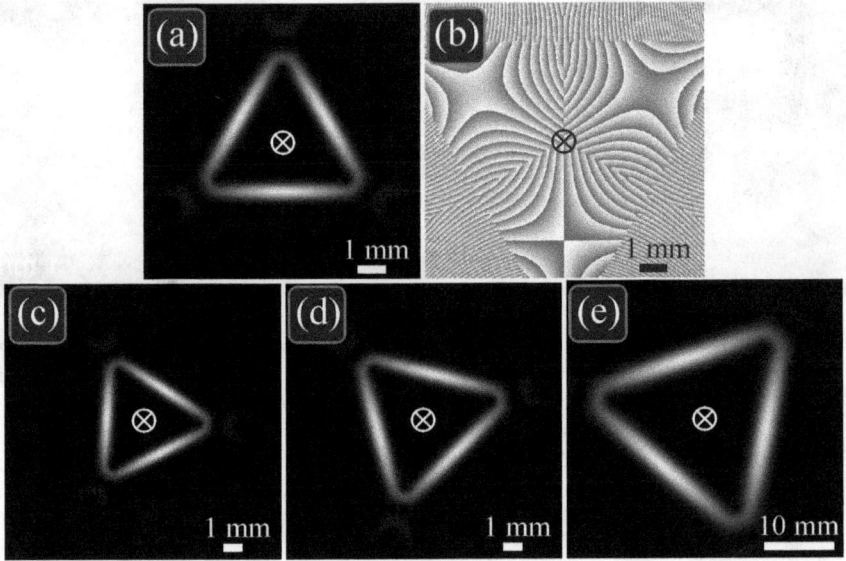

FIGURE 6.10 Intensity (a, c–e) and phase (b) distributions of the triangular propagation-invariant beam in the initial plane (a,b), in the near field $z = z_0/2$ (c), at the Rayleigh distance $z = z_0$ (d), and in the far field $z = 5z_0$ (e) for the following parameters: wavelength $\lambda = 532$ nm, waist radii of all three elliptic Gaussian beams $\sigma_x = 3102$ μm and $\sigma_y = 356$ μm ($w_0 = 500$ μm), coordinates of the beam centers $(x_0 p, y_0 p) = (r_0 \cos \varphi_p, r_0 \sin \varphi_p)$ ($r_0 = 1.768$ mm, $\varphi_p = \pi/6 + 2\pi p/3$, $p = 0, 1, 2$), tilt angles of the major axes of the ellipses to the x-axis $\alpha_p = \varphi_p + \pi/2$, computation domain |x|, |y| ≤ R with R = 5 mm (a, b), R = 10 mm (c, d), and R = 20 mm (e). The cross in the center shows the optical axis, around which the diffraction pattern rotates on propagation. The initial field (a, b) was obtained by Equation (6.67) (with a phase shift $2\pi/3$ between the beams to avoid the destructive interference in the corners), whereas distributions in the other planes (c–e) were obtained by the numerical Fresnel transform.

6.3.3 BEAM POWER AND THE ORBITAL ANGULAR MOMENTUM

For a paraxial light beam, its power and orbital angular momentum are obtained by the following formulae [162, 319]:

$$W = \int\limits_{-\infty}^{\infty} \int\limits_{-\infty}^{\infty} E^*(x,y) E(x,y) \, dx dy, \qquad (6.68)$$

$$J_z = \text{Im} \int\limits_{-\infty}^{\infty} \int\limits_{-\infty}^{\infty} E^*(x,y) \left[x \frac{\partial E(x,y)}{\partial y} - y \frac{\partial E(x,y)}{\partial x} \right] dx dy. \qquad (6.69)$$

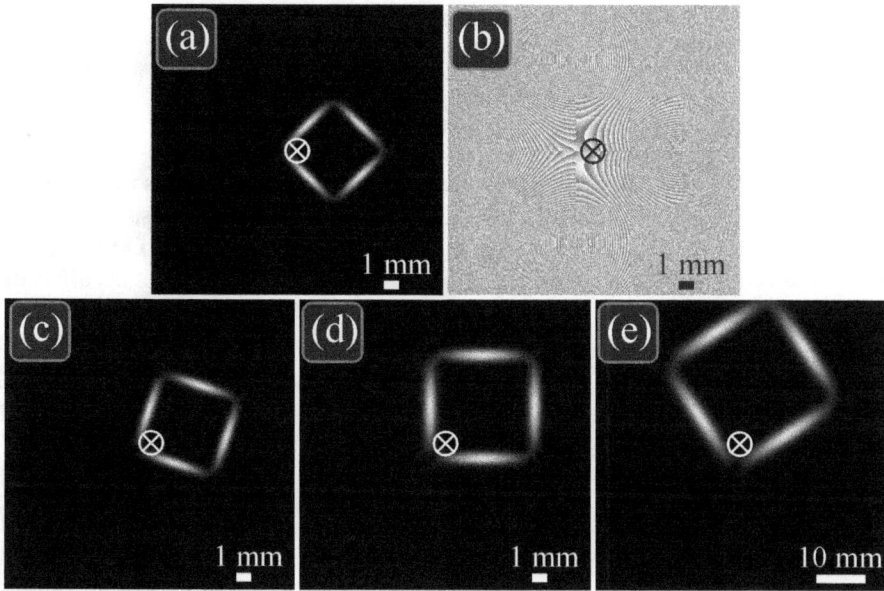

FIGURE 6.11 Intensity (a, c–e) and phase (b) distributions of the off-axis square propagation-invariant beam in the initial plane (a, b), in the near field $z = z_0/2$ (c), at the Rayleigh distance $z = z_0$ (d), and in the far field $z = 5z_0$ (e) for the following parameters: wavelength λ = 532 nm, waist radii of all four elliptic Gaussian beams σ_x = 2152 µm and σ_y = 358 µm (w_0 = 500 µm), coordinates of the beam centers $(x_{0p}, y_{0p}) = (r_0 + r_0 \cos \varphi_p, r0 \sin \varphi_p)$ ($r_0 = 2.5$ mm, $\varphi_p = \pi/4 + \pi p/2$, $p = 0, 1, 2, 3$), tilt angles of the major axes of the ellipses to the x-axis $\alpha_p = \varphi_p + \pi/2$, computation domain $|x|, |y| \leq R$ with $R = 10$ mm (a–d) and $R = 30$ mm (e). The cross in the center shows the optical axis, around which the diffraction pattern rotates on propagation. The initial field (a, b) was obtained by Equation (6.67), whereas distributions in the other planes (c–e) were obtained by the numerical Fresnel transform.

Substituting the complex amplitude from Equation (6.67) into these formulae, we get expressions for the power and OAM of the off-axis elliptic propagation-invariant Gaussian beam:

$$W = \frac{\pi w_0^2}{2\sqrt{1-\gamma^2}} \exp\left\{2\frac{r_0^2}{w_0^2}\left[1-\gamma \cos\left(2\varphi_0 - 2\alpha\right)\right]\right\}, \tag{6.70}$$

$$J_z = 2\frac{r_0^2}{w_0^2}\left(1-\gamma \cos 2\beta\right)W$$

$$+ \frac{\pi w_0^2 \gamma}{2\sqrt{1-\gamma^2}}\left(\frac{\gamma}{1-\gamma^2} + 2\frac{r_0^2}{w_0^2}\cos 2\beta\right)\exp\left\{2\frac{r_0^2}{w_0^2}\left[\left(1+\gamma\right)\sin^2\beta + \left(1-\gamma\right)\cos^2\beta\right]\right\}. \tag{6.71}$$

Dividing the OAM by power, we get the normalized OAM:

$$\frac{J_z}{W} = 2\frac{r_0^2}{w_0^2} + \frac{\gamma^2}{1-\gamma^2} = 2\frac{r_0^2}{w_0^2} + \frac{1}{4}\left(\frac{\sigma_y}{\sigma_x} - \frac{\sigma_x}{\sigma_y}\right)^2. \tag{6.72}$$

Equation (6.72) shows that shifting the beam from the optical axis leads to a parabolic growth of the normalized OAM with the shift distance r_0, which is consistent with the Steiner theorem in mechanics. It is also seen from Equation (6.72) that the OAM depends on the beam ellipticity, but the tilt of the ellipse to the coordinate axes does not affect the normalized OAM. At $r_0 = 0$, the obtained expression coincides with that given in [190].

Numerical computation confirms the expression from Equation (6.72). For instance, the theoretical normalized OAM value of the beam from Figure 6.8(a–c) is 35.328. Numerical computation by Equations (6.68) and (6.69) yields the values 35.022 in the initial plane (Figure 6.8(a)) and 34.881 at the Rayleigh distance (Figure 6.8(c)). For a narrower beam from Figure 6.8(d–f), the theoretical OAM is 491. By numerical computation (but over a wider domain, |x|, |y| $\leq R$ with $R = 30$ mm, and with finer sampling, 8192 × 8192 pixels), we obtained the values 473 in the initial plane (Figure 6.8(d)) and 471 at the Rayleigh distance (Figure 6.8(f)). Thus, the error between the numerical and theoretical normalized OAM is 4%.

Constructing a continuous superposition of the elementary spiral beams [189] in the transverse plane and solving a system of five nonlinear equations to obtain the weight coefficients of this superposition, we obtained an analytical expression describing monochromatic paraxial propagation-invariant elliptic Gaussian beams with a transverse shift from the optical axis (Equation (6.67)). The theory is also valid for partially coherent beams, but for distributions of the cross-spectral density function, instead of intensity. Qualitatively, the patterns are the same, but distorted near the edges [320]. On free-space propagation, such a beam is rotated, but around the optical axis, rather than around its center. It turns out, that both the shift of such an elliptic beam and its orientation (tilt angle) in the transverse plane can be arbitrary, but the waist radii are related to each other: the –2th mean power [318] of these two waist radii should be equal to the waist radius of a circular Gaussian beam propagating with the same phase velocity (Gouy phase).

We also derived a formula for the orbital angular momentum of such beams (Equation (6.72)). Similarly, to the Steiner theorem in mechanics, it is a sum of two terms. One of them describes the intrinsic OAM relative to the "mass center" (center of the ellipse) and increases with the beam ellipticity. The second term is proportional to the squared distance from the ellipse center to the optical axis. It turns out that the ellipse orientation (tilt angle) in the transverse plane does not affect the normalized orbital angular momentum.

The studied beam is of finite energy and is thus realizable by using a spatial light modulator and, possibly, some encoding technique [321,322]. The limitation is that the elliptic spot should fit within the modulator area in order to avoid the edge-diffraction effects [323].

7 Topological Charge of Polarization Singularities

7.1 TIGHTLY FOCUSING VECTOR BEAMS CONTAINING V-POINT POLARIZATION SINGULARITIES

In recent years, high-order vector light fields, whose linear polarization vector varies across the beam cross-section, have been the focus of research [324,325,326,327,328,329]. Such beams can be produced with a variety of techniques, including components with optical metasurfaces [330]. The vector beams feature a robust intensity profile on propagation through turbulence [331] and polarization singularity points [332,333,334] that, in many respects, are similar to phase singularity points of vortex fields [19]. Polarization singularity points (V-points) are intensity nulls in a vector field where the linear polarization vector is indefinite. The V-points are characterized [333] by a Poincaré–Hopf index denoted by η, which equals the number of integer phase steps by 2π when making a full circle around the V-point. The phase is understood as the argument of a complex field composed of transverse E-field components, $E_x + iE_y$. This definition is similar to a relationship utilized in [19] to calculate the topological charge (TC) of a scalar vortex field with complex amplitude $E(x,y)$. V-points can also be characterized using a Stokes index σ, which is defined through the Poincare-Hopf index η as $\sigma = 2\eta$ and also equals the number of integer phase steps by 2π of a complex Stokes field when making a full circle around the V-point. With the unit Stokes vector $\mathbf{S} = (S_1, S_2, S_3)$ [335] having three components, the complex Stokes field is composed of the first two components: $S_c = S_1 + iS_2$. The phase of the complex Stokes field is the argument of a complex number S_c.

In this section, we derive the Poincaré–Hopf and Stokes indices η and σ for nth-order cylindrical vector beams. We show that in the source plane of the beams (where the on-axis field component is zero), fields of linear polarization vectors are formed centered at the V-points, which look like a "flower" or a "web", with the number of petals depending on the vector field order n. Using Richards–Wolf formulae, we derive expressions for E-vector components at the tight focus for three types of vector fields, namely, for nth-order radial polarization (n is positive), $-n$th-order radial polarization ($-n$ is negative), and nth-order azimuthal polarization. Relying on the expressions derived for the complex E-field amplitudes, we deduce expressions for transverse intensity profiles of the fields of interest. Based on the expressions derived, we obtain a major finding of this work, showing that the number of petals of the "polarization flower" of the initial vector field equals the number of local intensity maxima at the focal plane. We also show that a V-point of an nth-order vector field is "disintegrated" at the tight focus into several 1st-order points with no petals around them.

DOI: 10.1201/9781003326304-7

7.1.1 VECTOR FIELD POLARIZATION INDEX IN THE SOURCE PLANE

Let us analyze an nth-order azimuthally polarized source field whose Jones vector takes the form [336,337]:

$$E_n(\varphi) = \begin{pmatrix} -\sin n\varphi \\ \cos n\varphi \end{pmatrix}, \tag{7.1}$$

where (r, φ) are the polar coordinates at the source plane. At the field center (at $r = 0$), there is a singular V-point, where the linear polarization vector is indefinite. According to [333], field from Equation (7.1) can be characterized by a singularity index similar to the TC of scalar optical vortices. V-points are described using a Poincaré–Hopf index η, which can be calculated for the field from Equation (7.1) similar to the TC of a complex field

$$E_{A,n}(\varphi) = E_x + iE_y = -\sin n\varphi + i\cos n\varphi = i\exp(in\varphi). \tag{7.2}$$

The index of field from Equation (7.1) and a V-point equal TC of field from Equation (7.2): $\eta = n$. On the other hand, vector field from Equation (7.1) can be characterized using Stokes parameters $\mathbf{S} = (S_1, S_2, S_3)$ [335], where:

$$S_1 = \frac{|E_x|^2 - |E_y|^2}{|E_x|^2 + |E_y|^2}, \; S_2 = \frac{2\,\mathrm{Re}\left(E_x^* E_y\right)}{|E_x|^2 + |E_y|^2}, \; S_3 = \frac{2\,\mathrm{Im}\left(E_x^* E_y\right)}{|E_x|^2 + |E_y|^2}, \tag{7.3}$$

with Re and Im stand for the real and imaginary parts of a number. From Equation (7.3), the Stokes vector is seen to be of unit length: $S_1^2 + S_2^2 + S_3^2 = 1$. For field from Equation (7.1), the Stokes parameters from Equation (7.3) are given by:

$$S_1 = -\cos(2n\varphi), \; S_2 = -\sin(2n\varphi), \; S_3 = 0. \tag{7.4}$$

Since $S_3 = 0$ in Equation (7.4), we can infer that the field from Equation (7.1) is linearly polarized at any point, except the V-point, where polarization is indefinite. The complex Stokes field for the vector in Equation (7.4) takes the form:

$$S_A = S_1 + iS_2 = -\cos(2n\varphi) - i\sin(2n\varphi) = -\exp(i2n\varphi). \tag{7.5}$$

The Stokes index for the field from Equation (7.1) equals TC of the field from Equation (7.5): $\sigma = 2\eta = 2n$. Thus, the Stokes index is twice as large as the Poincaré–Hopf index.

For a radially polarized nth-order field with the Jones vector:

$$E_{1,n}(\varphi) = \begin{pmatrix} \cos n\varphi \\ \sin n\varphi \end{pmatrix}, \tag{7.6}$$

the Poincaré–Hopf index of the central V-point ($r=0$) also equals $\eta=n$. The V-point singularity index has the opposite sign ($\eta=-n$) for a vector field:

$$E_{2,n}(\varphi) = \begin{pmatrix} \cos n\varphi \\ -\sin n\varphi \end{pmatrix}. \tag{7.7}$$

7.1.2   NUMBER OF LOCAL INTENSITY MAXIMA AT THE FOCUS OF A VECTOR FIELD

Interestingly, vector field from Equation (7.6) produces a "flower"-shaped pattern of linear polarization vectors composed of $2(n-1)$ petals. Actually, a petal is inscribed between the vector found at an angle $\varphi=0$ and the vector rotated by an angle $\varphi=\pi+\varphi_0$. From the first to the second angle, the phase of the field from Equation (7.6) changes by $n\varphi_0$ rad. Equating $\pi+\varphi_0=n\varphi_0$, we find the angle for a single petal to be $\varphi_0 = \pi/(n-1)$. In total, there are N petals: $2\pi = N\varphi_0$. Hence, we find that $N=2(n-1)$. A similar reasoning suggests that a polarization "web" composed of linear polarization vectors around the V-point of field (7.7) has $N = 2(n+1)$ cells.

Next, we demonstrate that a "flower" of linear polarization vectors composed of $2(n-1)$ petals formed by the field from Equation (7.6) in the source plane is transformed at the tight focus into a "flower"-shaped intensity pattern with $2(n-1)$ local maxima.

Actually, using Richards–Wolf formulae [338], which describe the electromagnetic field components in the tight focus neighborhood, the E-field components can be derived in the form:

$$E_x = -i^{n+1}\left(I_{0,n}\cos n\varphi + I_{2,n-2}\cos(n-2)\varphi\right),$$

$$E_y = -i^{n+1}\left(I_{0,n}\sin n\varphi - I_{2,n-2}\sin(n-2)\varphi\right), \tag{7.8}$$

$$E_z = 2i^n I_{1,n-1}\sin(n-1)\varphi,$$

where:

$$I_{v,\mu} = \left(\frac{\pi f}{\lambda}\right)\int_0^{\theta_0} \sin^{v+1}\left(\frac{\theta}{2}\right)\cos^{3-v}\left(\frac{\theta}{2}\right)\cos^{1/2}(\theta)A(\theta)e^{ikz\cos\theta}J_\mu(x)d\theta, \tag{7.9}$$

where λ is the wavelength of light, f is the focal length of an aplanatic optical system, $x = kr\sin\theta$, $J_\mu(x)$ is the first-kind Bessel function, and $NA = \sin\theta_0$ is the numerical aperture. The initial amplitude function $A(\theta)$ (herein assumed to be real) may be either constant (a plane wave) or in the form of a Gaussian beam:

$$A(\theta) = \exp\left(\frac{-\gamma^2 \sin^2\theta}{\sin^2\theta_0}\right), \tag{7.10}$$

where γ is constant. The transverse intensity (without regard for the longitudinal component of the field from Equation (7.8)) is given by:

$$I_t = |E_x|^2 + |E_y|^2 = I_{0,n}^2 + I_{2,n-2}^2 + 2I_{0,n}I_{2,n-2}\cos\left(2(n-1)\varphi\right). \qquad (7.11)$$

From Equation (7.11), the transverse intensity profile is seen to have $2(n-1)$ local intensity maxima centered on the optical axis, each being located on a ray $\varphi = 2\pi p/(2n-2), p = 1, 2, 3, \ldots, 2(n-1)$. Now we will determine an index of the V-point at the focus of the vector field from Equation (7.8). For this purpose, an equivalent complex field and its amplitude can be expressed as:

$$E_{c,n} = \left(I_{0,n}\cos n\varphi + I_{2,n-2}\cos(n-2)\varphi\right)$$

$$+ i\left(I_{0,n}\sin n\varphi - I_{2,n-2}\sin(n-2)\varphi\right) \qquad (7.12)$$

$$= I_{0,n}\exp\left(in\varphi\right) + I_{2,n-2}\exp\left(-i(n-2)\varphi\right).$$

In the general case, the index of the field from Equation (7.8) is undefined, because while at certain radii r coefficients in one exponential function can be larger than those in another one, the situation may be opposite at other radii. In the complex field of Equation (7.12), TC depends on the asymptotic properties of integrals in Equation (7.9). For instance, putting $A(\theta) = \delta(\theta - \theta_0)$, the integrals in Equation (7.9) are replaced by Bessel functions, so that Equation (7.12) is rearranged to:

$$E_{c,n} = AJ_n(\alpha r)\exp\left(in\varphi\right) + BJ_{n-2}(\alpha r)\exp\left(-i(n-2)\varphi\right), \qquad (7.13)$$

with $\alpha = kr\sin\theta_0$ and:

$$A = \left(\frac{\pi f}{\lambda}\right)\sin\left(\frac{\theta_0}{2}\right)\cos^3\left(\frac{\theta_0}{2}\right)\cos^{1/2}\theta_0,$$

$$B = \left(\frac{\pi f}{\lambda}\right)\sin^3\left(\frac{\theta_0}{2}\right)\cos\left(\frac{\theta_0}{2}\right)\cos^{1/2}\theta_0.$$

While from Equation (7.13), the index is still seen to be undefined, near the optical axis the amplitude of a lower-order Bessel function is larger than that of a higher-order Bessel function, which means that, similar to the TC of a superposition of two optical vortices [55], the near-axis index equals $\eta = -(n-2)$. In a particular case of $n = 1$ (conventional radial polarization) Equation (7.12) suggests that:

$$E_{c,1} = (I_{0,1} - I_{2,1})\exp\left(i\varphi\right). \qquad (7.14)$$

In this case, the V-point index is unit ($\eta = 1$) and, considering that $n = 1$, the source field index remains the same at the focus. This clearly follows from the fact that a singular point with a unit index is unable to disintegrate into a number of V-points with smaller indices. In a similar way, a scalar optical vortex with TC = 1 remains robust following stochastic amplitude and phase distortions.

For an nth-order azimuthally polarized vector source field of Equation (7.1), $2(n-1)$ local intensity maxima will also occur at the focus, though being located on other rays. Hence, a focal "flower" composed of local intensity maxima will be rotated by an angle of $\pi/(2n-2)$. Using the angle magnitude, it becomes possible to distinguish nth-order radial polarization from nth-order azimuthal one. Meanwhile, the number of "flower petals" enables a cylindrical polarization order to be determined. Actually, for a source field from Equation (7.1), E-vector components in the focal plane take a form similar to Equation (7.8):

$$E_x = i^{n+1}\left(I_{0,n}\sin n\varphi + I_{2,n-2}\sin(n-2)\varphi\right),$$

$$E_y = i^{n+1}\left(-I_{0,n}\cos n\varphi + I_{2,n-2}\cos(n-2)\varphi\right), \qquad (7.15)$$

$$E_z = -2i^n I_{1,n-1}\sin(n-1)\varphi.$$

For the source field from Equation (7.1), the transverse intensity distribution in the focus is:

$$I_t = |E_x|^2 + |E_y|^2 = I_{0,n}^2 + I_{2,n-2}^2 - 2I_{0,n}I_{2,n-2}\cos\left(2(n-1)\varphi\right). \qquad (7.16)$$

From Equation (7.16), $2(n-1)$ local maxima are seen to reside on a circle centered at the optical axis and on the rays outgoing from the center at angles $\varphi = (\pi+2\pi p)/(2n-2)$, $p = 0, 1, 2, \ldots, 2(n-2)$. To find indices of V-points at the focal spot of the vector field from Equation (7.1), we can express an equivalent complex field with the amplitude:

$$E_{c,n} = \left(I_{0,n}\sin n\varphi + I_{2,n-2}\sin(n-2)\varphi\right)$$

$$+ i\left(-I_{0,n}\cos n\varphi + I_{2,n-2}\cos(n-2)\varphi\right) \qquad (7.17)$$

$$= -iI_{0,n}\exp\left(in\varphi\right) + iI_{2,n-2}\exp\left(-i(n-2)\varphi\right).$$

In the general case, the index of the field from Equation (7.17) is undefined, because while at certain radii r coefficients in one exponential function can be larger than those in another one, the situation may be opposite at other radii. However, at $n=1$ (ordinary azimuthal polarization), from Equation (7.17) it follows that:

$$E_{c,1} = -i(I_{2,1}+I_{0,1})\exp\left(i\varphi\right). \qquad (7.18)$$

In this case, the V-point index is unit ($\eta=1$), meaning that the index of the initial field (7.1) remains unchanged at the focus.

A vector "web" of source field from Equation (7.7) with $2(n+1)$ cells, centered on the V-point polarization singularity is transformed at the focus into an intensity pattern with $2(n+1)$ local maxima. Actually, for the source field in Equation (7.7), projections of the E-vector are given by ($n > 0$):

$$E_x = i^{n-1}\left(I_{0,n}\sin n\varphi + I_{2,n+2}\sin(n+2)\varphi\right),$$

$$E_y = i^{n-1}\left(I_{0,n}\cos n\varphi - I_{2,n+2}\cos(n+2)\varphi\right), \qquad (7.19)$$

$$E_z = -2i^n I_{1,n+1}\sin(n+1)\varphi.$$

For the field from Equation (7.19), the transverse intensity distribution at the focus is given by:

$$I_t = |E_x|^2 + |E_y|^2 = I_{0,n}^2 + I_{2,n+2}^2 - 2I_{0,n}I_{2,n+2}\cos\left(2(n+1)\varphi\right). \qquad (7.20)$$

From Equation (7.20), the intensity distribution is seen to have $2(n+1)$ local intensity maxima at the focus on an axis-centered circle of a certain radius. Hence, the vector "web" in the source field of Equation (7.7) can be identified based on the number of petals of an nth-order vector "flower".

Putting $n = -1$ in Equation (7.7) for the source field, we may infer from Equation (7.14) that the V-point index changes sign at the focus, because based on Equation (7.19) for the E-vectors at the focus, we find that:

$$E_{c,1} = -i(I_{2,1} + I_{0,1})\exp\left(i\varphi\right). \qquad (7.21)$$

Aiming to determine the V-point index at the focus of the vector field from Equation (7.7) and using Equation (7.19), we form an equivalent complex field with the amplitude:

$$E_{c,n} = \left(I_{0,n}\sin n\varphi + I_{2,n+2}\sin(n+2)\varphi\right)$$

$$+ i\left(I_{0,n}\cos n\varphi - I_{2,n+2}\cos(n+2)\varphi\right) \qquad (7.22)$$

$$= iI_{0,n}\exp\left(-in\varphi\right) - iI_{2,n+2}\exp\left(i(n+2)\varphi\right).$$

Just like in Equation (7.17), the index of field (7.19) is undefined, but like in Equation (7.13), it can be asserted that at the focus the near-axis V-point index is equal to a lesser number of the Bessel function, i.e. $\eta = -n$. That is, given the source field of Equation (7.7), the near-axis V-point index at the focus is the same as in the source plane.

7.1.3 POLARIZATION SINGULARITY INDEX FOR A GENERALIZED VECTOR FIELD

Obviously, the above reasoning cannot be automatically applied to a generalized vector field as it has different orders on the different axes. For such a field, the Jones vector is [333]:

$$E_{2,n}(\varphi) = \begin{pmatrix} \cos n\varphi \\ \sin m\varphi \end{pmatrix}. \qquad (7.23)$$

Although the field from Equation (7.23) may also be said to have a central V-point, its index can be defined analytically only in some cases. Actually, the complex field equivalent to the field from Equation (7.23) is given by:

$$E_{A,n}(\varphi) = E_x + iE_y = \cos n\varphi + i \sin m\varphi. \tag{7.24}$$

In the topic-related work [333], it was not specified in which way the index of such a field could be determined if $n \neq m$. In this work, we propose that the V-point index of the vector field from Equation (7.23) should be calculated in a similar way to calculating the TC of scalar optical vortices using Berry's formula [19]:

$$TC = \frac{1}{2\pi} \lim_{r \to \infty} \mathrm{Im} \int_0^{2\pi} d\varphi \frac{\partial E(r,\varphi)/\partial \varphi}{E(r,\varphi)}. \tag{7.25}$$

Then, according to Equation (7.25), the Poincaré–Hopf index for vector field from Equation (7.24) is given by:

$$\eta = \frac{1}{2\pi} \lim_{r \to \infty} \mathrm{Im} \int_0^{2\pi} d\varphi \frac{-n \sin n\varphi + im \cos m\varphi}{\cos n\varphi + i \sin m\varphi} =$$

$$\frac{1}{2\pi} \int_0^{2\pi} d\varphi \frac{n \sin n\varphi \sin m\varphi + m \cos m\varphi \cos n\varphi}{\cos^2 n\varphi + \sin^2 m\varphi}. \tag{7.26}$$

From Equation (7.26), it follows that at $m = n$, $\eta = n$, whereas at $m = -n$, $\eta = -n$. However, at $n \neq \pm m$, the integral in Equation (7.26) is not reduced to reference integrals. In separate cases, Equation (7.26) can be calculated analytically, but in other cases, it needs to be calculated numerically.

Calculating polarization singularity index of a generalized vector field

Below, we deduce some properties of polarization singularity index from Equation (7.26), including properties of parity, symmetry, reciprocity, and multiplicity. The parity property is expressed in the fact that for different-parity m and n (i.e. $m + n$ is odd), polarization singularity index from Equation (7.26) equals zero. Actually, the first integral in from Equation (7.26) can be broken down into two (with the range of integration in the second integral shifted from $(\pi, 2\pi)$ to $(0, \pi)$):

$$\eta_{n,m} = \frac{1}{2\pi} \mathrm{Im} \left\{ \int_0^{\pi} \frac{-n \sin n\varphi + im \cos m\varphi}{\cos n\varphi + i \sin m\varphi} d\varphi \right.$$

$$\left. + \int_0^{\pi} \frac{-n(-1)^n \sin n\varphi + im(-1)^m \cos m\varphi}{(-1)^n \cos n\varphi + i(-1)^m \sin m\varphi} d\varphi \right\}. \tag{7.27}$$

Multiplying the numerator and denominator of the first integral by $(-1)^n$ and taking into account that $(-1)^{m+n} = -1$, we get a sum of two complex conjugated numbers whose imaginary part equals zero. Thus, we obtain a symmetry property of the Poincaré–Hopf index for vector field from Equation (7.23). In this way, it stands to reason that when n changes sign, the first integral in Equation (7.26) does not change:

$$\eta_{-n,m} = \frac{1}{2\pi} \operatorname{Im} \int_0^{2\pi} d\varphi \, \frac{-(-n)\sin(-n\varphi) + im\cos m\varphi}{\cos(-n\varphi) + i\sin m\varphi} = \eta_{n,m}. \tag{7.28}$$

On the contrary, when m changes sign, the integrand becomes complex conjugated and, hence, the imaginary part changes sign:

$$\eta_{n,-m} = -\eta_{n,m}. \tag{7.29}$$

Shifting the range of integration by $\pi/2$, we get the following relationships between the indices:

$$\eta_{n,m} = \begin{cases} \dfrac{1}{2\pi} \operatorname{Im} \displaystyle\int_0^{2\pi} d\varphi \, \dfrac{-n(-1)^{\frac{n}{2}} \sin n\varphi + im(-1)^{\frac{m}{2}} \cos m\varphi}{(-1)^{\frac{n}{2}} \cos n\varphi + i(-1)^{\frac{m}{2}} \sin m\varphi}, \\[6pt] \qquad \text{if } n, m \text{ are even,} \\[12pt] \dfrac{1}{2\pi} \operatorname{Im} \displaystyle\int_0^{2\pi} d\varphi \, \dfrac{-n(-1)^{\frac{n-1}{2}} \cos n\varphi + im(-1)^{\frac{m+1}{2}} \sin m\varphi}{(-1)^{\frac{n+1}{2}} \sin n\varphi + i(-1)^{\frac{m-1}{2}} \cos m\varphi}, \\[6pt] \qquad \text{if } n, m \text{ are odd,} \end{cases} \tag{7.30}$$

$$= \begin{cases} (-1)^{(m-n)/2} \eta_{n,m}, \text{ if } n, m \text{ are even,} \\[6pt] (-1)^{(m-n)/2} \eta_{m,n}, \text{ if } n, m \text{ are odd.} \end{cases}$$

This can be termed as a reciprocity property because it enables the indices to be swapped if they are odd.

From Equation (7.30), it also follows that if n and m are even, but $(m-n)/2$ is odd, then $\eta = 0$.

If the orders m and n have a common divisor, i.e. $m = p\mu$ and $n = p\nu$, then, performing a change of variables $\varphi = \theta/p$ in Equation (7.26), we obtain a multiplicity property:

$$\eta_{p\nu,p\mu} = p\frac{1}{2\pi} \operatorname{Im} \int_0^{2\pi p} \frac{-\nu \sin \nu\theta + i\mu \cos \mu\theta}{\cos \nu\theta + i\sin \mu\theta} \frac{d\theta}{p} = p\eta_{\nu,\mu}. \tag{7.31}$$

For instance, at $m = 2n$, the polarization singularity index equals zero thanks to the multiplicity and parity properties: $\eta_{n,2n} = n\eta_{1,2} = 0$.

In a simple case, we determine the index $\eta_{1,2p+1}$ analytically through the use of residues. If we denote $\zeta = \cos\varphi + i(2p+1)\sin\varphi$, the integral in Equation (7.26) can be written as (at $n = 1$)

$$\eta = \frac{1}{2\pi} \operatorname{Im} \oint_{\Gamma} \frac{d\zeta}{\zeta}, \tag{7.32}$$

where Γ is the oriented closed contour in the complex plane drawn by the variable ζ, when $0 \leq \varphi \leq 2\pi$. Figure 7.1 illustrates this contour for $p = 0, 1, 2$.

If $p = 0$, this contour is a simple unit-radius circle. Otherwise, Γ has self-intersections and the integral over Γ can be replaced by a sum of the integrals over several simple contours without the self-intersections [dashed contours in Figure 7.1(b, c)]. The only pole of the integrand in Equation (7.32) is $\zeta = 0$. If $p = 0$, this pole is within the unit-radius circle and, according to the residues theorem, applied to the integral in Equation (7.32), $\eta_{11} = 1$. For $p > 0$, only one simple contour contains the pole (Figure 7.1(b, c)). Thus, integration over the other simple contours yields 0. If $p = 1$ (and at other odd p), the pole is bypassed clockwise and, therefore, the integration yields $\eta_{13} = -1$. Similarly, if $p = 2$ (and at other even p), the pole is bypassed counterclockwise and, therefore, the integration yields $\eta_{15} = +1$. Thus, we can write a general rule for the Poincaré–Hopf index η_{nm} at $n = 1$ and odd m:

$$\eta_{1m} = (-1)^{(m-1)/2}, \tag{7.33}$$

or, using the reciprocity property:

$$\eta_{n,1} = 1. \tag{7.34}$$

All the properties of index from Equation (7.26) for field from Equation (7.23) derived herein can be verified using the Table 7.1.

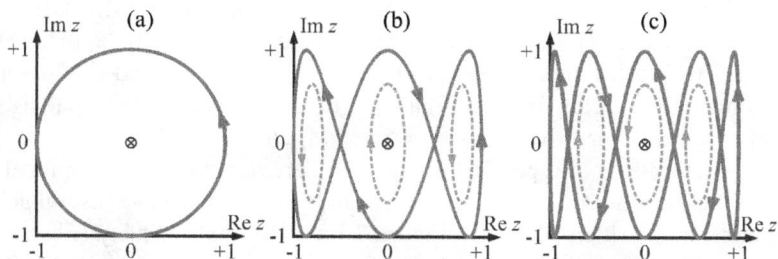

FIGURE 7.1 Calculation of the Poincaré–Hopf index. Contours Γ (solid curves) in the complex plane defined as $\zeta = \cos\varphi + i(2p+1)\sin\varphi$ $(0 \leq \varphi \leq 2\pi)$ for $p = 0$ (a), $p = 1$ (b), $p = 2$ (c). Dashed ellipses show the simple contours without self-intersections, into which the contour Γ can be split. The cross in the center denotes $\zeta = 0$, the only pole of the integrand in Equation (7.32).

TABLE 7.1

Poincaré–Hopf index η of a vector field from Equation (7.23): n shown on the horizontal lines and m – on the vertical

m	n										
	0	1	2	3	4	5	6	7	8	9	10
0	0	0	0	0	0	0	0	0	0	0	0
1	0	1	0	1	0	1	0	1	0	1	0
2	0	0	2	0	0	0	2	0	0	0	2
3	0	−1	0	3	0	−1	0	−1	0	3	0
4	0	0	0	0	4	0	0	0	0	0	0
5	0	1	0	1	0	5	0	1	0	1	0
6	0	0	−2	0	0	0	6	0	0	0	−2
7	0	−1	0	−1	0	−1	0	7	0	−1	0
8	0	0	0	0	0	0	0	0	8	0	0
9	0	1	0	−3	0	1	0	1	0	9	0
10	0	0	2	0	0	0	2	0	0	0	10

Table 7.1 gives values of η, which were calculated using Equation (7.26) for vector field from Equation (7.23), with the orders m and n being varied from 0 to +10 (for negative m and n, symmetry rules can be used, as is derived in Appendix A: $\eta_{-n,m} = \eta_{n,m}$ and $\eta_{n,-m} = -\eta_{n,m}$). From Table 7.1, the polarization singularity index can be an only integer. It is also interesting that at $n = 1$, -1, and any m, the η index is equal to either 1, or 0, or −1. Also, at $n = 8$, -8, and any m, the η index equals either 8, or 0, or −8. The same holds for $n = 4$ and $n = 2$.

7.1.4 NUMERICAL MODELING

Shown in Figure 7.2 are source vector fields with polarization singularity (V-point) at the center for the nth-order vector field (7.6): (a) 3, (b) 4, (c) −3, and (d) −4. In compliance with the theoretical predictions, the vector fields in Figure 7.2 (a, b) are shaped as "flowers" with the number of petals equal to (a) $2(n-1) = 4$ and (b) $2(n-1) = 6$. Whereas two other vector fields in Figure 7.2 (c, d) produce "lattice" patterns with the number of cells equal to (c) $2(n+1) = 8$ and (d) $2(n+1) = 10$.

Source vector fields of type (7.6) in Figure 7.2 are transformed at the focal plane into vector fields from Equations (7.8), (7.15), and (7.19), which have several points of polarization singularities. Shown in Figure 7.3 are the total intensity (Figure 7.3(a)) and the transverse intensity (Figure 7.3(b)) for a source vector field with the index $n = 3$ of Figure 7.2(a). The numerical modeling of focusing vector fields was conducted using Richards–Wolf formulae [338] for wavelength 532 nm and numerical aperture NA = 0.95.

In accordance with theoretical predictions (Equation (7.11)), there occur $2(n-1) = 4$ local maxima of the total and transverse intensities at the focus. Due to the

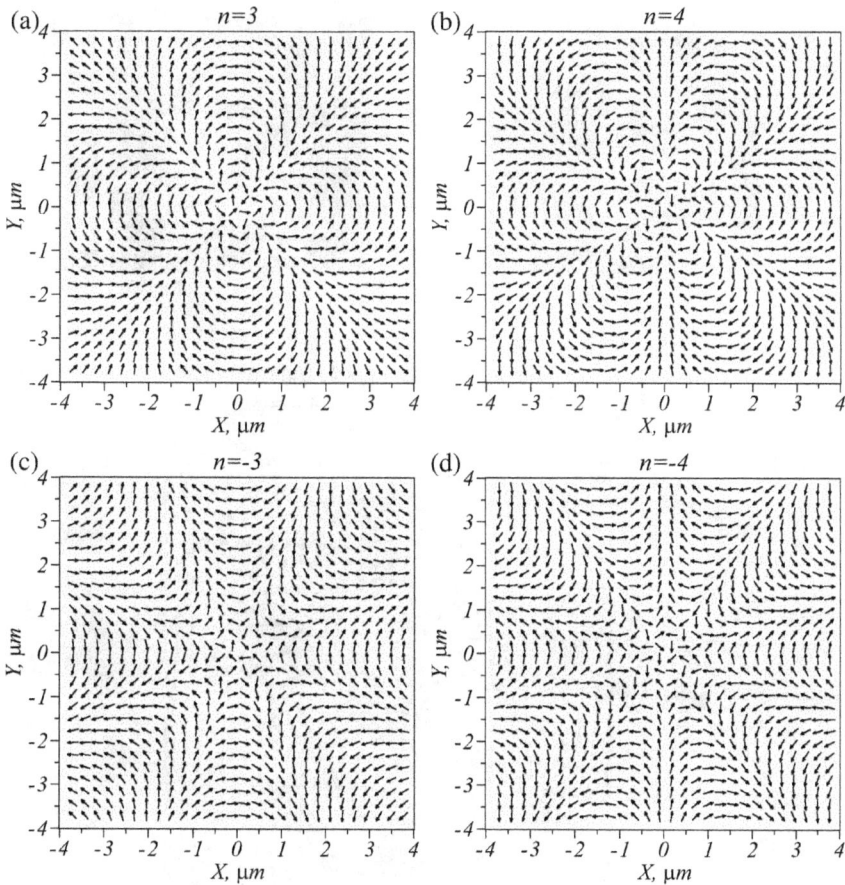

FIGURE 7.2   Vector field (7.6) (arrows mark linear polarization vectors at particular points), whose order n coincides with the index of the V-point polarization singularity (Poincare-Hopf index η) at the field center and equals:(a) 3, (b) 4,(c) –3, and (d) –4.

longitudinal intensity components, the coordinates of four local maxima in Figure 7.3(a) are different from those of the transverse intensity in Figure 7.3(b).

Shown in Figure 7.4 is a distribution of linear polarization vectors at the focus from the source vector field in Figure 7.2(a) ($n = 3$).

From Figure 7.4, four polarization singularity centers are seen to be located at the corners of the dark cross of Figure 7.3, with an on-axis V-point with the index $\eta = -1$ located at the center. The indices of the four V-points at the corners of the dark cross (Figure 7.3) are the same in magnitude but of different signs, with two vertical V-points having $\eta = +1$, and two horizontal V-points $\eta = -1$. Hence, the total near-axis index of the vector field of Figure 7.4 equals that of the central V-point, i.e. $\eta = -1$. This conclusion agrees well with Equations (7.12) and (7.13): $\eta = -(n-2) = -1$.

Figure 7.5 depicts numerically simulated patterns for the (a) total intensity and (b) transverse intensity from the source vector field with $n = 4$ (Figure 7.2(b)). From

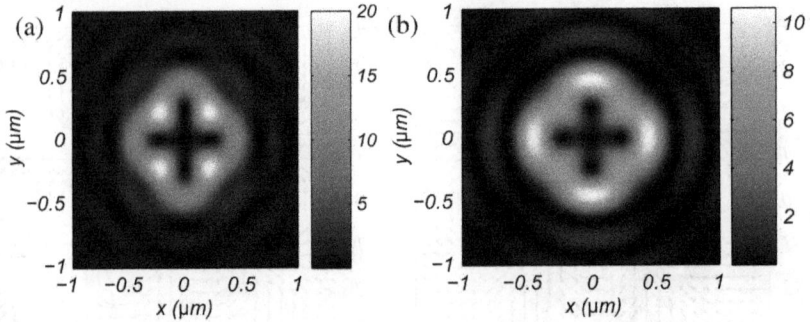

FIGURE 7.3 Patterns of the (a) total intensity $I_x + I_y + I_z$ and (b) transverse intensity components $I_x + I_y$ from the source vector field of Figure 7.2(a) at n = 3.

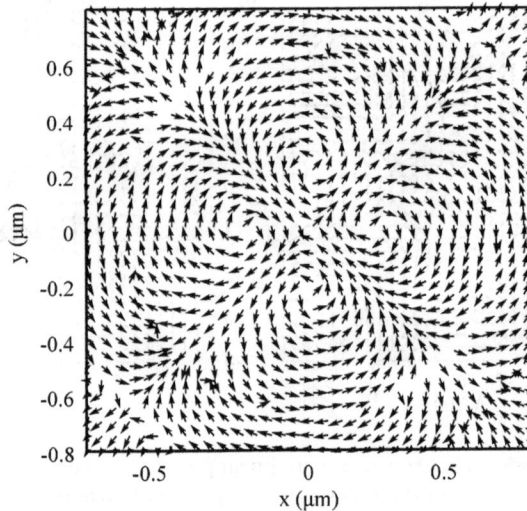

FIGURE 7.4 Pattern of linear polarization vectors at the focal plane from the source vector field in Figure 7.2(a) (n = 3).

Figure 7.5, the theoretical relation from Equation (7.11) is again seen to be corroborated, with $2(n-1) = 6$ local maxima in the intensity pattern found symmetrically to the optical axis being observed.

Figure 7.6 shows a pattern of linear polarization vectors at the focus from a source vector field with n = 4 (Figure 7.2(b)). From Figure 7.6, a set of V-points with indices $\eta = +1, -1$ are seen to form at the vertices of a "dark six-point star" of Figure 7.5. Equation (7.12) suggests that an on-axis V-point with $\eta = -2$ is found at the center.

Figure 7.7 depicts patterns for the total (Figure 7.7(a)) and transverse (Figure 7.7(b)) component of the intensity at the focal plane (NA = 0.95) from the source vector field with n = -3 of Figure 7.2(c). Figure 7.7 shows that in compliance with

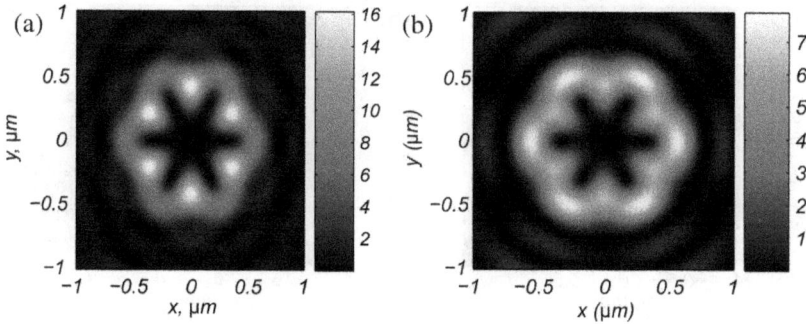

FIGURE 7.5 Patterns of the (a) total intensity $I_x + I_y + I_z$ and (b) transverse intensity $I_x + I_y$ component at the focal plane (NA = 0.95) from the source vector field with the index $n = 4$ (Figure 7.2(b)).

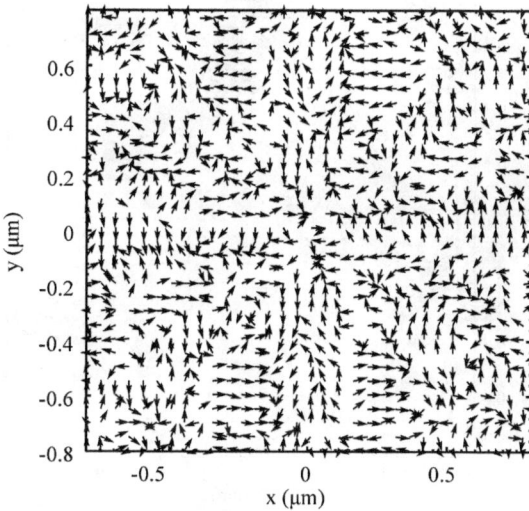

FIGURE 7.6 Pattern of linear polarization vectors for the source vector field of Figure 7.2(b) with the index $n = 4$.

theoretical predictions, there are $2(n + 1) = 8$ local intensity maxima in the intensity distribution.

Shown in Figure 7.8 is a pattern of linear polarization vectors at the focus from the source vector field of Figure 7.2(c) at $n = -3$. From Figure 7.8, eight V-points are seen to be located on a circle (at the vertices of a "dark eight-point star"), with four of them having the index $\eta = +1$ and four having the index $\eta = -1$. Equation (7.22) suggests that at the center of the focal spot there is a V-point with $\eta = 3$.

Figure 7.9 presents patterns of linear polarization vectors for the source field of Equation (7.23) at different values of (n, m): (a) (2,1), (b) (3,−7), (c) (9,−3), and (d) (6,2). Using Table 7.1, the Poincaré–Hopf indices η for the said vector fields can be found

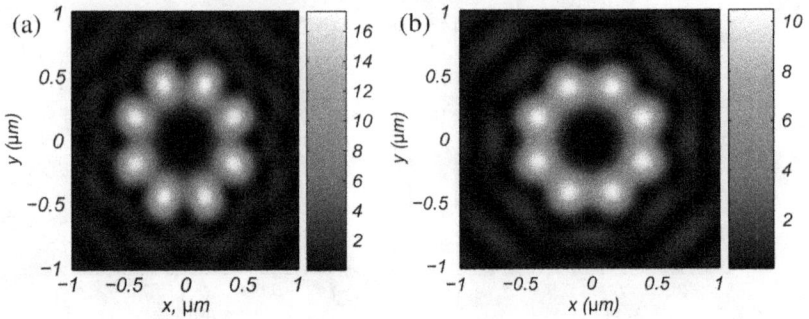

FIGURE 7.7 Patterns of the (a) total intensity $I_x + I_y + I_z$ and (b) transverse intensity $I_x + I_y$ component at the focal plane (NA = 0.95) for the source vector field with the index $n = -3$ of Figure 7.2(c).

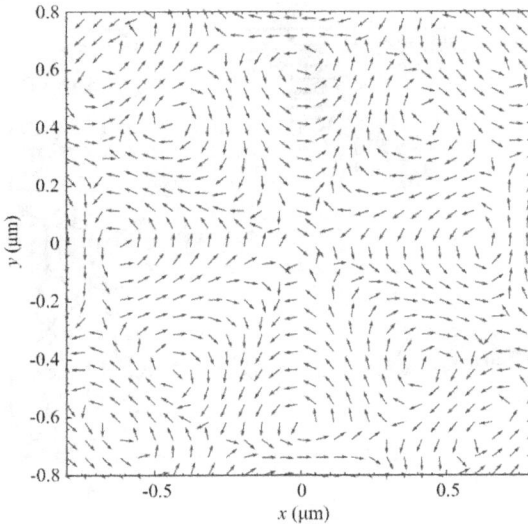

FIGURE 7.8 Pattern of linear polarization vectors at the focus from the source field with the index $n = -3$ of Figure 7.2(c).

to be (a) 0, (b) 1, (c) –3, and (d) 2. By looking at Figure 7.9, indices of the V-points of such complex vector fields would be difficult to determine. The pattern for linear polarization vectors at the focus would be even more complicated (not presented here). Shown in Figure 7.10 is an intensity pattern at the focus of an aplanatic objective with NA = 0.95 when focusing vector beams with $n = 2$, $m = 1$ (Figure 7.10(a)) and $n = 3$, $m = -7$ (Figure 7.10(b)).

Figure 7.10 suggests that a source field with $\eta = 0$ (Figure 7.9(a)) produces neither an intensity null nor a V-point at the center of the focal spot (Figure 7.10(a)), whereas a source vector field with $\eta = 1$ (Figure 7.10(b)) produces at the center an intensity null and a V-point.

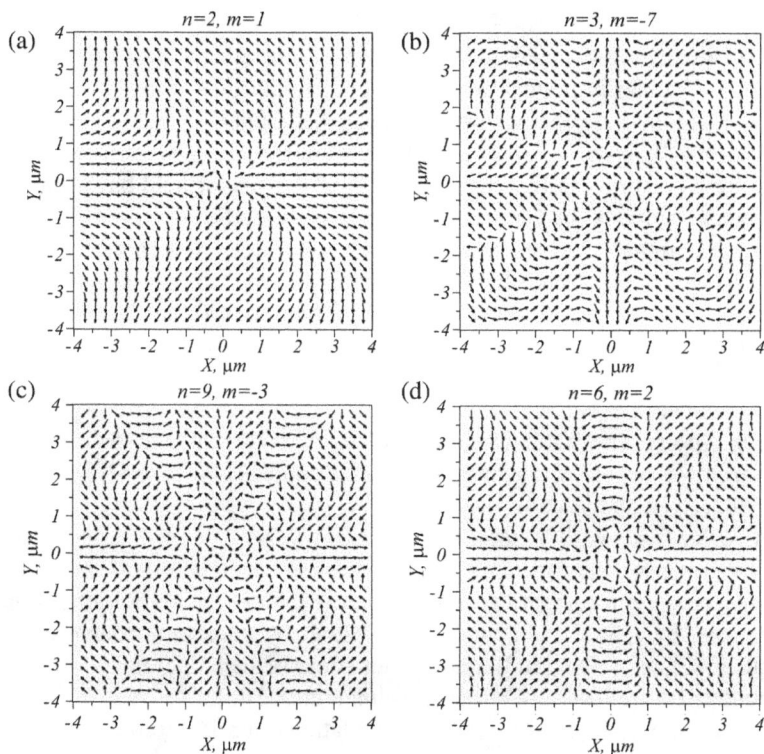

FIGURE 7.9 Source vector fields from Equation (7.23) at different n and m.

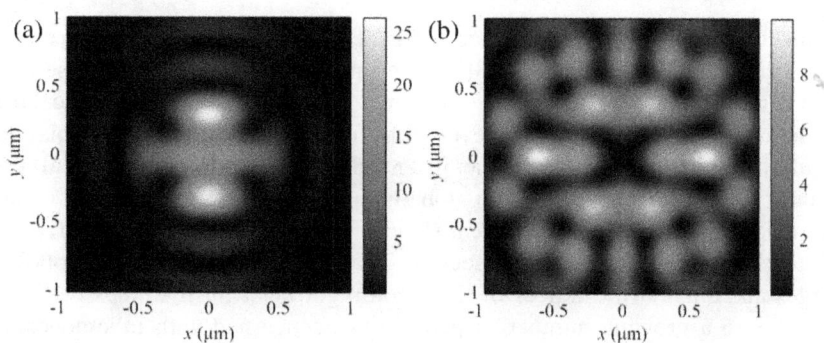

FIGURE 7.10 Intensity patterns at the focus of vector beams with (a) $n = 2$, $m = 1$ (a "butterfly") and (b) $n = 3$, $m = -7$ (a "dragon mouth").

Summing up, we have shown both theoretically and numerically that an nth-order source vector field has a central V-point with the Poincaré–Hopf index $\eta = n$ and the Stokes index $2n$ [339]. Such a vector field is "flower"-shaped with $2(n-1)$ petals. When tightly focused, this field produces at the focus an intensity pattern with

$2(n-1)$ local maxima located on a circle of a certain radius, centered on the optical axis. Near those intensity maxima, $2(n-1)$ local minima are found (intensity nulls, polarization singularity points), where V-point singularities with alternating indices $+1$ and -1 (the total index being zero) are located. An intensity null, or a V-point with the index $-(n-2)$, has also been shown to occur at the center of the focus.

It has also been shown that an $-n$th-order source vector field has at the center a V-point with the index $-n$. Such a vector field is in the form of a "web" with $2(n+1)$ cells. At the tight focus, this field produces an intensity pattern with $2(n+1)$ local maxima located on a circle of a certain radius centered on the optical axis. Near those intensity maxima, $2(n+1)$ local minima are found (intensity nulls, polarization singularity points), where V-points with alternating indices $+1$ and -1 (the total index being zero) are located. An intensity null, or a V-point with the index $-n$, has also been shown to occur at the center of the focus. For an (n, m)-order vector field, indices of V-points (see Table 7.1) have been numerically calculated for the numbers varying from -10 to $+10$. For a number of cases, indices of a generalized vector field have been derived analytically.

Such vector fields with V-point singularities can be generated experimentally by using either q-plates, i.e. especially transversely patterned liquid crystal cells inducing an integer or semi-integer topological charge [340,341], or, for higher Poincaré–Hopf indices, by spatial light modulators: either by one, with double modulation technique [342], or by two [343]. Application areas of such light fields with polarization singularities are laser information technologies [344], laser material processing [345], microscopy [346], and particle manipulation or optical trapping [347].

7.2   SHARP FOCUSING OF A HYBRID VECTOR BEAMS CONTAINING C-POINT POLARIZATION SINGULARITY

For the first time, vector singularities as a generalization of scalar singularities were proposed in 1983 by J. F. Nye [348], where lines of zero-valued transverse components of the E-field were called "disclinations" (to distinguish them from scalar edge and screw dislocations [21]). However, both in [348,21] and [333] the polarization singularities were studied locally, i.e. in a neighborhood of singular (critical) points. It would be of significant interest to globally investigate inhomogeneously polarized light fields characterized by different (linear, elliptical, or circular) polarization at different points of the beam cross-section. That is, we aim to determine topological charges and singularity indices of the whole light field. Such studies become relevant due to a growing number of publications concerned with inhomogeneously polarized vector fields [336]. Inhomogeneously polarized beams can be generated by interferometry [349], inside a cavity [350], as well as with q-plates [237], metasurfaces [351,352], polarization prisms [353], and spatial light modulators [354]. Points of intensity nulls at which the linear polarization vector is not defined are called V-points [333]. In a similar way, points of a light field with inhomogeneous elliptical polarization where the direction of the major axis of the polarization ellipse is undefined are called "C-points", with the light being circularly polarized at such points. If the C-points are arranged on a line, the line is called a C-line [333]. Polarization singularities are described by singularity indices, which are calculated similarly to the topological charges of scalar light fields [19]. The polarization singularity index of V-points is called a Poincaré–Hopf index [333] and is calculated using Stokes parameters [355,328,356]. Meanwhile, the C-points are described by an index equal to the number of turns by π the major axis of the polarization ellipse makes around the C-point. The index of a C-point can take a fractional (half-integer) value if on a complete turn the polarization ellipse makes an odd number of turns by π. When intersecting C-lines, the polarization ellipse axis makes a jump by $\pi/2$.

In this section, we look into a hybrid nth-order vector light field whose polarization varies from linear to elliptical, to circular depending on the polar angle. This field contains just C-lines with their number being equal to n. For this field, we find components of the Stokes vector and show the polarization index to be half-integer, $n/2$. Using a Richards–Wolf formalism [338], we derive analytical expressions for projections of the E-vector at the tight focus for a source hybrid nth-order vector field and analytical relations for the field intensity at the focus. We find that at an even number n, the intensity has nth-order symmetry and C-points at the focus. Thus, we numerically demonstrate that C-lines in the source field "disintegrate" into C-points at the focus, which are located on the same C-lines. We also derive analytical relationships for the projection of the Stokes vector at the focus, which suggest that for an odd number n, the field at the focus is purely vector, consists of vectors of linear polarization, has n V-points, and has no C-points.

7.2.1 SOURCE HYBRID VECTOR FIELD WITH POLARIZATION SINGULARITY POINTS

Let us analyze a new hybrid nth-order vector field defined in the original plane by two transverse projections of the E-vector and a Jones vector in the form:

$$E_n(\varphi) = \frac{1}{\sqrt{2}} \begin{pmatrix} \cos n\varphi \\ i\alpha + \sin n\varphi \end{pmatrix}, \tag{7.35}$$

where n is integer and $0 \le |\alpha| \le 1$. From Equation (7.35) it follows that at $n = 0$, the light field from Equation (7.35) is elliptically polarized, while at $|\alpha| = 1$ it is circularly polarized. At $\alpha = 0$, field from Equation (7.35) has inhomogeneous nth-order vector polarization.

Field (7.35) has points of linear, elliptical, and circular polarization. Points of circular polarization are called C-points of polarization singularity because the direction of the major axis of the ellipse polarization is undefined at such points [333]. The topology of the polarization ellipses around a C-point is described by an index Ic, which shows how many (integer) times the major axis of the polarization ellipse changes its direction by an angle of π while making a full circle around the C-point. To find the index Ic of field from Equation (7.35), let us find all projections of the Stokes vector [355] $\mathbf{S} = (S_1, S_2, S_3)$, where:

$$S_1 = \frac{|E_x|^2 - |E_y|^2}{|E_x|^2 + |E_y|^2}, \quad S_2 = \frac{2\operatorname{Re}\left(E_x^* E_y\right)}{|E_x|^2 + |E_y|^2}, \quad S_3 = \frac{2\operatorname{Im}\left(E_x^* E_y\right)}{|E_x|^2 + |E_y|^2}, \tag{7.36}$$

where Re and Im denote the real and imaginary parts of the number. From Equation (7.36), the Stokes vector is seen to be of unit length: $S_1^2 + S_2^2 + S_3^2 = 1$. For field from Equation (7.35), the Stokes vector components in Equation (7.36) take the form:

$$S_1 = 2\frac{\cos 2n\varphi - \alpha^2}{1+\alpha^2}, \quad S_2 = \frac{2\sin 2n\varphi}{1+\alpha^2}, \quad S_3 = \frac{2\alpha \cos n\varphi}{1+\alpha^2}. \tag{7.37}$$

From Equation (7.37) it follows that polarization of light is linear on the rays outgoing from the center at angles defined by the equation $S_3 = \cos n\varphi = 0$. At angles φ that satisfy the equation $S_3 = 1$ or $\cos n\varphi = \pm 1$ and $\alpha = +1, -1$ the light is circularly polarized. Elsewhere, the light is elliptically polarized. Thus, we can infer that field (7.35) has no isolated C-points but has C-lines, with the direction of the major axis of a polarization ellipse jumping by $\pi/2$ on crossing the line. A single C-point is equivalent to a screw dislocation and a C-line is equivalent to an edge dislocation. The number of C-lines in the source field from Equation (7.35) equals the field order n, with the lines found on $2n$ rays outgoing from the center at angles $\pi m/n$, $m = 0, 1, 2, \ldots, 2n-1$.

In [333], a local index of hybrid vector fields for polarization singularities (C-points) was calculated and the hybrid vector field itself was locally defined near the singularity. Hereinafter, we shall calculate the topological index of the whole hybrid vector field from Equation (7.35) in a global way, in a similar way to calculating the topological charge of the whole scalar complex vortex field using Berry's formula [19]. To these ends, let us form a complex Stokes field by the rule:

$$S_c = S_1 + iS_2. \tag{7.38}$$

For the source vector field from Equation (7.35), the complex Stokes field is given by:

$$S_c = 2\frac{\exp(2in\varphi) - \alpha^2}{1+\alpha^2}. \tag{7.39}$$

The Stokes index σ for the field from Equation (7.39) can be calculated using Berry's formula [19]:

$$\sigma = \frac{1}{2\pi} \text{Im} \int_0^{2\pi} d\varphi \frac{\partial S_c(\varphi)/\partial\varphi}{S_c(\varphi)}. \tag{7.40}$$

Substituting Stokes field in Equation (7.39) into Equation (7.40) yields:

$$\sigma = \frac{1}{2\pi} \text{Im} \int_0^{2\pi} d\varphi \frac{2in\exp(2in\varphi)}{\exp(2in\varphi) - \alpha^2} = \frac{n}{\pi} \int_0^{2\pi} d\varphi \frac{(1 - \alpha^2 \cos 2n\varphi)}{(1+\alpha^4) - 2\alpha^2 \cos 2n\varphi}. \tag{7.41}$$

Putting in Equation (7.41) $\alpha^2 = 1$, we find that $\sigma = n$ and the index of the C-points and the whole field from Equation (7.35) equals $Ic = \sigma/2 = n/2$. The index Ic can be half-integer owing to the tilt of the major axis of the polarization ellipse varying from 0 to π, rather than to 2π. Putting in Equation (7.41) $\alpha = 0$, Equation (7.35) will describe an inhomogeneous linearly polarized field ($S_3 = 0$), containing just V-points (where the linear polarization vector is undefined), where the Stokes index of Equation (7.41) equals $\sigma = 2n$, meanwhile the Poincaré–Hopf index [333] of field from Equation (7.35) is half as large: $\eta = n$. At $0 < |\alpha| < 1$, the Stokes index in Equation (7.41) can be calculated using a reference integral [275]:

$$\int_0^{2\pi} \frac{\cos mx}{a + b\cos x}\,dx = \frac{2\pi}{\sqrt{a^2 - b^2}}\left(\frac{\sqrt{a^2 - b^2} - a}{b}\right)^m. \tag{7.42}$$

In view of Equation (7.42) and at $0 < |\alpha| < 1$, the Stokes index of field from Equation (7.35) equals $\sigma = 2n$, whereas the Poincaré–Hopf index is $\eta = \sigma/2 = n$. In this case, there are no points where the light is circularly polarized.

7.2.2 Vector Field with Polarization Singularity Points in the Plane of the Tight Focus

In this subsection, using a Richards–Wolf formalism [338] we derive projections of the E-vector in the focal plane from source field from Equation (7.35). Thus, we obtain:

$$E_x = -\frac{i^{n+1}}{\sqrt{2}}\left(I_{0,n}\cos n\varphi + I_{2,n-2}\cos(n-2)\varphi\right) + \frac{\alpha}{\sqrt{2}}I_{2,2}\sin 2\varphi,$$

$$E_y = -\frac{i^{n+1}}{\sqrt{2}}\left(I_{0,n}\sin n\varphi - I_{2,n-2}\sin(n-2)\varphi\right) + \frac{\alpha}{\sqrt{2}}\left(I_{0,0} - I_{2,2}\cos 2\varphi\right), \tag{7.43}$$

$$E_z = \sqrt{2}i^n I_{1,n-1}\cos(n-1)\varphi - i\alpha\sqrt{2}I_{1,1}\sin\varphi,$$

where the integrals in Equation (7.43) take the form:

$$I_{v,\mu} = \left(\frac{\pi f}{\lambda}\right)\int_0^{\theta_0}\sin^{v+1}(\frac{\theta}{2})\cos^{3-v}(\frac{\theta}{2})\cos^{1/2}(\theta)A(\theta)e^{ikz\cos\theta}J_\mu(x)\,d\theta, \tag{7.44}$$

where λ is the wavelength of light, f is the focal length of an aplanatic system, $x = kr\sin\theta$, $J_\mu(x)$ is a Bessel function of the first kind, and $NA = \sin\theta_0$ is the numerical aperture. The original amplitude function $A(\theta)$ (here, assumed to be real) may be constant (for a plane wave) or described by a Gaussian beam:

$$A(\theta) = \exp\left(\frac{-\gamma^2\sin^2\theta}{\sin^2\theta_0}\right),$$

where γ is constant. At $\alpha = 0$, the field at the focus described by Equation (7.43) is identical (up to a constant $1/\sqrt{2}$) to the field at the focus from an nth order radially polarized wave [337]:

$$E_x = -\frac{i^{n+1}}{\sqrt{2}}\left(I_{0,n}\cos n\varphi + I_{2,n-2}\cos(n-2)\varphi\right),$$

$$E_y = -\frac{i^{n+1}}{\sqrt{2}}\left(I_{0,n}\sin n\varphi - I_{2,n-2}\sin(n-2)\varphi\right), \qquad (7.45)$$

$$E_z = \sqrt{2}i^n I_{1,n-1}\cos(n-1)\varphi.$$

Field from Equation (7.45) contains just V-points of polarization singularity while having neither C-points nor C-lines. At $n = 0$ and $\alpha = 1$, field (7.43) is fully identical to the field at the focus from an incident wave with right-handed circular polarization [357]:

$$E_x = -\frac{i}{\sqrt{2}}\left(I_{0,0} + e^{i2\varphi}I_{2,2}\right),$$

$$E_y = \frac{1}{\sqrt{2}}\left(I_{0,0} - e^{i2\varphi}I_{2,2}\right), \qquad (7.46)$$

$$E_z = -\sqrt{2}e^{i\varphi}I_{1,1}.$$

Because of this, the source field from Equation (7.35) and the field in the focus in Equation (7.43) can be called hybrid, as at some points they have linear, elliptical or circular polarization. For field from Equation (7.43), the intensity at the focus is given by:

$$I = \frac{1}{2}\left\{I_{0,n}^2 + I_{2,n-2}^2 + 2I_{0,n}I_{2,n-2}\cos 2(n-1)\varphi + \right.$$

$$\alpha^2 I_{0,0}^2 + \alpha^2 I_{2,2}^2 - 2\alpha I_{0,0}I_{2,2}\cos 2\varphi +$$

$$4I_{1,n-1}^2\cos^2(n-1)\varphi + 4\alpha^2 I_{1,1}^2\sin^2\varphi -$$

$$2\alpha\cos\left(\frac{n+1}{2}\right)\pi\left[\sin n\varphi\left(I_{0,0}I_{0,n} + I_{2,2}I_{2,n-2}\right) - \right. \qquad (7.47)$$

$$\sin(n-2)\varphi\left(I_{0,0}I_{2,n-2} + I_{2,2}I_{0,n}\right) -$$

$$\left.\left.\sin\varphi\sin(n-1)\varphi I_{1,1}I_{1,n-1}\right]\right\}.$$

Equation (7.47) is rather cumbersome, but putting $n = 2p$ (even) yields $\cos(n+1)\pi/2 = 0$, leading to a simpler relationship of the intensity:

$$I_{n=2p} = \frac{1}{2} \left\{ I_{0,n}^2 + I_{2,n-2}^2 + 2I_{0,n}I_{2,n-2} \cos 2(n-1)\varphi + \right.$$

$$\alpha^2 I_{0,0}^2 + \alpha^2 I_{2,2}^2 - 2\alpha I_{0,0}I_{2,2} \cos 2\varphi + \tag{7.48}$$

$$\left. 4I_{1,n-1}^2 \cos^2(n-1)\varphi + 4\alpha^2 I_{1,1}^2 \sin^2 \varphi \right\}.$$

From Equation (7.48), the intensity at the center of the focal plane is seen to be non-zero because the term $\alpha^2 I_{0,0}^2$ is non-zero. The intensity pattern has central symmetry as Equation (7.48) contains cosines of the double angle 2φ, as well as squared cosine and sine functions, meaning that replacing φ with $\varphi + \pi$ introduces no changes to the intensity pattern. From Equation (7.48), the intensity pattern is also seen to have $2(n-1)$ local intensity peaks (not considering central intensity maximum) because the term $\cos 2(n-1)\varphi$ changes sign $2(n-1)$ times per full circle. At odd numbers $n = 2p+1$, we obtain $\cos(n+1)\pi/2 = \pm 1$, which means that the intensity in Equation (7.47) has no central symmetry due to different intensity values at φ and $\varphi + \pi$, but has a central intensity peak, similar to the previous case.

Let us derive formulae for projections of the Stokes vector at the focus. Since these formulae are rather cumbersome, below, we give only relationships for projections of symmetrical fields at the focus for an even number $n = 2p$. The Stokes vector can be defined in a different way using four projections, rather than three use in definition from Equation (7.37):

$$s_0 = |E_x|^2 + |E_y|^2, s_1 = |E_x|^2 - |E_y|^2,$$

$$s_2 = 2\operatorname{Re}\left(E_x^* E_y\right), s_3 = 2\operatorname{Im}\left(E_x^* E_y\right). \tag{7.49}$$

Based on Equation (7.49), for field from Equation (7.34) (at $n = 2p$) we obtain:

$$s_0 = \frac{1}{2}\left(I_{0,n}^2 + I_{2,n-2}^2 + 2I_{0,n}I_{2,n-2} \cos 2(n-1)\varphi + \right.$$

$$\left. \alpha^2 I_{0,0}^2 + \alpha^2 I_{2,2}^2 - 2\alpha^2 I_{0,0}I_{2,2} \cos 2\varphi \right),$$

$$s_1 = \frac{1}{2}\left(I_{0,n}^2 \cos 2n\varphi + I_{2,n-2}^2 \cos 2(n-2)\varphi + \right.$$

$$\left. 2I_{0,n}I_{2,n-2} \cos 2\varphi - \alpha^2 I_{0,0}^2 - \alpha^2 I_{2,2}^2 \cos 4\varphi + 2\alpha^2 I_{0,0}I_{2,2} \cos 2\varphi \right),$$

$$s_2 = \frac{1}{2}\left(I_{0,n}^2 \sin 2n\varphi + I_{2,n-2}^2 \sin 2(n-2)\varphi + \right. \tag{7.50}$$

$$\left. 2I_{0,n}I_{2,n-2} \sin 2\varphi - \alpha^2 I_{2,2}^2 \cos 4\varphi + 2\alpha^2 I_{0,0}I_{2,2} \cos 2\varphi \right),$$

$$s_3 = \alpha \sin\left(\frac{n+1}{2}\right)\pi\left[\cos n\varphi\left(I_{0,0}I_{0,n} - I_{2,2}I_{2,n-2}\right) + \right.$$

$$\left. \cos(n-2)\varphi\left(I_{0,0}I_{2,n-2} - I_{2,2}I_{0,n}\right)\right].$$

In Equation (7.50), the relations for s_0, s_1, s_2 are given for even numbers $n = 2p$, except for s_3 which holds at any n. The purpose is to demonstrate that at odd $n = 2p+1$ we have $s_3 = S_3 = 0$, hence we can infer that the field at the focus has no C-points, being purely vector and composed of linear polarization vectors.

Using two components of the field from Equation (7.50), the complex Stokes field can be expressed as:

$$S_A = s_1 + is_2 = \frac{1}{2}\left[I_{0,n}^2 e^{2in\varphi} + I_{2,n-2}^2 e^{2i(n-2)\varphi} \right.$$

$$\left. - \alpha^2 I_{2,2}^2 e^{4i\varphi} + 2e^{2i\varphi}\left(\alpha^2 I_{0,0} I_{2,2} + I_{0,n} I_{2,n-2} \right) - \alpha^2 I_{0,0}^2 \right]. \quad (7.51)$$

From Equation (7.51), it follows that the topological charge of the vortex Stokes field is undefined, varying over the entire focal plane, for at large radii r, the amplitudes by the exponents vary in magnitudes, making it impossible to determine which term in the sum from Equation (7.51) is larger in the absolute value at each particular case. For instance, at some radii, the Stokes index of field from Equation (7.51) can be $\sigma = 2n$, being $\sigma = 2(n-2)$ at other radii and taking values of 4, 2, or 0 elsewhere. What may be said definitely is that near the optical axis only the last term in Equation (7.51) remains non-zero, which has no vortex phase. Hence, at any n, the Stokes index at the center of the focus is zero ($\sigma = 0$). Conclusions arrived at in this section are validated by numerical modeling.

7.2.3 NUMERICAL MODELING

Figure 7.11 depicts a distribution of polarization ellipses in the source field from Equation (7.35) at different n: 3(a), 2(b), 1(c), and 4(d). Indices for C-lines of the fields in Figure 7.11, derived from Equation (7.41) using a complex Stokes field, equal $Ic = \sigma/2 = n/2$: 3/2(a), 1(b), 1/2(c), and 2(d). From Figure 7.11(a), field from Equation (7.35) with $n = 3$ is seen to have three C-lines located at angles $\varphi = \pi m/3$, $m = 0,1,2$. The tilt of the major axis of the polarization ellipses changes by $\pi/2$ at each of the six sectors between the adjacent C-lines. Thus, after a full circle, the tilt of the major axis changes by $6\pi/2 = 3\pi$, meaning that the index of the field in Figure 7.11(a) is $Ic = 3\pi/(2\pi) = 3/2$. In a similar way, in Figure 7.11(b), field from Equation (7.35) with $n = 2$ has two C-lines located on the Cartesian axes. With the angle φ in a sector between C-lines changing from 0 to $\pi/2$, the tilt of the major axis of the polarization ellipse is rotated by an angle of $\pi/2$, hence after a full circle around the center, the tilt of the major axis changed by $4\pi/2$. Hence, the index of the field in Figure 7.11(b) equals $Ic = 2\pi/(2\pi) = 1$. In Figure 7.11(c), the C-line is found on the horizontal Cartesian axis. With the angle φ in one of the sectors between C-lines changing from 0 to π (in the upper semi-plane), the tilt of the polarization ellipse major axis is rotated by $\pi/2$, and in the bottom semi-plane, the tilt of the major axis is also rotated by $\pi/2$. Hence, after a full circle, the polarization ellipse makes a turn by π and the singularity index equals $Ic = \pi/(2\pi) = 1/2$. Finally, in Figure 7.11(d), a change in the tilt of the major axis of polarization ellipses can be analyzed in a similar way.

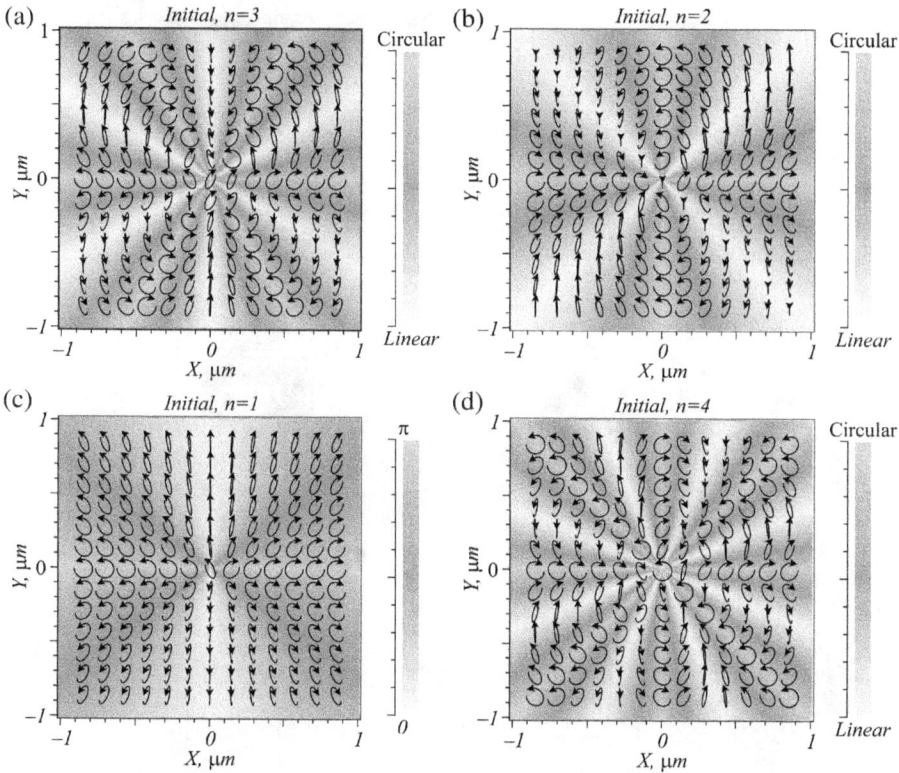

FIGURE 7.11 Polarization patterns (shown in half-tones and with arrows) in the source field from Equation (7.35) at $\alpha = 1$ at different orders n: 3(a), 2(b), 1 (c), 4(d). The arrows show the handedness of the E-vector; the origin of the ellipse is determined based on the phase of the field at this point.

Shown in Figure 7.12(a) is a total intensity at the focus of field from Equation (7.35) at $\alpha = 1$ and $n = 2$. The numerical modeling was conducted using a Richards-Wolf formalism [338] for a wavelength of $\lambda = 532$ nm and numerical aperture NA = 0.95. Shown in Figure 7.12(b, c) are an amplitude and phase of the complex Stokes field $S_A = s_1 + is_2$, which was calculated with the aid of the Stokes vector components in Equation (7.50). From Figure 7.12(a), it is seen that according to theoretical predictions in Equations (7.47) and (7.48), the intensity pattern at the focus remains unchanged after replacing φ by $\varphi + \pi$, with an intensity peak located at the center. From Figure 7.12(c), it is seen that there is no singular point at the center of the phase pattern for the Stokes field from Equation (7.51), because there is no isolated intensity null. Two isolated intensity nulls (singularity points) in Figure 7.12(c), each having the topological charge 1, are seen on the vertical axis (1 and 2 points in Figure 7.12(d)). In Figure 7.12(d), the arrows specify a pattern of the polarization ellipses at the focus. Figure 7.12(e) depicts C-points at the focus, which are all located on the Cartesian axes, it is where the C-lines are located in the source plane (Figure 7.11(b)).

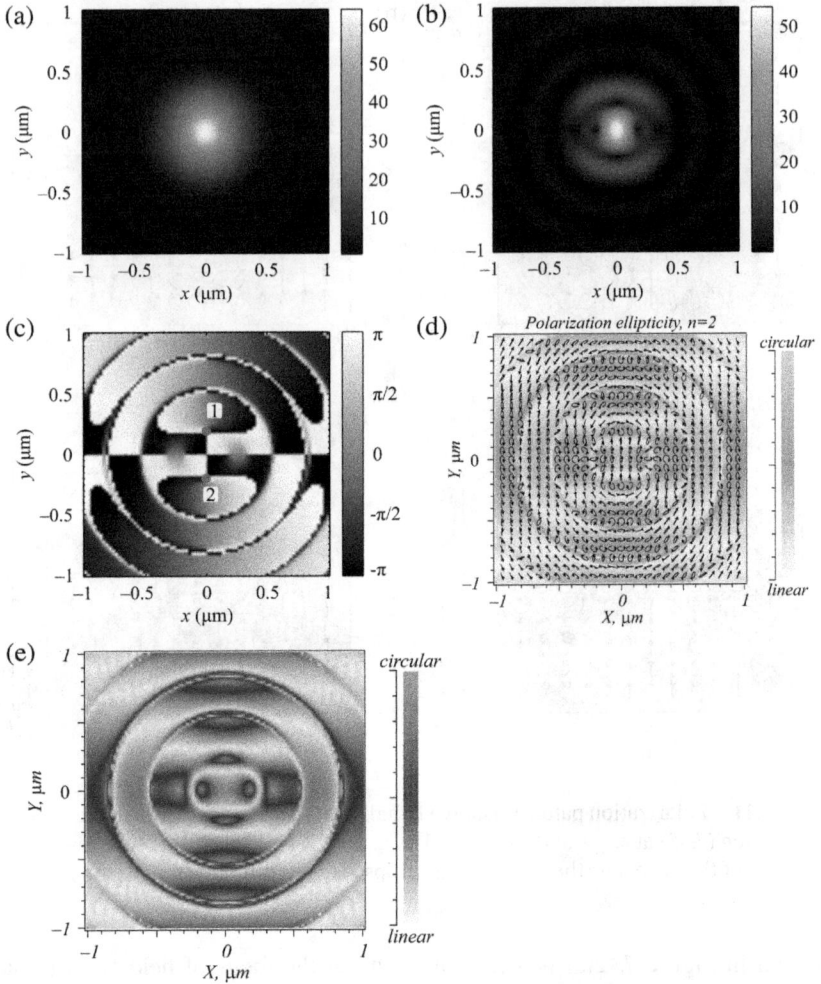

FIGURE 7.12 (a) An intensity $I = I_x + I_y + I_z$, (b) amplitude, and (c) phase of a complex Stokes field from Equation (7.51) when focusing field from Equation (7.35) at $n = 2$. (d) Pattern of elliptic polarization at the focus, and (e) pattern of points with circular, elliptic, and linear polarization.

Thus, the numerical modeling has shown that as a result of tightly focusing, C-lines "disintegrate" into a number of C-points arranged on the same lines. This effect is analogous to an effect of astigmatic edge-to-screw dislocation conversion of a wavefront in scalar paraxial optics [162]. Indices of two symmetrical and closest to the center C-points on the horizontal Cartesian axis are $Ic = \pm 1/2$ (1 and 2 points in Figure 7.12(c)), with the indices of the next two neighboring C-points located farther from the center on the horizontal axis being $Ic = \mp 1/2$ (3 and 4 points in Figure 7.12(c)).

In a similar way, Figure 7.13 depicts numerical simulation results at the focus for $n = 3$ (the rest parameters are the same as in Figure 7.12). Figure 7.13(a) suggests that in agreement with the theoretical prediction in Equation (7.47), the intensity pattern at the focus at odd $n = 3$ is asymmetric. As can be inferred from Equation (7.50), there is a vector field with purely inhomogeneous linear polarization at the focus (Figure 7.13(d)), as putting $n = 2p+1$, we obtain $s_3=S_3=0$. In a phase pattern of the complex Stokes field in Figure 7.13(c), there occur three phase singularity points with the topological charge +2 (1, 2, and 3 points in Figure 7.13(c)). In total, the Stokes index is $\sigma = 6$, and the V-points singularity index is $Ic = \sigma/2 = 3$. The pattern in Figure 7.13(d) is seen to contain three V-points (2 points of the "center" type and 1 point of the "knot" type). Thus, when field from Equation (7.35) has an odd order n, the C-lines (Figure 7.11(a)) of the original plane are transformed into a number of V-points (Figure 7.13(d)) and the whole field becomes vectorial (with no points of elliptical polarization).

Shown in Figure 7.14(a) is an intensity pattern at the focus, which has a fourth-order symmetry relative to the Cartesian coordinates. The amplitude and phase of

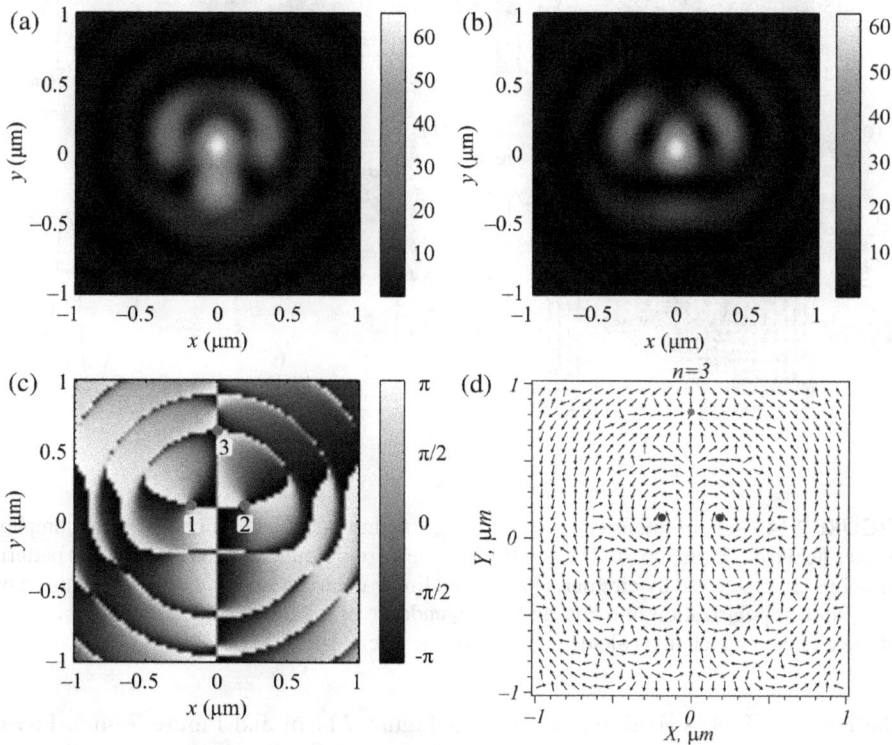

FIGURE 7.13   (a) An intensity $I = I_x + I_y + I_z$, (b) amplitude, and (c) phase of the complex Stokes field from Equation (7.51) when focus in field from Equation (7.35) at $n = 3$. (d) A pattern of linear polarization vectors at the focus: (two V-points of the "center" type and one V-point of the "knot" type).

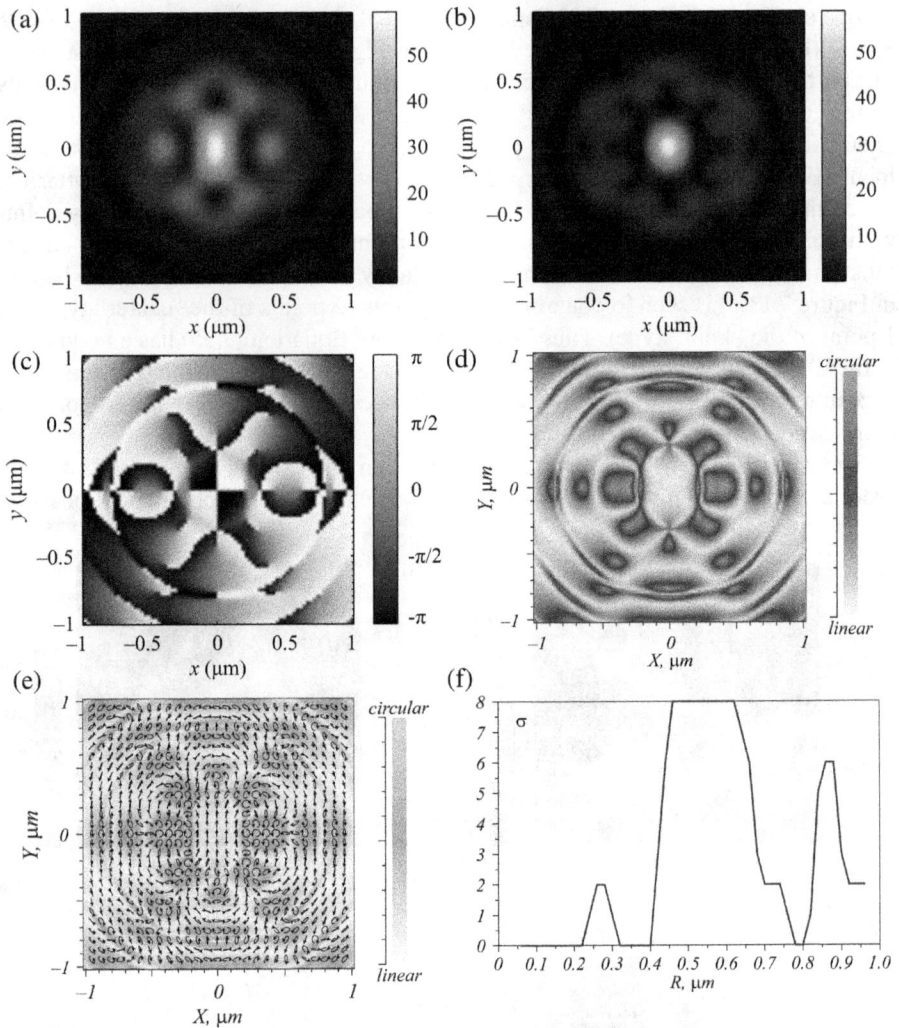

FIGURE 7.14 (a) An intensity $I = I_x + I_y + I_z$, (b) amplitude, and (c) phase of the complex Stokes field from Equation (7.51) when focusing field from Equation (7.35) at $n = 4$. (d) pattern of points characterized by circular, elliptic, and linear polarization at the focus, (e) pattern of the elliptical polarization in focus, and (f) dependence of the Stokes index σ on the radius R of the circle on which it is calculated (1 μm frame size).

the complex Stokes field are depicted in Figure 7.14(b) and Figure 7.14(c). Phase singularities points linked with the C-points are observed in the phase pattern of the Stokes field (Figure 7.14(c)). Finally, Figure 7.14(f) depicts a plot for the Stokes index σ against the radius R of an origin-centered circle along which the phase delay of the Stokes field from Equation (7.51) of Figure 7.14(c) is calculated. From the plot it is

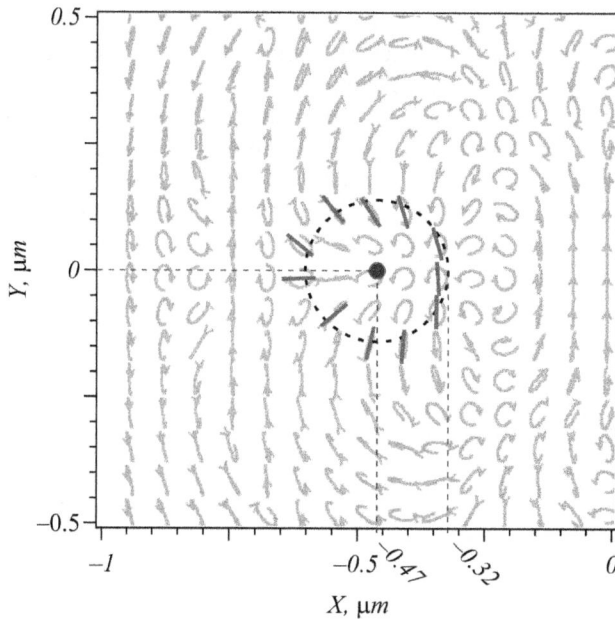

FIGURE 7.15 Polarization ellipses for a field fragment at the focus depicted in Fig. 7.15e limited by (–1 μ <x<0; -0.5 μm< y<0.5 μm) and a C-point with Ic=1/2 found at (–0.47 μm; 0) marked with a black point around which the major axes of the polarization ellipses centered on a circle of radius 0.15 μm are depicted as black lines.

seen that at different radii R, the Stokes index takes values of 8, 6, 2, and 0, which is in good agreement with the theory in Equation (7.51).

From Figure 7.14(e) and (d), the C-points are seen to lie on the Cartesian axes and two diagonal lines, where C-lines were located in the original plane (Figure 7.11(d)). Two center-symmetrical C-points located on the horizontal axis (Figure 7.14(e, f)) have a singularity index of ±1/2, producing a "lemon"-shaped topology. Remarkably, when making a circle around such a C-point, the surface of the polarization ellipses produces a Mobius strip in the 3D space [358,359,360].

Figure 7.15 depicts the neighborhood of a C-point (marked with bold black) in more detail, showing a characteristic tilt of the major axis of polarization ellipses ("lemon"-type topology) lying on a circle centered on the C-point. The axes of the polarization ellipses are seen to turn by an angle of π when making a full circle, i.e. the index of the C-point equals 1/2.

Summing up, we have theoretically and numerically studied a new type of nth order hybrid vector light field that is tightly focused with an aplanatic system [361]. The polarization of the source hybrid vector field varies in the original plane from linear, to elliptical and circular with the polar angle. The polarization pattern of the field has n C-lines of circular polarization that pass through the center. Components for the Stokes vector of such a field have been analytically derived and the field polarization singularity index has been shown to equal $n/2$. Based on a Richards–Wolf formalism, analytical relationships for projections of the E-vector, and intensity of light at the tight focus have been deduced. At even n, the intensity at the focus has been shown to possess a center symmetry and have C-points lying on lines coincident with the C-lines of the source field. Analytical relationships have been deduced to describe projections of the Stokes vector at the focus, which suggest that at odd n, the field at the focus is purely vector and has several V-points, while having no C-points.

Conclusion

In this book, we have calculated the topological charge (T C) for a number of paraxial scalar vortex beams. The peculiarity of this book is that the TC was calculated not according to the usual formula for calculating TC, but according to a more general formula. As a rule, the TC of an optical vortex is understood as an integer 2π, which changes the phase of the light field when approaching a closed loop of phase singularity (phase uncertainty points). But in the cross-section of the light beam there can be several phase singularities (screw dislocations). Therefore, in order to calculate the TC of such a beam, it is necessary to draw a closed contour around each phase singularity and calculate an integer number of phase jumps by 2π for each phase singularity. This approach is non-constructive. The book uses a more general approach suggested by M. Berry. The topological charge is calculated as an integer number of phase jumps by 2π along a circle of infinite radius. The complex amplitude of the light field at infinity is found as a limit as the radial variable tends to infinity. In this case, there is no need to find all the centers of phase singularities in the beam and go around them in a closed loop. All possible singularities in the beam cross-section are automatically taken into account when traversing a circle of infinite radius.

The second feature of this book is that the generalized definition of TC is applied to the calculation of the indices of polarization singularities in vector beams. It is well-known that light has, in addition to amplitude and phase, also polarization, and is described by three projections of the electric field strength vector. Vector light fields have their own points of polarization singularity, which, similarly to the topological charge, which characterizes the topology of the wavefront around the point of phase singularity for a scalar field, are described by the corresponding indices. For example, vector light beams with inhomogeneous polarization, at each cross-section point of which the linear polarization vector changes its direction, have points of polarization singularity (V-points) in which the direction of the polarization vector is not determined. Such points are characterized by the Poincaré–Hopf index, which, similar to the topological charge for scalar fields, is equal to an integer number of complete rotations of the linear polarization vector when traversing a V-point along a closed contour. Similarly, for light fields with inhomogeneous elliptical polarization, when the angle at which the major axis of the polarization ellipse is directed changes over the cross-section of the beam at each point, there are points at which the angle of inclination of the polarization ellipse (C-points) cannot be determined. At these points, the polarization is circular. Such points of polarization singularity are characterized by an index equal to an integer number of rotations by an angle π of the polarization ellipses along a closed contour around the C-point. Therefore, the methods considered in this book for calculating the topological charge of paraxial scalar light fields in the last chapter were applied to calculate the topological indices, with the help of which the points of polarization singularity of non-paraxial vector light fields are characterized.

References

1. J. Durnin, J.J. Micely, and J.H. Eberly, "Diffraction-free beams," *Phys. Rev. Lett.* 58(15), 1499–1501 (1987).
2. M.A. Bandres, J.C. Gutiérrez-Vega, and S. Chávez-Cerda, "Parabolic nondiffracting optical wave fields," *Opt. Lett.* 29(1), 44–46 (2004).
3. J.C. Gutiérrez-Vega, M.D. Iturbe-Castillo, and S. Chávez-Cerda, "Alternative formulation for invariant optical fields: Mathieu beams," *Opt. Lett.* 25(20), 1493–1495 (2000).
4. A.E. Siegman, *Lasers* (University Science Book, Mill Valley, 1986).
5. N. Efremidis, Z. Chen, M. Segev, and D. Christodoulides, "Airy beams and accelerating waves: an overview of recent advances," *Optica* 6(5), 686 (2019).
6. P. Zhang, Y. Hu, T. Li, D. Cannan, X. Yin, R. Morandotti, Z. Chen, and X. Zhang, "Nonparaxial Mathieu and Weber Accelerating Beams," *Phys. Rev. Lett.* 109(19), 193901 (2012).
7. J. Turunen, and A.T. Friberg, "Propagation-invariant optical fields," *Prog. Opt.* 54, 1–88 (2010).
8. U. Levy, S. Derevyanko, and Y. Silberberg, "Light modes of free space," *Prog. Opt.* 61, 237–281 (2016).
9. H. Xiong, Y. Huang, and Y. Wu, "Laguerre-Gaussian optical sum-sideband generation via orbital angular momentum exchange," *Phys. Rev. A* 103(4), 043506 (2021).
10. G. Gbur, and R.K. Tuson, "Vortex beam propagation through atmospheric turbulence and topological charge conservation," *J. Opt. Soc. Am. A* 25(1), 225–230 (2008).
11. A.V. Volyar, M.V. Brezko, Ya.E. Akimova, Yu.A. Egorov, and V.V. Milyukov, "Sector perturbation of a vortex beam: Shannon entropy, orbital angular momentum and topological charge," *Comput. Opt.* 43(5), 723–734 (2019).
12. S.N. Alperin, R.D. Niederriter, J.T. Gopinath, and M.E. Siemens, "Quantitative measurement of the orbital angular momentum of light with a single, stationary lens," *Opt. Lett.* 41(21), 5019–5022 (2016).
13. V.V. Kotlyar, A.A. Kovalev, and A.P. Porfirev, "Calculation of fractional orbital angular momentum of superpositions of optical vortices by intensity moments," *Opt. Express* 27(8), 11236–11251 (2019).
14. L.A. Melo, A.J. Jesus-Silva, S. Chávez-Cerda, P.H. Souto Ribeiro, and W.C. Soares, "Direct measurement of the topological charge in elliptical beams using diffraction by a triangular aperture," *Sci. Rep.* 8(1), 6370 (2018).
15. V.V. Kotlyar, A.A. Kovalev, and A.P. Pofirev, "Astigmatic transforms of an optical vortex for measurement of its topological charge," *Appl. Opt.* 56(14), 4095–4104 (2017).
16. L. Allen, S.M. Barnett, and M.J. Padgett, *Orbital angular momentum* (CRC Press, Boca Raton, 2003).
17. J. Courtial, K. Dholakia, L. Allen, and M.J. Padgett, "Gaussian beams with very high orbital angular momentum," *Opt. Commun.* 144(4–6), 210–213 (1997).
18. I. Martinez-Castellanos and J. Gutiérrez-Vega, "Shaping optical beams with non-integer orbital-angular momentum: A generalized differential operator approach," *Opt. Lett.* 40(8), 1764–1767 (2015).
19. M.V. Berry, "Optical vortices evolving from helicoidal integer and fractional phase steps," *J. Opt. A: Pure Appl. Opt.* 6(2), 259–268 (2004).
20. L. Allen, M.J. Padgett, and M. Babiker, "IV The orbital angular momentum of light," *Prog. Opt.* 39, 291–372 (1999).

21. J.F. Nye, and M.V. Berry, "Dislocations in wave trains," *Proc. R. Soc. Lond. A* 336, 165–190 (1974).

22. M.S. Soskin, V.N. Gorshkov, M.V. Vastnetsov, J.T. Malos, and N.R. Heckenberg, "Topological charge and angular momentum of light beams carrying optical vortex," *Phys. Rev. A* 56(5), 4064–4075 (1997).

23. V.V. Kotlyar, A.A. Kovalev, and A.G. Nalimov, "Optical phase singularities 'going to' infinity with a higher-than-light speed," *J. Opt.* 23(10), 105702 (2021).

24. F. Gori, in *Current trends in optics*, ed. by J. C. Dainty (Academic Press, London, 1994), p. 140.

25. J. Humblet, "Sur le moment d'impulsion d'une onde electromagnetique," *Physica* 10(7), 585 (1943).

26. Sommerfeld, *Lectures on theoretical physics* (Academic Press, New York, 1954), pp. 361–373.

27. Z. Sacks, D. Rozas, and G. Swartzlander, "Holographic formation of optical-vortex filaments," *J. Opt. Soc. Am. B* 15(8), 2226 (1998).

28. V.V. Kotlyar, A.A. Kovalev, R.V. Skidanov, O.Yu. Moiseev, and V.A. Soifer, "Diffraction of a finite-radius plane wave and a Gaussian beam by a helical axicon and a spiral phase plate," *J. Opt. Soc. Am. A* 24(7), 1955 (2007).

29. M.R. Dennis, "Rows of optical vortices from elliptically pertubing a high-order beam," *Opt. Lett.* 31(9), 1325–1327 (2006).

30. A.A. Kovalev, V.V. Kotlyar, and A.G. Nalimov, "Topological charge and asymptotic phase invariants of vortex laser beams," *Photonics* 8(10), 445 (2021).

31. B. Paroli, M. Siano, and M. Potenza, "Measuring the topological charge of orbital angular momentum radiation in single-shot by means of the wavefront intrinsic curvature," *Appl. Opt.* 59(17), 5258 (2020).

32. D. Wang, H. Huang, H. Toyoda, and H. Liu, "Topological Charge Detection Using Generalized Contour-Sum Method from Distorted Donut-Shaped Optical Vortex Beams: Experimental Comparison of Closed Path Determination Methods," *Appl. Sci.* 9(19), 3956 (2019).

33. L. Allen, M. Beijersbergen, R. Spreeuw, and J. Woerdman, "Orbital angular momentum of light and the transformation of Laguerre-Gaussian laser modes," *Phys. Rev. A* 45(11), 8185 (1992).

34. A. Volyar, M. Bretsko, Ya. Akimova, and Yu. Egorov, "Vortex avalanche in the perturbed singular beams," *J. Opt. Soc. Am. A* 36(6), 1064–1071 (2019).

35. Y. Zhang, X. Yang, and J. Gao, "Orbital angular momentum transformation of optical vortex with aluminium metasurfaces," *Sci. Rep.* 9(1), 9133 (2019).

36. H. Zhang, X. Li, H. Ma, M. Tang, H. Li, J. Tang, and Y. Cai, "Grafted optical vortex with controllable orbital angular momentum distribution," *Opt. Express* 27(16), 22930–22938 (2019).

37. A. Volyar, M. Bretsko, Ya. Akimova, and Yu. Egorov, "Measurement of the vortex and orbital angular momentum spectra with a single cylindrical lens," *Appl. Opt.* 58(21), 5748–5755 (2019).

38. V.V. Kotlyar, A.A. Kovalev, A.P. Porfirev, and E.S. Kozlova, "Orbital angular momentum of a laser beam behind an off-axis spiral phase plate," *Opt. Lett.* 44(15), 3673–3676 (2019).

39. S. Maji, P. Jacob, and M.M. Brundavanam, "Geometric phase and intensity-controlled extrinsic orbital angular momentum of off-axis vortex beams," *Phys. Rev. Appl.* 12(5), 054053 (2019).

40. F. Gori, G. Guattary, and C. Padovani, "Bessel-Gauss beams," *Opt. Commun.* 64(6), 491–495 (1987).

41. V.V. Kotlyar, R.V. Skidanov, S.N. Khonina, and V.A. Soifer, "Hypergeometric modes," *Opt. Lett.* 32(7), 742–744 (2007).

42. M.A. Bandres, and J.C. Gutierrez-Vega, "Circular beams," *Opt. Lett.* 33(2), 177–179 (2008).
43. V.V. Kotlyar, A.A. Kovalev, and V.A. Soifer, "Asymmetric Bessel modes," *Opt. Lett.* 39(8), 2395–2398 (2014).
44. A.A. Kovalev, V.V. Kotlyar, and A.P. Porfirev, "Asymmetric Laguerre-Gaussian beams," *Phys. Rev. A* 93(6), 063858 (2016).
45. M.A. Garcia-March, A. Ferrando, M. Zacares, S. Sahu, and D.E. Ceballos-Herrera, "Symmetry, winding number, and topological charge of vortex solitons in discrete-symmetry media," *Phys. Rev. A* 79(5), 053820 (2009).
46. A. Volyar, M. Bretsko, Ya. Akimova, and Yu. Egorov, "Orbital angular momentum and informational entropy in perturbed vortex beams," *Opt. Lett.* 44(22), 2687–2690 (2019).
47. V.V. Kotlyar, A.A. Kovalev, and A.P. Porfirev, "Asymmetric Gaussian optical vortex," *Opt. Lett.* 42(1), 139–142 (2017).
48. V.V. Kotlyar, A.A. Kovalev, A.P. Porfirev, and E.G. Abramochkin, "Fractional orbital angular momentum of a Gaussian beam with an embedded off-axis optical vortex," *Comput. Opt.* 41(1), 22–29 (2017).
49. G. Indebetouw, "Optical vortices and their propagation," *J. Mod. Opt.* 40(1), 73–87 (1993).
50. E. Abramochkin, and V. Volostnikov, "Spiral-type beams," *Opt. Commun.* 102(3–4), 336–350 (1993).
51. I.S. Gradshteyn, and I.M. Ryzhik, Table *of integrals, series, and products* (Academic, New York, 1965).
52. J.B. Gotte, S. Franke-Arnold, R. Zambrini, and S.M. Barnett, "Quantum formulation of fractional orbital angular momentum," *J. Mod. Opt.* 54(12), 1723–1738 (2007).
53. C.N. Alexeyev, Yu.A. Egorov, and A.V. Volyar, "Mutual transformations of fractional-order and integer-order optical vortices," *Phys. Rev. A* 96(6), 063807 (2017).
54. G. Liang and W. Cheng, "Splitting and rotating of optical vortices due to non-circular symmetry in amplitude and phase distributions of the host beams," *Phys. Lett. A* 384, 126046 (2020).
55. V.V. Kotlyar, A.A. Kovalev, and A.V. Volyar, "Topological charge of a linear combination of optical vortices: Topological competition," *Opt. Express* 28(6), 8266–8281 (2020).
56. V.V. Kotlyar, A.A. Kovalev, and A.P. Porfirev, *Vortex laser beams* (CRC Press, Boca Raton, 2018).
57. S. Li, X. Pan, Y. Ren, H. Liu, S. Yu, and J. Jing, "Deterministic generation of orbital-angular-momentum multiplexed tripartite entanglement," *Phys. Rev. Lett.* 124(8), 083605 (2020).
58. M. Hiekkamaki, S. Prabhakar, and R. Fickler, "Near-perfect measuring of full-field transverse-spatial modes of light," *Opt. Express* 27(22), 31456–31464 (2019).
59. S.Li, X. Li, L. Zhang, G. Wang, L. Zhang, M. Liu, C. Zeng, L. Wang, Q. Sun, W. Zhao, and W. Zhang, "Efficient optical angular momentum manipulation for compact multiplexing and demultiplexing using a dielectric metasurface," *Adv. Opt. Mater.* 2020, 1901666 (2020).
60. A. Pryamikov, G. Alagashev, G. Falkovich, and S. Turitsyn, "Light transport and vortex-supported wave-guiding in micro-strustured optical fibers," *Sci. Rep.* 10(1), 2507 (2020).
61. K. Dai, W. Li, K.S. Morgan, Y. Li, J.K. Miller, R.J. Watkins, and E.G. Johnson, "Second-harmonic generation of asymmetric Bessel-Gaussian beams carrying orbital angular momentum," *Opt. Express* 28(2), 2536–2546 (2020).
62. N. Dimitrov, M. Zhekova, G.G. Paulus, and A. Dreischuh, "Inverted field interferometer for measuring the topological charges of optical vortices carried by short pulses," *Opt. Commun.* 456, 124530 (2020).

63. R.J. Watkins, K. Dai, G. White, W. Li, J.K. Miller, K.S. Morgan, and E.G. Johnson, "Experimental probing of turbulence using a continuous spectrum of asymmetric OAM beams," *Opt. Express* 28(2), 924–935 (2020).

64. J.M. Hickmann, E.J.S. Fonseca, W.C. Soares, and S.Chavez-Cerda, "Unveiling a truncated optical lattice associated with a triangular aperture using lights orbital angular momentum," *Phys. Rev. Lett.* 105(5), 053904 (2010).

65. V.V. Kotlyar, S.N. Khonina, and V.A. Soifer, "Light field decomposition in angular harmonics by means of diffractive optics," *J. Mod. Opt.* 45(7), 1495–1506 (1998).

66. A.V. Volyar, M.V. Brezko, Ya.E. Akimova, and Yu.A. Egorov, "Beyond the intensity or intensity moments and measuring the spectrum of optical vortices in complex beams," *Comput. Opt.* 42(5), 736–743 (2018).

67. V.V. Kotlyar, A.A. Kovalev, R.V. Skidanov, and V.A. Soifer, "Asymmetric Bessel-Gauss beams," *J. Opt. Soc. Am. A* 31(9), 1977–1983 (2014).

68. V.V. Kotlyar, A.A. Kovalev, and E.G. Abramochkin, "Kummer laser beams with a transverse complex shift," *J. Opt.* 22(1), 015606 (2020). https://doi.org/10.1088/2040-8986/ab5ef1.

69. V.V. Kotlyar, and A.A. Kovalev, "Hermite-Gaussian modal laser beams with orbital angular momentum," *J. Opt. Soc. Am. A* 31(2), 274–282 (2014).

70. V.V. Kotlyar, A.A. Kovalev, and A.P. Porfirev, "Vortex Hermite-Gaussian laser beams," *Opt. Lett.* 40(5), 701–704 (2015).

71. A.Y. Bekshaev, M.S. Soskin, and M.V. Vasnetsov, "Optical vortex symmetry breakdown and decomposition of the orbital angular momentum of light beams," *J. Opt. Soc. Am. A* 20(8), 1635–1643 (2003).

72. V.G. Volostnikov, and E.G.Abramochkin, *The modern optics of the Gaussian beams* [In Russian] (Fizmatlit Publisher, Moscow, 2010).

73. V.V. Kotlyar, A.A. Kovalev, "Topological charge of asymmetric optical vortices," *Opt. Express* 28(14), 20449–20460 (2020).

74. H. Wang, L. Liu, C. Zhou, J. Xu, M. Zhang, S. Teng, and Y. Cai, "Vortex beam generation with variable topological charge based on a spiral slit," *Nanophotonics* 8(2), 317–324 (2019).

75. J. Leach, E. Yao, and M.J. Padgett, "Observation of the vortex structure of a non-integer vortex beam," *New J. Phys.* 6, 71 (2004).

76. A.J. Jesus-Silva, E.J.S. Fonseca, and J.M. Hickmann, "Study of the birth of a vortex at Fraunhofer zone," *Opt. Lett.* 37(21), 4552–4554 (2012).

77. J. Wen, L. Wang, X. Yang, J. Zhang, and S. Zhu, "Vortex strength and beam propagation factor of fractional vortex beams," *Opt. Express* 27(4), 5893–5904 (2019).

78. A. Mourka, J. Baumgartl, C. Shanor, K. Dholakia, and E.M. Wright, "Visualization of the birth of an optical vortex using diffraction from a triangular aperture," *Opt. Express* 19(7), 5760–5771 (2011).

79. G. Gibson, J. Courtial, M.J. Padgett, M. Vasnetsov, V. Pasko, S.M. Barnett, and S. Franke-Arnold, "Free-space information transfer using light beams carrying orbital angular momentum," *Opt. Express* 12(22), 5448 (2004).

80. V.V. Kotlyar, A.A. Almazov, S.N. Khonina, V.A. Soifer, H. Elfstrom, and J. Turunen, "Generation of phase singularity through diffracting a plane or Gaussian beam by a spiral phase plate," *J. Opt. Soc. Am. A* 22(5), 849–861 (2005).

81. "Handbook of mathematical functions," Ed. by M. Abramowitz, and I.A. Stegun (National bureau of standards, Applied math, series, 55, Issue June (1964).

82. G.R. Salla, C. Perumangattu, S. Prabhakar, A. Anwar, and R.P. Singh, "Recovering the vorticity of a light beam after scattering," *Appl. Phys. Lett.* 107(2), 021104 (2015).

83. E.G. Abramochkin, and V.G. Volostnikov, "Beam transformations and nontransformed beams," *Opt. Commun.* 83(1–2), 123–135 (1991).

84. V.V. Kotlyar, A.A. Kovalev, and V.A. Soifer, "Optical beams: Vortices with fractional orbital angular momentum," in *Chapter 8 in book "compendium of electromagnetic analysis"* (World Scientific, NJ, 2020). 5, pp. 279–340. https://doi.org/10.1142/9789813270343.0008. ISBN 978-981-3270-16-9.

85. V.V. Kotlyar, A.A. Kovalev, A.G. Nalimov, and A.P. Porfirev, "Evolution of an optical vortex with an initial fractional topological charge," *Phys. Rev. A*. 102, 023516 (2020).

86. A.V. Volyar, M.V. Brecko, Y.E. Akimova, and Y.A. Egorov, "Shaping and processing the vortex spectra of singular beams with anomalous orbital angular momentum," *Comput. Opt.* 43(4), 517–527 (2019).

87. A.G. White, C.P. Smith, N.R. Heckenberg, H. Rubinsztein-Dunlop, R. McDuff, C.O. Weiss, and C. Tamm, "Interferometric measurements of phase singularities in the output of a visible laser," *J. Mod. Opt.* 38(12), 2531–2541 (1991).

88. A.G. Nalimov, V.V. Kotlyar, and V.A. Soifer, "Modeling of an image forming by a zone plate in X-ray," *Comput. Opt.* 35(3), 290–296 (2011).

89. G.E. Volovik, and V.P. Mineev, "Line and point singularitie in superfluid He3," *Pis'ma Zh. Eksp. Teor. Fiz.* 24, 605–608 (1976).

90. J. Jang, D.G. Ferguson, V. Vakarynk, R. Budakian, S.B. Chung, P.M. Goldbart, and Y. Maeno, "Observation of half-height magnetization spets in S_2RuO_4," *Science* 331(6014), 186–188 (2011).

91. Y.G. Rubo, "Half vortices in exciton polariton condensates," *Phys. Rev. Lett.* 99(10), 106401 (2007).

92. H. Flayac, I.A. Shelukh, D.D. Solnyshkov, and G. Malpuech, "Topological stability of the half-vortices in spinor exciton-polariton condensates," *Phys. Rev. B* 81(4), 045318 (2010).

93. M. Krenn, R. Fickler, M. Fink, J. Handsteiner, M. Malik, T. Scheidl, R. Ursin, and A. Zeilinger, "Communication with spatially modulated light through turbulent air across Vienna," *New J. Phys.* 16(11), 113028 (2014).

94. J. Durnin, "Exact solutions for nondiffracting beams. I. The scalar theory," *J. Opt. Soc. Am. A* 4(4), 651–654 (1987).

95. V.S. Vasilyev, A.I. Kapustin, R.V. Skidanov, V.V. Podlipnov, N.A. Ivliev, and S.V. Ganchevskaya, "Experimental investigation of the stability of Bessel beams in the atmosphere," *Comput. Opt.* 43(3), 376–384 (2019).

96. F. Wang, Y. Cai, H.T. Eyyuboglu, and Y. Baykal, "Average intensity and spreading of partially coherent standard and elegant Laguerre-Gaussian beams in turbulent atmosphere," *Prog. Electromagn. Res.* 103, 33–55 (2010).

97. Y. Chen, F. Wang, C. Zhao, and Y. Cai, "Experimental demonstration of a Laguerre-Gaussian correlated Schell-model vortex beam," *Opt. Express* 22(5), 5826–5838 (2014).

98. V.P. Lukin, P.A. Konyaev, and V.A. Sennikov, "Beam spreading of vortex beams propagating in turbulent atmosphere," *Appl. Opt.* 51(10), C84–C87 (2012).

99. K. Zhu, G. Zhou, X. Li, X. Zheng, and H. Tang, "Propagation of Bessel-Gaussian beams with optical vortices in turbulent atmosphere," *Opt. Express* 16(26), 21315–21320 (2008).

100. S. Avramov-Zamurovic, C. Nelson, S. Guth, O. Korotkova, and R. Malek-Madani, "Experimental study of electromagnetic Bessel-Gaussian Schell model beams propagating in a turbulent channel," *Opt. Commun.* 359, 207–215 (2016).

101. I.P. Lukin, "Integral momenta of vortex Bessel-Gaussian beams in turbulent atmosphere," *Appl. Opt.* 55(12), B61–B66 (2016).

102. L.G. Wang, and W.W. Zheng, "The effect of atmospheric turbulence on the propagation properties of optical vortices formed by using coherent laser beam arrays," *J. Opt. A: Pure Appl. Opt.* 11(6), 065703 (2009).

103. M.V. Berry, "A note on superoscillations associated with Bessel beams," *J. Opt.* 15(4), 044006 (2013).

104. E. Ostrovsky, K. Cohen, S. Tsesses, B. Gjona, and G. Bartal, "Nanoscale control over optical singularities," *Optica* 5(3), 283–288 (2018).

105. J.F. Nye, J.V. Hajnal, and J.H. Hannay, "Phase saddles and dislocations in two-dimensional waves such as the tides," *Proc. R. Soc. Lond. A* 417(1852), 7–20 (1988).

106. J.F. Nye, *Natural focusing and fine structure of light* (Institute of Physics Publishing, Bristol, 1999).

107. M.V. Berry, "Wave dislocation reactions in non-paraxial Gaussian beams," *J. Mod. Opt.* 45(9), 1845–1858 (1998).

108. F. Flossmann, U.T. Schwarz, and M. Maier, "Optical vortices in a Laguerre-Gaussian LG (1,0) beam," *J. Mod. Opt.* 52(7), 1009–1017 (2005).

109. I.V. Basistiy, V.Yu. Bazhenov, M.S. Soskin, and M.V. Vasnetsov, "Optics of light beams with screw dislocations," *Opt. Commun.* 103(5–6), 422–428 (1993).

110. A.Ya. Bekshaev, M.S. Soskin, and M.V. Vasnetsov, "Centrifugal transformation of the transverse structure of freely propagating paraxial light beams," *Opt. Lett.* 31(6), 694–696 (2006).

111. A.Ya. Bekshaev, and A.I. Karamoch, "Spatial characteristics of vortex light beams produced by diffraction gratings with embedded phase singularity," *Opt. Commun.* 28(6), 1366–1374 (2008).

112. D. Rozas, C.T. Law, and G.A. Shwartzlander Jr, "Propagation dynamics of optical vortices," *J. Opt. Soc. Am. B* 14(11), 3054–3065 (1997).

113. M. Abramowitz, and I.A. Stegun, *Handbook of mathematical functions: With formulas, graphs, and mathematical tables* (Dover Publications Inc, New York, 1979).

114. A. Longman, and R. Fedosejevs, "Optimal Laguerre-Gaussian modes for high-intensity optical vortices," *J. Opt. Soc. Am. A* 37(5), 841–848 (2020).

115. V.V. Kotlyar, A.A. Kovalev, and A.P. Porfirev, "Topological stability of optical vortices diffracted by a random phase screen," *Comput. Opt.* 43(4), 917–925 (2019).

116. A.A. Kovalev, V.V. Kotlyar, and A.P. Porfirev, "Orbital angular momentum and topological charge of a multi-vortex Gaussian beam," *J. Opt. Soc. Am. A* 37(11), 1740–1747 (2020).

117. G. Ruffato, "OAM-inspired new optics: The angular metalens," *Light Sci. Appl.* 10(1), 96 (2021).

118. Y. Guo, S. Zhang, M. Pu, Q. He, J. Jin, M. Xu, Y. Zhang, P. Gao, and X. Luo, "Spin-decoupled metasurface for simultaneous detection of spin and orbital angular momenta via momentum transformation," *Light Sci. Appl.* 10(1), 63 (2021).

119. Z. Jin, D. Janoschka, J. Deng, L. Ge, P. Dreher, B. Frank, G. Hu, J. Ni, J. Yang, J. Li, G. Yu et al, "Phyllotaxis-inspired nanosieves with multiplexed orbital angular momentum," *eLight* 1, 1 (2021).

120. D. Wei, Y. Cheng, R. Ni, Y. Zhang, X. Hu, S. Zhu, and M. Xiao, "Generating controllable Laguerre-Gaussian laser modes through intracavity spin-orbital angular momentum conversion of light," *Phys. Rev. Appl.* 11(1), 014038 (2019).

121. V. Stella, T. Grogjeon, N. De Leo, L. Boarino, P. Munzerd, J.R. Lakowicz, and E. Descrovi, "Vortex beam generation by spin-orbit interaction with Bloch surface waves," *ACS Photonics* 7(3), 774–783 (2020).

122. T. Arikawa, T. Hiraoka, S. Morimoto, F. Blanchard et al, "Transfer of optical angular momentum of light to plasmoni excitations in metamaterials," *Sci. Adv.* 6(24), 253 (2020).

123. V.V. Kotlyar, S.S. Stafeev, A.G. Nalimov, L. O'Faolain, and M.V. Kotlyar, "A dual-functionality metalens to shape a circularly polarized optical vortex or a second-order cylindrical vector *beam*," *Phot. Nanostr Fund. Appl.* 43, 100898 (2021).

124. L. Zhu, M. Tang, H. Li, Y. Tai, and X. Li, "Optical vortex lattice: An explotation of orbital angular momentum," *Nanophotonics* 10(9), 0139 (2021).
125. S. Fu, Y. Zhai, J. Zhang, X. Liu, R. Song, H. Zhou, and C. Gao, "Universal orbital angular momentum spectrum analyser for beams," *PhotoniX* 1(1), 19 (2020).
126. D.M. Fatkhiev, M.A. Butt, E.P. Grakhova, R.V. Katluyarov, I.V. Stepanov, N.L. Kazanskiy, S.N. Khonina, V.S. Luubopytov, and A.K. Sultanov, "Recent advances in generation and detection of orbital angular momentum optical vortices: A review," *Sensors* 21(15), 4988 (2021).
127. L. Zhu, and J. Wang, "A review of multiple optical vortices generation: Methods and applications," *Front. Optoelectron.* 12(1), 52–68 (2019).
128. D.L. Andrews, "Symmetry and quantum features in optical vortices," *Symmetry* 13(8), 1368 (2021).
129. X. Wang, Z. Nie, Y. Liang, J. Wang, T. Li, and B. Jia, "Recent advances on optical vortex generation," *Nanophotonics* 7, 1533–1556 (2018).
130. R. Chen, H. Zhou, M. Moretti, X. Wang, and J. Li, "Orbital angular momentum waves: Generation, detection and emerging applications," *IEEE Commun. Surv. Tutor.* 22(2), 840–868 (2020).
131. A. Forbes, M. de Oliveira, and M.R. Dennis, "Structured light," *Nat. Photon.* 15(4), 253–262 (2021).
132. Y. Shen, X. Wang, Z. Xie, C. Min, X. Fu, Q. Liu, M. Gong, and X. Yuan, "Optical vortices 30 years on: OAM manipulation from topological charge to multiple singularities," *Light Sci. Appl.* 8, 90 (2019).
133. V.V. Kotlyar, and A.A. Kovalev, *Accelerating and vortex laser beams.* (CRC Press, Boca Raton, 2019), p. 298. https://doi.org/10.1201/9780429321610. ISBN: 9780429321610.
134. *Frontires in optics and photonics*, ed. by F. Capasso, and D. Couwenberg (de Gruyter GmbH, Berlin, 2021), p. 783.
135. G. Gbur, "Fractional vortex Hilbert's hotel," *Optica* 3(3), 222–225 (2016).
136. I.V. Basistiy, M.S. Soskin, and M.V. Vasnetsov, "Optical wavefront dislocations and their properties," *Opt. Commun.* 119(5–6), 604–612 (1995).
137. B.Ya. Zel'dovich, N.D. Kundikova, F.V. Podgornov, and L.F. Rogacheva, "Formation of a 'light flash' moving at an arbitrary velocity in vacuum," *Quantum Electron.* 26(12), 1097–1099 (1996).
138. P. Saari, and I. Besieris, "Backward energy flow in simple 4-wave electromagnetic fields," *Eur. J. Phys.* 42(5), 055301 (2021).
139. V.V. Kotlyar, S.N. Khonina, R.V. Skidanov, and V.A. Soifer, "Rotation of laser beams with zero of the orbital angular momentum," *Opt. Commun.* 274(1), 8 (2007).
140. A.V. Volyar, M.V. Bretsko, Ya.E. Akimova, and Yu.A. Egorov, "Avalanche instability of the orbital angular momentum higher order optical vortices," *Comput. Opt.* 43(1), 14 (2019).
141. V.V. Kotlyar, A.A. Kovalev, A.P. Porfirev, and E.S. Kozlova, "Orbital angular momentum of laser beam behied an off-axis spiral phase plate," *Opt. Lett.* 44(15), 3673 (2019).
142. G. Ruffalo, M. Massari, and F. Romanato, "Multiplication and division of the orbital angular momentum of light with diffractive transformation optics," *Light Sci. Appl.* 8, 113 (2019).
143. L. Gong, Q. Zhao, H. Zheng, X. Hu, K. Huang, J. Yang, and Y. Li, "Optical orbital-angular-momentum multiplexed data transmission under high scattering," *Light Sci. Appl.* 8, 27 (2019).
144. S.H. Kazemi, and M. Mahmoudi, "Identifying orbital angular momentum of light in quantum wells," *Laser Phys. Lett.* 16(7), 076001 (2019).
145. A.M. Konzelmann, S.O. Kruger, and H. Giessen, "Interaction of orbital angular momentum light with Rydberg excitons: Modifying dipole selection rules," *Phys. Rev. B* 100(11), 115308 (2019).

146. F.N. Rybakov, and N.S. Kiselev, "Chiral magnetic skyrmions with arbitrary topological charge," *Phys. Rev. B* 99(6), 064437 (2019).

147. L. Zhang, L. Zhang, and X. Liu, "Dynamical detection of topological charges," *Phys. Rev. A* 99(5), 053606 (2019).

148. V.V. Kotlyar, and A.A. Kovalev, "Hermite–Gaussian modal laser beams with orbital angular momentum," *J. Opt. Soc. Am. A* 31(2), 274 (2014).

149. S. Franke-Arnold, J. Leach, M.J. Padgett, V.E. Lembessis, D. Ellinas, A.J. Wright, J.M. Girkin, P. Ohberg, and A.S. Arnold, "Optical Ferris wheel for ultracold atoms," *Opt. Express* 15(14), 8619 (2007).

150. T. Ando, N. Matsumoto, Y. Ohtake, Y. Takiguchi, and T. Inoue, "Structure of optical singularities in coaxial superpositions of Laguerre-Gaussian modes," *J. Opt. Soc. Am. A* 27(12), 2602 (2010).

151. E.J. Galvez, N. Smiley, and N. Fernandes, "Composite optical vortices formed by collinear Laguerre-Gauss beams," *Proc. SPIE* 6131, 613105 (2006).

152. C. Schulze, F.S. Roux, A. Dudley, R. Rop, M. Duparre, and A. Forbes, "Accelerated rotation with orbital angular momentum modes," *Phys. Rev. A* 91(4), 043821 (2015).

153. L. Yang, D. Qian, C. Xin, Z. Hu, S. Ji, D. Wu, Y. Hu, J. Li, W. Huang, and J. Chu, "Two-photon polymerization of microstructures by a non-diffraction multifoci pattern generated from a superposed Bessel beam," *Opt. Lett.* 42(4), 743 (2017).

154. S. Supp, and J. Jahns, "Coaxial superposition of Bessel beams by discretized spiral axicons," *J. Eur. Opt. Soc. Rapid Publ.* 14(1), 18 (2018).

155. S. Orlov, and A. Stabinis, "Propagation of superpositions of coaxial optical Bessel beams carrying vortices," *J. Opt. A: Pure Appl. Opt.* 6(5), S259 (2004).

156. M.V. Berry, "Quantal phase factors accompanying adiabatic changes," *Proc. R. Soc. Lond. A* 392(1802), 45 (1984).

157. A.P. Prudnikov, Y.A. Brychkov, and O.I. Marichev, *Integrals and series, special functions* (Gordon and Breach, New York, 1981).

158. J.W. Goodman, *Introduction to fourier optics*, 2nd ed. (McGraw-Hill, New York, 1996).

159. V.V. Kotlyar, A.A. Kovalev, and A.G. Nalimov, "Conservation of the half-integer topological charge on propagation of a superposition of two Bessel-Gaussian beams," *Phys. Rev. A.* 104(3), 033507 (2021).

160. A.J. Jesus-Silva, E.J.S. Fonseca, and J.M. Hickman, "Study of the birth of a vortex at Frauhofer zone," *Opt. Lett.* 37(12), 4552–4554 (2012).

161. J. Zeng, H. Zhang, Z. Xu, C. Zhao, Y. Cai, and G. Gbur, "Anomalous multi-ramp fractional vortex beams with arbitrary topological charge jumps," *Appl. Phys. Lett.* 117(24), 241103 (2020).

162. A.A. Kovalev, and V.V. Kotlyar, "Optical vortex beams with the infinite topological charge," *J. Opt.* 23(5), 055601 (2021).

163. A.A. Kovalev, and V.V. Kotlyar, "Propagation-invariant laser beams with an array of phase singularities," *Phys. Rev. A* 103(6), 063502 (2021).

164. E.G. Abramochkin, and V.G. Volostnikov, "Spiral-type beams: Optical and quantum aspects," *Opt. Commun.* 125(4–6), 302–323 (1996).

165. S. Rasouli, P. Amiri, V.V. Kotlyar, and A.A. Kovalev, "Characterization of a pair of superposed vortex beams having different winding numbers via diffraction from a quadratic curved-line grating," *J. Opt. Soc. Am. B* 38(8), 2267–2276 (2021).

166. A.A. Kovalev, and V.V. Kotlyar, "Orbital angular momentum of superposition of identical shifted vortex beams," *J. Opt. Soc. Am. A* 32(10), 1805–1810 (2015).

167. V.V. Kotlyar, A.A. Kovalev, P. Amiri, P. Soltani, and S. Rasouli, "Topological charge of two parallel Laguerre-Gaussian beams," *Opt. Express* 29(26), 42962–42977 (2021).

168. V.V. Kotlyar, and A.A. Kovalev, *Topological charge of optical vortices* (Novaya Tekhnika, Samara, 2021), 185 pages, ISBN 978-5-88940-157-5.

169. O.O. Arkhelyuk, P.V. Polyanskii, A.A. Ivanovskii, and M.S. Soskin, "Creation and diagnostics of stable rainbow optical vortices," *Opt. Appl.* XXXIV(3), 419–426 (2004).

170. V. Denisenko, V. Shvedov, A.S. Desyatnikov, D.N. Neshev, W. Krolikovski, A. Volyar, M. Soskin, and Y.S. Kivshar, "Determination of topological charges of polychromatic optical vortices," *Opt. Express* 17(26), 23374–23379 (2009).

171. D. Hakobyan, H. Magallanes, G. Seniutinas, S. Juodkazis, and E. Brasselet, "Tailoring orbital angular momentum of light in the visible domain with metallic metasurfaces," *Adv. Opt. Mater.* 4(2), 306–312 (2015).

172. J. Kobashi, H. Yoshida, and M. Ozaki, "Polychromatic optical vortex generation from patterned cholesteric liquid crystals," *Phys. Rev. Lett.* 116(25), 253903 (2016).

173. Y. Zhang, H. Guo, X. Qiu, X. Lu, X. Ren, and L. Chen, "LED-based chromatic and white-light vortices of fractional topological charges," *Opt. Commun.* 485, 126732 (2021).

174. G. Vallone, "On the properties of circular beams: Normalization, Laguerre-Gauss expansion, and free-space divergence," *Opt. Lett.* 40(8), 1717–1720 (2015).

175. G. Vallone, "Role of beam waist in Laguerre-Gauss expansion of vortex beams," *Opt. Lett.* 42(6), 1097–1100 (2017).

176. A.P. Prudnikov, Yu.A. Brychkov, and O.I. Marichev, *Integrals and series, special functions* (Taylor&Francis, London, 1992), p. 444.

177. G. Campbell, B. Hage, B. Buchler, and P. Lam, "Generation of high-order optical vortices using directly machined spiral phase mirrors," *Appl. Opt.* 51(7), 873–876 (2012).

178. Y. Chen, Z. Fang, Y. Ren, L. Gong, and R. Lu, "Generation and characterization of a perfect vortex beam with a large topological charge through a digital micromirror device," *Appl. Opt.* 54(27), 8030–8035 (2015).

179. C. Wang, Y. Ren, T. Liu, C. Luo, S. Qiu, Z. Li, and H. Wu, "Generation and measurement of high-order optical vortices by using the cross phase," *Appl. Opt.* 59(13), 4040–4047 (2020).

180. D. Chen, Y. Miao, H. Fu, H. He, J. Tong, and J. Dong, "High-order cylindrical vector beams with tunable topological charge up to 14 directly generated from a microchip laser with high beam quality and high efficiency," *APL Photonics* 4(10), 106106 (2019). https://doi.org/10.1063/1.5119789.

181. P. Kumar, and N. Nishchal, "Modified Mach-Zehnder interferometer for determining the high-order topological charge of Laguerre-Gaussian vortex beams," *J. Opt. Soc. Am. A* 36(8), 1447–1455 (2019).

182. Y. Li, Y. Han, and Z. Cui, "Measuring the topological charge of vortex beams with gradually changing-period spiral spoke grating," *IEEE Photonics Technol. Lett.* 32(2), 101–104 (2020).

183. I. Nape, B. Sephton, Y.-W. Huang, A. Vallés, C.-W. Qiu, A. Ambrosio, F. Capasso, and A. Forbes, "Enhancing the modal purity of orbital angular momentum photons," *APL Photon* 5, 070802 (2020).

184. S. Hong, Y.S. Lee, H. Choi, C. Quan, Y. Li, S. Kim, and K. Oh, "Hollow silica photonic crystal fiber guiding 101 orbital angular momentum modes without phase distortion in C+ L band," *J. Lightw. Technol.* 38(5), 1010–1018 (2020).

185. R. Fickler, G. Campbell, B. Buchler, P.K. Lam, and A. Zeilinger, "Quantum entanglement of angular momentum states with quantum numbers up to 10010," *Proc. Natl. Acad. Sci. U. S. A.* 113(48), 13642–13647 (2016).

186. S. Fu, Y. Zhai, H. Zhou, J. Zhang, T. Wang, C. Yin, and C. Gao, "Demonstration of free-space one-to-many multicasting link from orbital angular momentum encoding," *Opt. Lett.* 44(19), 4753–4756 (2019).

187. Z. Qiao, Z. Wan, G. Xie, J. Wang, L. Qian, and D. Fan, "Multi-vortex laser enabling spatial and temporal encoding," *PhotoniX* 1, 13 (2020).

188. J. Serna, and J. Movilla, "Orbital angular momentum of partially coherent beams," *Opt. Lett.* 26(7), 405–407 (2001).

189. E.G. Abramochkin, and V.G. Volostnikov, *Modern optics of Gaussian beams* ("Fizmatlit" Publisher, Moscow, 2010, in Russian).

190. V.V. Kotlyar, A.A. Kovalev, and A.P. Porfirev, "Vortex astigmatic Fourier-invariant Gaussian beams," *Opt. Express* 27(2), 657–666 (2019).

191. P. Vaity, A. Aadhi, and R. Singh, "Formation of optical vortices through superposition of two Gaussian beams," *Appl. Opt.* 52(27), 6652–6656 (2013).

192. M.P. Backlund, M.D. Lew, A.S. Backer, S.J. Sahl, G. Grover, A. Agrawal, R. Piestun, and W.E. Moerner, "The double-helix point spread function enables precise and accurate measurement of 3D single-molecule localization and orientation," *Proc. SPIE Int. Soc. Opt. Eng.* 8590, 85900L (2013).

193. J.M. Hickmann, E.J.S. Fonseca, W.C. Soares, and S. Chávez-Cerda, "Unveiling a truncated optical lattice associated with triangular aperture using light's orbital angular momentum," *Phys. Rev. Lett.* 105(5), 053904 (2010).

194. L.E.E. de Araujo, and M.E. Anderson, "Measuring vortex charge with a triangular aperture," *Opt. Lett.* 36(6), 787–789 (2011).

195. P. Vaity, J. Banerji, and R.P. Singh, "Measuring the topological charge of an optical vortex by using a titled convex lens," *Phys. Lett. A* 377(15), 1154–1156 (2013).

196. D. Shen, and D. Zhao, "Measuring the topological charge of optical vortices with a twisting phase," *Opt. Lett.* 44(9), 2334–2337 (2019).

197. G. Liu, K. Wang, Y. Lee, D. Wang, P. Li, F. Gou, Y. Li, C. Tu, S. Wu, and H. Wang, "Measurement of the topological charge and index of vortex vector optical fields with a space-variant half-wave plate," *Opt. Lett.* 43(4), 823–826 (2018).

198. B. Lan, C. Liu, D. Rui, M. Chen, F. Shen, and H. Xian, "The topological charge measurement of the vortex beam based on dislocation self-reference interferometry," *Phys. Scr.* 94(5), 055502 (2019).

199. B. Kodatskii, A. Sevryugin, E. Shalimov, I. Tursunov, and V. Venediktov, "Comparative study of reference wave lacking measurement of topological charge of the incoming optical vortex," *Proc. SPIE* 11153, 111530G (2019).

200. S. Fu, Y. Zhai, J. Zhang, X. Liu, R. Song, H. Zhou, and C. Gao, "Universal orbital angular momentum spectrum analyzer for beams," *PhotoniX (Springer)* 1, 19 (2020).

201. Z. Zhang, X. Qiao, B. Midya, K. Liu, J. Sun, T. Wu, W. Liu, R. Agarwal, J.M. Jornet, S. Longhi, N.M. Litchinitser, and L. Feng, "Tunable topological charge vortex microlaser," *Science* 368(6492), 760–763 (2020).

202. K. Zhang, Y. Wang, Y. Yuan, and S.N. Burokur, "A review of orbital angular momentum vortex beams generation: from traditional methods to metasurfaces," *Appl. Sci.* 10, 1015 (2020).

203. K. Ciesielski, *Set theory for the working mathematician* (Cambridge University Press, Cambridge, 1997), https://doi.org/10.1017/CBO9781139173131.

204. A.A. Kovalev, and V.V. Kotlyar, "Orbital angular momentum of generalized cosine Gaussian beams with an infinite number of screw dislocations," *Optik* 242, 166863 (2021).

205. S.N. Khonina, V.V. Kotlyar, V.A. Soifer, P. Paakkonen, and J. Turunen, "Measuring the light field orbital angular momentum using DOE," *Opt. Mem. Neural Netw.* 10, 241–255 (2001).

206. T. Kaiser, D. Flamm, S. Schroter, and M. Duppare, "Complete modal decomposition of optical fibers using CGH-based correlation filter," *Opt. Express* 17(11), 9347–9356 (2009).

207. M.V. Vasnetsov, J.P. Torres, D.V. Petrov, and L. Torner, "Observation of the orbital angular momentum spectrum of a light beam," *Opt. Lett.* 28(23), 2285–2287 (2003).

208. A. Forbes, A. Dudley, and M. McLaren, "Creation and detection of optical modes with spatial light modulators," *Adv. Opt. Phot.* 8(2), 200 (2016).
209. A. D'Errico, R. D'Amelio, B. Piccirillo, F. Cardano, and L. Marrucci, "Measuring the complex orbital angular momentum spectrum and spatial mode decomposition of structured light beams," *Optica* 4(11), 1350–1357 (2017).
210. J. Pinnell, V. Rodriguez-Fajardo, and A. Forbes, "Single-step shaping of the orbital angular momentum spectrum of light," *Opt. Express* 27(20), 28009–28021 (2019).
211. Y. Yang, Q. Zhao, L. Liu, Y. Liu, C. Rosales-Guzman, and C. Qiu, "Manipulation of orbital angular momentum spectrum using pinhole plates," *Phys. Rev. Appl.* 12(6), 064007 (2019).
212. J. Leach, M.J. Padgett, S.M. Barnett, S. Franke-Arnold, and J. Courtial, "Measuring the orbital angular momentum of a single photon," *Phys. Rev. Lett.* 88(25 Pt 1), 257901 (2002).
213. A. Schulze, A. Dadley, D. Flamm, M. Duparre, and A. Forbes, "Measurements of the orbital angular momentum density of light by modal decomposition," *New J. Phys.* 15(7), 073025 (2013).
214. P. Bierdz, M. Kwon, C. Roncaioli, and H. Deng, "High fidelity detection of the orbital angular momentum of light by time mapping," *New J. Phys.* 15(11), 113062 (2013).
215. S. Li, P. Zhao, X. Feng, K. Cui, F. Liu, W. Zhang, and Y. Huang, "Measuring the orbital angular momentum spectrum with a single point detector," *Opt. Lett.* 43(19), 4607–4610 (2018).
216. E. Karimi, B. Piccirillo, E. Nagali, L. Marucci, and E. Santamato, "Efficient generation and sorting of orbital angular momentum eigenmodes of light by thermally tuned q-plates," *Appl. Phys. Lett.* 94(23), 231124 (2009).
217. G.C.G. Berghout, M.P.J. Lavery, J. Cortial, M.W. Beijersbergen, and M.J. Padgett, "Efficient sorting of orbital angular momenyum states of light," *Phys. Rev. Lett.* 105(15), 153601 (2010).
218. M. Mirhosseini, M. Malik, Z. Shi, and R.W. Boyd, "Efficient separation of the orbital angular momentum eigenstates of light," *Nat. Commun.* 4, 1–6 (2013).
219. H. Zucker, "Optical resonators with variable reflectivity mirrors," *Bell Syst. Tech. J.* 49(9), 2349–2376 (1970).
220. E. Karimi, G. Zito, B. Piccirillo, L. Marrucci, and E. Santamato, "Hypergeometric-Gaussian modes," *Opt. Lett.* 32(21), 3053–3055 (2007).
221. V.V. Kotlyar, A.A. Kovalev, and A.P. Porfirev, "Elliptic Gaussian optical vortices," *Phys. Rev. A* 95(5), 053805 (2017).
222. E.G. Abramochkin, and V.G. Volostnikov, "Beam transformation and nontransformed beams," *Opt. Commun.* 83(1–2), 123–135 (1991).
223. S.N. Khonina, V.V. Kotlyar, V.A.Soifer, P. Paakkonen, J. Simonen, and J. Turunen , "An analysis of the angular momentum of a light field in terms of angular harmonics," *J. Mod. Opt.* 48(10), 1543–1557 (2001).
224. Y. Yang, L. Wu, Y. Liu, D. Xie, Z.Jin, J. Li, G. Hu, and C. Qiu, "Deuterogenic plasmonic vortices," *Nano Lett.* 20(9), 6774–6779 (2020).
225. D.V. Petrov, "Vortex-edge dislocation interaction in a linear medium," *Opt. Commun.* 188(5–6), 307–312 (2001).
226. D.V. Petrov, "Splitting of an edge dislocation by an optical vortex," *Opt. Quant. Electr.* 34(8), 759–773 (2002).
227. M.V. Berry, "Optical currents," *J. Opt. A: Pure Appl. Opt.* 11(9), 094001 (2009).
228. V.V. Kotlyar, and A.A. Kovalev, "Optical vortex beams with a symmetric and almost symmetric OAM spectrum," *J. Opt. Soc. Am. A* 38(9), 1276–1283 (2021).
229. I. Hebri, and S. Rasouli, "Combined half-integer Bessel-like beams: A set of solutions of the wave equation," *Phys. Rev. A* 98(4), 043826 (2018).

230. A. Vasara, J. Turunen, and A.T. Friberg, "Realization of general nondiffracting beams with computer-generated holograms," *J. Opt. Soc. Am. A* 6(11), 1748–1754 (1989).

231. N.R. Heckenberg, R. McDaff, C.P. Smith, and A.G. White, "Generation of optical phase singularities by computer-generated holograms," *Opt. Lett.* 17(3), 221–223 (1992).

232. V.Yu. Bazhenov, M.S. Soskin, and M.V. Vasnetsov, "Screw dislocations in light wavefronts," *J. Mod. Opt.* 39(5), 985–990 (1992).

233. V.V. Kotlyar, S.N. Khonina, G.V. Uspleniev, M.V. Shinkarev, and V.A. Soifer, "The phase rotor filter," *J. Mod. Opt.* 39(5), 1147–1154 (1992).

234. M.V. Beijersbergen, R.P.C. Coerwinkel, M. Kristensen, and J.P. Woerdman, "Helical-wavefront laser beams produced with a spiral phase plate," *Opt. Commun.* 112(5–6), 321–327 (1994).

235. A. Fedotowsky, and K. Lehovec, "Optimal filter design for annular imaging," *Appl. Opt.* 13(12), 2919–2923 (1974).

236. S.N. Khonina, V.V. Kotlyar, V.A. Soifer, G.V. Uspleniev, and M.V. Shinkarev, "Trochoson," *Opt. Commun.* 91(3–4), 158–162 (1992).

237. L. Marrucci, C. Manzo, and D. Paparo, "Optical spin-to-orbital angular momentum conversion in inhomogeneous anisotropic media," *Phys. Rev. Lett.* 96(16), 163905 (2006).

238. D.N. Naik, and N.K. Viswanahan, "Generation of singular optical beams from fundamental Gaussian beam using Sagnac interferometer," *J. Opt.* 18(9), 095601 (2016).

239. L. Zhu, and J. Wang, "Arbitrary manipulation of spatial amplitude and phase using phase-only spatial light modulators," *Sci. Rep.* 4, 7441 (2015).

240. R.C. Devlin, A. Ambrosio, A. Rubin, J.B. Mueller, and F. Capasso, "Arbitrary spin-to-orbital angular momentum conversion of light," *Science* 358(6365), 896–901 (2017).

241. P. Miao, Z. Zhang, J. Sun, W. Walasik, S. Longhi, N.M. Litchintser, and L. Feng, "Orbital angular momentum microlaser," *Science* 353(6298), 464–467 (2016).

242. J. Zhang, Z. Guo, R. Li, W. Wang, A. Zhang, J. Liu, S. Qu, and J. Gao, "Circular polarization analyzer based on the combined coaxial Archimedes' spiral structure," *Plasmonics* 10(6), 1256–1261 (2015).

243. J. Zhang, Z. Guo, K. Zhou, L. Ran, L. Zhu, W. Wang, Y. Sun, F. Shen, J. Gao, and S. Liu, "Circular polarization analyzer based on an Archimedean nano-pinholes array," *Opt. Express* 23(23), 30523–30531 (2015).

244. H. Zhan, J. Li, K. Guo, and Z. Guo, "Generation of acoustic vortex beams with designed Fermat's spiral diffraction grating," *J. Acoust. Soc. Am.* 146(6), 4237–4243 (2019).

245. D. He, H. Yan, and B. Lu, "Interaction of the vortex and edge dislocation embedded in a cosh-Gaussian beam," *Opt. Commun.* 282(20), 4035–4044 (2009).

246. H. Chen, W. Wang, Z. Gao, and W. Li, "Splitting of an edge dislocation by a vortex emergent from a nonparaxial beam," *J. Opt. Soc. Am. B* 36(10), 2804–2809 (2019).

247. A.T. O'Neil, and J.Courtial, "Mode transformations in terms of the constituent Hermite-Gaussian or Laguerre-Gaussian modes and the variable-phase mode converter," *Opt. Commun.* 181(1–3), 35–45 (2000).

248. M.J. Padgett, and L. Allen, "Orbital angular momentum exchange in cylindrical-lens mode converters," *J. Opt. B: Quant. Semicl* 4(2), S17–S19 (2002).

249. R. Zeng, and Y. Yang, "Generation of an asymmetric optical vortex array with tunable singularity distribution," *J. Opt. Soc. Am. A* 38(3), 313–320 (2021).

250. H. Fan, H. Zhang, C. Cai, M. Tang, H. Li, J. Tang, and X. Li, "Flower-shaped optical vortex array," *Ann. Phys.* 533, 2000573 (2021).

251. J. Jin, X. Li, M. Pu, Y. Guo, P. Gao, M. Xu, Z. Zhang, and X. Luo, "Angular-multiplexed nultichannel optical vortex arrays generators based on geometric metasurface," *iScince* 24, 102107 (2021).

252. V.V. Kotlyar, A.A. Kovalev, A.P. Porfirev, and E.S. Kozlova, "Three different types of astigmatic Hermite-Gaussian beams with orbital angular momentum," *J. Opt.* 21(11), 115601 (2019).

253. J. Lin, J. Dellinger, P. Genevet, B. Cluzel, F. de Fornel, and F. Capasso, "Cosine-Gauss plasmon beam: A localized long-range nondiffracting surface wave," *Phys. Rev. Lett.* 109(9), 093904 (2012).

254. V. Kotlyar, and A. Kovalev, "Converting an array of edge dislocations into a multivortex beam," *J. Opt. Soc. Am. A* 38(5), 719–726 (2021).

255. B. Lu, and P. Wu, "Analytical propagation equation of astigmatic Hermite-Gaussian beams through a 4x4 paraxial optical systems and their symmetrizing transformation," *Opt. Las. Technol.* 35, 497–504 (2003).

256. Y.F. Chen, C.C. Chay, C.Y. Lee, J.C. Tung, H.C. Liang, and K.T. Huang, "Characterizing the propagation evolution of wave patterns and vortex structures in astigmatic transformations of Hermite-Gaussian beams," *Las. Phys.* 28, 015002 (2017).

257. E.G. Abramochkin, E.U. Razueva, and V.G. Volostnikov, "Hermite-Laguerre-Gaussian beams in astigmatic optical systems," *Proc. SPIE* 7009, 70090M (2008).

258. A.Y. Bekshaev, M.S. Soskin, and M.V. Vasnetsov, "Transformation of higher-order optical vortices upon focusing by an astigmatic lens," *Opt. Commun.* 241(4–6), 237–247 (2004).

259. A.Y. Bekshaev, and A.I. Karamoch, "Astigmatic telescopic transformation of a high-order optical vortex," *Opt. Commun.* 281(23), 5687–5696 (2008).

260. K. Zhu, J. Zhu, Q. Su, and H. Tang, "Propagation properties of an astigmatic sin-Gaussian beam in strongly nonlocal nonlinear media," *Appl. Sci.* 9(1), 71 (2019).

261. T.D. Huang, and T.H. Lu, "Large astigmatic laser cavity modes and astigmatic compensation," *Appl. Phys. B* 124(5), 72 (2018).

262. J. Pan, Y. Shen, Z. Wan, X. Fu, H. Zhang, and Q. Liu, "Index-tunable structured-light beams from a laser with a intracavity astigmatic mode converter," *Phys. Rev. Appl.* 14(4), 044048 (2020).

263. H. Chen, Z. Gao, H. Yang, S. Xiao, F. Wang, X. Huang, and X. Liu, "Evolution behavior of two edge dislocations passing through an astigmatic lens," *J. Mod. Opt.* 59(21), 1863–1872 (2012).

264. H. Yan, and B. Lu, "Vortex-edge dislocation interaction in the presence of an astigmatic lens," *Opt. Commun.* 282(5), 717–726 (2009).

265. V.V. Kotlyar, A.A. Kovalev, and A.G. Nalimov, "Converting an nth-order edge dislocation to a set of optical vortices," *Optik* 243, 167453 (2021).

266. A.M. Sedletsky, "Asymptotics of nulls of a degenerate hypergeometric function," *Mathematicheskii Zametki* 82(2), 262–271 (2007).

267. J.H. McLeod, "The axicon: A new type of optical element," *J. Opt. Soc. Am.* 44(8), 592–597 (1954).

268. G. Indebetouw, "Nondiffracting optical fields: Some remarks on their analysis and synthesis," *J. Opt. Soc. Am. A* 6(1), 150–152 (1989).

269. V.V. Kotlyar, A.A. Kovalev, and A.P. Porfirev, "Optimal phase element for generating a perfect optical vortex," *J. Opt. Soc. Am. A* 33(12), 2376–2384 (2016).

270. C.E.R. Caron, and R.M. Potuliege, "Bessel-modulated Gaussian beams with quadratic radial dependence," *Opt. Commun.* 164(1–3), 83–93 (1999).

271. G. Li, H. Lee, and E. Wolf, "New generalized Bessel-Gauss laser beams," *J. Opt. Soc. Am. A* 21, 640–646 (2004).

272. V.V. Kotlyar, A.A. Kovalev, and V.A. Soifer, "Hankel-Bessel laser beams," *J. Opt. Soc. Am. A* 29(5), 741–747 (2012).

273. Y. Ismail, N. Khilo, V. Belyi, and A. Forbes, "Shape invariant higher-order Bessel-like beams carrying orbital angular momentum," *J. Opt.* 14(8), 085703 (2012).

274. S.C. Pei, and J.J. Ding, "Eigenfunctions of linear canonical transform," *IEEE Trans. Signal Process.* 50(1), 11–26 (2002).
275. A.P. Prudnikov, Y.A. Brychkov, and O.I. Marichev, *Integrals and series, volume 2, special functions* (Gordon and Breach, New York, 1986).
276. V.V. Kotlyar, A.A. Kovalev, and D.S. Kalinkina, "Fractional-order-Bessel Fourier-invariant optical vortices," *Opt. Commun.* 492, 126974 (2021).
277. V.V. Kotlyar, and A.A. Kovalev, "Family of hypergeometric laser beams," *J. Opt. Soc. Amer. A* 25(1), 262–270 (2008). https://doi.org/10.1364/JOSAA.25.000262.
278. E. Karimi, B. Piccirillo, L. Marrucci, and E. Santamato, "Improved focusing with hypergeometric-Gaussian type-II optical modes," *Opt. Express* 16(25), 21069–21075 (2008). https://doi.org/10.1364/OE.16.021069.
279. V.V. Kotlyar, A.A. Kovalev, R.V. Skidanov, S.N. Khonina, and J. Turunen, "Generating hypergeometric laser beams with a diffractive optical element," *Appl. Opt.* 47(32), 6124–6133 (2008). https://doi.org/10.1364/AO.47.006124.
280. S.N. Khonina, S.A. Balalaev, R.V. Skidanov, V.V. Kotlyar, B. Paivanranta, and J. Turunen, "Encoded binary diffractive element to form hypergeometric laser beams," *J.Opt. A: Pure Appl. Opt.* 11(6), 065702 (2009). https://doi.org/10.1088/1464-4258/11/6/065702.
281. V.V. Kotlyar, and A.A. Kovalev, "Nonparaxial hypergeometric beams," *J. Opt. A: Pure Appl. Opt.* 11(4), 045711 (2009). https://doi.org/10.1088/1464-4258/11/4/045711.
282. V.V. Kotlyar, A.A. Kovalev, and V.A. Soifer, "Lensless focusing of hypergeometric laser beams," *J. Opt.* 13(7), 075703 (2011). https://doi.org/10.1088/2040-8978/13/7/075703.
283. B. de Lima Bernardo, and F. Moraes, "Data transmission by hypergeometric modes through a hyperbolic-index medium," *Opt. Express* 19(12), 11264–11270 (2011). https://doi.org/10.1364/OE.19.011264.
284. V.V. Kotlyar, A.A. Kovalev, and A.G. Nalimov, "Propagation of hypergeometric laser beams in a medium with a parabolic refractive index," *J. Opt.* 15(12), 125706 (2013). https://doi.org/10.1088/2040-8978/15/12/125706.
285. L. Bian, and B. Tang, "Propagation properties of hypergeometric-Gaussian type-II beams through the quadratic-index medium," *Appl. Opt.* 57(17), 4735–4742 (2018). https://doi.org/10.1364/AO.57.004735.
286. B. Tang, C. Jiang, and H. Zhu, "Fractional Fourier transform for confluent hypergeometric beams," *Phys. Lett. A* 376(38–39), 2627–2631 (2012). https://doi.org/10.1016/j.physleta.2012.07.017.
287. J. Peng, Z. Shan, Y. Yuan, Z. Cui, W. Huang, and J. Qu, "Focusing properties of hypergeometric Gaussian beam through a high numerical-aperture objective," *Prog. Electromagn. Res.* 51, 21–26 (2015). https://doi.org/10.2528/pierl14101304.
288. T. Bin, J. Chun, Z. Haibin, Z. Xin, and W. Shuai, "The propagation of hypergeometric beams through an annular apertured paraxial ABCD optical system, *Laser Phys.* 24(12), 125002 (2014). https://doi.org/10.1088/1054-660x/24/12/125002.
289. J. Li, and Y. Chen, "Propagation of confluent hypergeometric beam through uniaxial crystals orthogonal to the optical axis," *Opt. Laser Technol.* 44(5), 1603–1610 (2012). https://doi.org/10.1016/j.optlastec.2011.11.041.
290. Y. Zhu, L. Zhang, Z. Hu, and Y. Zhang, "Effects of non-Kolmogorov turbulence on the spiral spectrum of hypergeometric-Gaussian laser beams," *Opt. Express* 23(7), 9137–9146 (2015). https://doi.org/10.1364/OE.23.009137.
291. X. Wang, L. Wang, B. Zheng, Z. Yang, and S. Zhao, "Effects of oceanic turbulence on the propagation of hypergeometric-Gaussian beam carrying orbital angular momentum." In *Proceedings of the 2020 IEEE international conference on communications*

workshops, Dublin, Ireland, 7–11 June 2020. https://doi.org/10.1109/iccworkshops49005 .2020.9145217.

292. X. Wang, L. Wang, and S. Zhao, "Research on hypergeometric-Gaussian vortex beam propagating under oceanic turbulence by theoretical derivation and numerical simulation," *J. Mar. Sci. Eng.* 9(4), 442 (2021). https://doi.org/10.3390/jmse9040442.

293. L. Bian, and B. Tang, "Evolution properties of hypergeometric-Gaussian type-II beams in strongly nonlocal nonlinear media," *J. Opt. Soc. Am. B* 35(6), 1362–1367 (2018). https://doi.org/10.1364/JOSAB.35.001362.

294. T. Bin, B. Lirong, Z. Xin, and C. Kai, "Propagation of hypergeometric Gaussian beams in strongly nonlocal nonlinear media," *Laser Phys.* 28(1), 015001 (2018). https://doi.org /10.1088/1555-6611/aa9628.

295. A.A. Kovalev, V.V. Kotlyar, and A.P. Porfirev, "Auto-focusing accelerating hyper-geometric laser beams," *J. Opt.* 18(2), 025610 (2016). https://doi.org/10.1088/2040-8978/18 /2/025610.

296. Y. Zhu, Y. Zhang, and G. Yang, "Evolution of orbital angular momentum mode of the autofocusing hypergeometric-Gaussian beams through moderate-to-strong anisotropic non-Kolmogorov turbulence," *Opt. Commun.* 405, 66–72 (2017). https://doi.org/10 .1016/j.optcom.2017.07.047.

297. A.A.A. Ebrahim, F. Saad, and L. Ez-zariy, "Theoretical conversion of the hypergeometric-Gaussian beams family into a high-order spiraling Bessel beams by a curved fork-shaped hologram," *Opt. Quant. Electron.* 49, 169 (2017). https://doi.org/10.1007/ s11082-017-0987-6.

298. G. Jin, L. Bian, L. Huang, and B. Tany, "Radiation forces of hypergeometric-Gaussian type-II beams acting on a Rayleigh dielectric sphere," *Opt. Las. Techn.* 126, 106124 (2020). https://doi.org/10.1016/j.optlastec.2020.106124.

299. R.L. Phillips, and L.C. Andrews, "Spot size and divergence for Laguerre-Gaussian beams of any order," *Appl. Opt.* 22(5), 643–644 (1983). https://doi.org/10.1364/AO.22 .000643.

300. C. Alpmann, C. Scholer, and C. Denz, "Elegant Gaussian beams for enhanced optical manipulation," *Appl. Phys. Lett.* 106(24), 241102 (2015). https://doi.org/10.1063/1 .4922743.

301. N. Nossir, L. Dalil-Essakali, and A. Belafhal, "Optical trapping of particles by radiation forces of doughnut laser beams in the Rayleigh regime," *Opt. Quantum Electron.* 53(2), 100 (2021). https://doi.org/10.1007/s11082-021-02752-y.

302. J. Su,, N. Li, J. Mou, Y. Liu, X. Chen, and H. Hu, "Simultaneous trapping of two types of particles with focused elegant third-order hermite–Gaussian beams," *Micromachines* 12(7), 769 (2021). https://doi.org/10.3390/mi12070769.

303. D. Wei, S. Li, J. Zeng, X. Zhu, T. Chen, Y. Cai, and J. Yu, "Comparative study of spiral spectrum of elegant and standard Laguerre–Gaussian beams in atmospheric turbulence," *J. Russ. Laser Res.* 41(4), 364–372 (2020). https://doi.org/10.1007/s10946-020 -09887-5.

304. M. Dong, and Y. Yang, "Coherent vortices properties of partially coherent elegant laguerre-Gaussian beams in the free space," *Opt. Photonics News* 10(6), 159–166 (2020). https://doi.org/10.4236/opj.2020.106017.

305. E.G. Abramochkin, and V.G. Volostnikov, "Generalized Gaussian beams," *J. Opt. A: Pure Appl. Opt.* 6(5), S157–S161 (2004).

306. V.V. Kotlyar, A.A. Kovalev, and A.P. Porfirev, "Vortex Hermite–Gaussian laser beams," *Opt. Lett.* 40(5), 701–704 (2015).

307. M. Bock, S. Das, and R. Grunwald, "Ultrashort highly localized wavepackets," *Opt. Express* 20(11), 12563–12578 (2012).

308. X. Liu, and J. Pu, "Investigation on the scintillation reduction of elliptical vortex beams propagating in atmospheric turbulence," *Opt. Express* 19(27), 26444–26450 (2011).

309. L.C. Andrews, and R.L. Phillips, *Laser beam propagation through random media* (SPIE Press, Bellingham, Washington, 1998).

310. X. Zhang, T. Xia, S. Cheng, and S. Tao, "Free-space information transfer using the elliptic vortex beam with fractional topological charge," *Opt. Commun.* 431, 238–244 (2019).

311. K. Wu, Y. Huai, T. Zhao, and Y. Jin, "Propagation of partially coherent four-petal elliptic Gaussian vortex beams in atmospheric turbulence," *Opt. Express* 26(23), 30061–30075 (2018).

312. R.V. Skidanov, and M.A. Rykov, "The modification of laser beam for optimization of optical trap force characteristics," *Comput. Opt.* 37(4), 431–435 (2013).

313. D.A. Belousov, A.V. Dostovalov, V.P. Korolkov, and S.L. Mikerin, "A microscope image processing method for analyzing TLIPSS structures," *Comput. Opt.* 43(6), 936–945 (2019).

314. A.V. Dostovalov, K.A. Okotrub, K.A. Bronnikov, V.S. Terentyev, V.P. Korolkov, and S.A. Babin, "Influence of femtosecond laser pulse repetition rate on thermochemical laser-induced periodic surface structures formation by focused astigmatic Gaussian beam," *Laser Phys. Lett.* 16(2), 026003 (2019).

315. A.V. Dostovalov, T.J.Y. Derrien, S.A. Lizunov, F. Přeučil, K.A. Okotrub, T. Mocek, V.P. Korolkov, S.A. Babin, and N.M. Bulgakova, "LIPSS on thin metallic films: New insights from multiplicity of laser-excited electromagnetic modes and efficiency of metal oxidation," *Appl. Surf. Sci.* 491, 650–658 (2019).

316. Y. Cai, and Q. Lin, "Decentered elliptical Gaussian beam," *Appl. Opt.* 41(21), 4336–4340 (2002).

317. V.V. Kotlyar, A.A. Kovalev, and A.P. Porfirev, "Astigmatic laser beams with a large orbital angular momentum," *Opt. Express* 26(1), 141–156 (2018).

318. P.S. Bullen, "The power means," in *Handbook of means and their inequalities* (Kluwer, Dordrecht, 2003), pp. 175–265. https://doi.org/10.1007/978-94-017-0399-4.

319. M.V. Berry, M.R. Jeffrey, and M. Mansuripur, "Orbital and spin angular momentum in conical diffraction," *J. Opt. A: Pure Appl. Opt.* 2005, 7(11), 685–690 (2005).

320. Z. Mei, O. Korotkova, D. Zhao, and Y. Mao, "Self-focusing vortex beams," *Opt. Lett.* 46(10), 2384–2387 (2021).

321. S. Goorden, J. Bertolotti, and A. Mosk, "Superpixel-based spatial amplitude and phase modulation using a digital micromirror device," *Opt. Express* 22(15), 17999–18009 (2014).

322. O. Mendoza-Yero, G. Mínguez-Vega, and J. Lancis, "Encoding complex fields by using a phase-only optical element," *Opt. Lett.* 39(7), 1740–1743 (2014).

323. A.A. Kovalev, V.V. Kotlyar, and D.S. Kalinkina, "Propagation-invariant off-axis elliptic Gaussian beams with the orbital angular momentum," *Photonics* 8(6), 190 (2021).

324. Z. Liu, Y. Liu, Y. Ke, Y. Liu, W. Shu, H. Luo, and S. Wen, "Generation of arbitrary vector vortex beams on hybrid-order Poincare sphere," *Photon. Res.* 5(1), 15–21 (2017). https://doi.org/10.1364/PRJ.5.000015.

325. S. Fu, Y. Zhai, T. Wang, C. Yin, and C. Gao, "Tailoring arbitrary hybrid Poincare beams through a single hologram," *Appl. Phys. Lett.* 111(21), 211101 (2017). https://doi.org/10.1063/1.5008954.

326. Y. Zhang, P. Chen, S. Ge, T. Wei, J. Tang, W. Hu, and Y. Lu, "Spin-controlled massive channels of hybrid-order Poincare sphere beams," *Appl. Phys. Lett.* 117(8), 081101 (2020). https://doi.org/10.1063/5.0020398.

327. J. Liu, X. Chen, Y. He, L. Lu, H. Ye, G. Chai, S. Chen, and D. Fan, "Generation of arbitrary cylindrical vector vortex beams with cross-polarized modulation," *Res. Phys.* 19, 103455 (2020). https://doi.org/10.1016/j.rinp.2020.103455.

328. G. Arora, S. Deepa, S.N. Khan, and P. Senthilkumaran, "Detection of degenerate Stokes index states," *Sci. Rep.* 10(1), 20759 (2020). https://doi.org/10.1038/s41598-020-77365-8.

329. G. Arora, Ruchi, and P. Senthilkumaran, "Hybrid order poincare spheres for stokes singularities," *Opt. Lett.* 45(18), 5136–5139 (2020). https://doi.org/10.1364/OL.400946.

330. S.S. Stafeev, V.V. Kotlyar, A.G. Nalimov, M.V. Kotlyar, and L. O'Faolain, "Subwavelength gratings for polarization conversion and focusing of laser light," *Photonics Nanostructures: Fundam. Appl.* 27, 32–41 (2017). https://doi.org/10.1016/j.photonics.2017.09.001.

331. P. Lochab, P. Senthilkumaran, and K. Khare, "Designer vector beams maintaining a robust intensity profile on propagation through turbulence," *Phys. Rev. A* 98(2), 023831 (2018). https://doi.org/10.1103/PhysRevA.98.023831.

332. M. Berry, "Geometry of phase and polarization singularities illustrated by edge diffraction and the fides," *Proc. SPIE* 4403 (2001). https://doi.org/10.1117/12.428252.

333. I. Freund, "Polarization singularity indices in Gaussian laser beams," *Opt. Commun.* 201(4–6), 251–270 (2002). https://doi.org/10.1016/S0030-4018(01)01725-4.

334. A.A. Kovalev, and V.V. Kotlyar, "Gaussian beams with multiple polarization singularities," *Opt. Commun.* 423, 111–120 (2018). https://doi.org/10.1016/j.optcom.2018.04.023.

335. M. Born, and E. Wolf, *Principles of optics* (Pergamon Press, Oxford, 1968).

336. Q. Zhan, "Cylindrical vector beams: From mathematical concepts to applications," *Adv. Opt. Photon.* 1(1), 1–57 (2009). https://doi.org/10.1364/AOP.1.000001.

337. V.V. Kotlyar, S.S. Stafeev, and A.A. Kovalev, "Sharp focusing of a light field with polarization and phase singularities of an arbitrary order," *Comput. Opt.* 43(3), 337–346 (2019). https://doi.org/10.18287/2412-6179-2019-43-3-337-346.

338. B. Richards, and E. Wolf, "Electromagnetic diffraction in optical systems. II. Structure of the image field in an aplanatic system," *Proc. R. Soc. Lond. A* 253(1274), 358–379 (1959). https://doi.org/10.1098/rspa.1959.0200.

339. V.V. Kotlyar, A.A. Kovalev, S.S. Stafeev, A.G. Nalimov, and S. Rasouli, "Tightly focusing vector beams containing V-points polarization singularities," *Opt. Las. Techn.* 145 107479 (2022).

340. F. Cardano, E. Karimi, S. Slussarenko, L. Marrucci, C. de Lisio, and E. Santamato, "Polarization pattern of vector vortex beams generated by q-plates with different topological charges," *Appl. Opt.* 51(10), C1–C6 (2012). https://doi.org/10.1364/AO.51.0000C1.

341. F. Cardano, E. Karimi, L. Marrucci, C. de Lisio, and E. Santamato, "Generation and dynamics of optical beams with polarization singularities," *Opt. Express* 21(7), 8815–8820 (2013). https://doi.org/10.1364/OE.21.008815.

342. P. Kumar, S. Pal, N. Nishchal, and P. Senthilkumaran, "Non-interferometric technique to realize vector beams embedded with polarization singularities," *J. Opt. Soc. Am. A* 37(6), 1043–1052 (2020). https://doi.org/10.1364/JOSAA.393027.

343. B. Khajavi, and E.J. Galvez, "High-order disclinations in space-variant polarization," *J. Opt.* 18(8), 084003 (2016). https://doi.org/10.1088/2040-8978/18/8/084003.

344. L. Lu, Z. Wang, and Y. Cai, "Propagation properties of phase-locked radially-polarized vector fields array in turbulent atmosphere," *Opt. Express* 29(11), 16833–16844 (2021). https://doi.org/10.1364/OE.427003.

345. M. Meier, V. Romano, and T. Feurer, "Material processing with pulsed radially and azimuthally polarized laser radiation," *Appl. Phys. A* 86(3), 329–334 (2007). https://doi.org/10.1007/s00339-006-3784-9.

346. P. Meng, S. Pereira, and P. Urbach, "Confocal microscopy with a radially polarized focused beam," *Opt. Express* 26(23), 29600–29613 (2018). https://doi.org/10.1364/OE.26.029600.

347. L. Carretero, P. Acebal, and S. Blaya, "Three-dimensional analysis of optical forces generated by an active tractor beam using radial polarization," *Opt. Express* 22(3), 3284–3295 (2014). https://doi.org/10.1364/OE.22.003284.

348. J.F. Nye, "Polarization effects in the diffraction of electromagnetic waves: The role of disclinations," *Proc. R. Soc. Lond.* 387, 105–132 (1983).

349. X. Wang, J. Ding, W. Ni, C. Guo, and H. Wang, "Generation of arbitrary vector beams with a spatial light modulator and a common path interferometric arrangement," *Opt. Lett.* 32(24), 3549–3551 (2007).

350. D. Naidoo, F.S. Roux, A. Dudley, I. Litvin, B. Piccirillo, and L. Marrucci, "Controlled generation of higher-order poincare sphere beams from a laser," *Nat. Phot.* 10(5), 327–333 (2016).

351. Z. Bomzon, G. Biener, V. Kleiner, and E. Hasman, "Radially and azimuthally polarized beams generated by space-variant dielectric subwavelength gratings," *Opt. Lett.* 27(5), 285–287 (2002).

352. V.V. Kotlyar, S.S. Stafeev, A.G. Nalimov, and L. O'Faolain, "Subwavelength grating-based metalens for focusing of laser light," *Appl. Phys. Lett.* 114(14), 141107 (2019).

353. Z. Ren, Z. Chen, X. Wang, J. Ding, and H. Wang, "Polarization interferometric prism: A versatile tool for generation of vector fields, measurement of topological charges, and implementation of a spin-orbit controlled-not gate," *Appl. Phys. Lett.* 118(1), 011105 (2021).

354. V. Kumar, and N.K. Viswanathan, "Topological structures in the poynting vector field: An experimental realization," *Opt. Lett.* 38(19), 3886–3889 (2013).

355. M. Born, and E. Wolf, "Principles of optics," *Pergamon Press, Oxford*, p. 836 (1980).

356. G. Arora, Ruchi, P. Senthilkumaran, "Hybrid order poincare spheres for stoks singularities," *Opt. Lett.* 45(18), 5136 (2020).

357. V.V. Kotlyar, A.G. Nalimov, and S.S. Stafeev, "Inversion of the axial projection of the spin angular momentum in the region of the backward energy flow in sharp focus," *Opt. Express* 28(23), 33830–33839 (2020).

358. T. Bauer, M. Neugebauer, G. Leuchs, and P. Banzer, "Optical polarization Möbius strips and points of purely transverse spin density," *Phys. Rev. Lett.* 117(1), 013601 (2016).

359. T. Bauer, P. Banser, E. Karimi, S. Orlov, A. Rubano, L. Marrucci, E. Santamato, R.W. Boyd, and G. Leuchs, "Observation of optical polarization Möbius strips," *Science* 347(6225), 964–966 (2015).

360. V.V. Kotlyar, A.G. Nalimov, A.A. Kovalev, A.P. Porfirev, and S.S. Stafeev, "Spin-orbit and orbit-spin conversion in the sharp focus of laser light: Theory and experiment," *Phys. Rev. A* 102(3), 033502 (2020).

361. V.V. Kotlyar, S.S. Stafeev, and A.G. Nalimov, "Sharp focusing of a hybrid vector beam with a polarization singularity," *Photonics* 8(6), 227 (2021).

Index

For Product Safety Concerns and Information please contact our EU
representative GPSR@taylorandfrancis.com
Taylor & Francis Verlag GmbH, Kaufingerstraße 24, 80331 München, Germany

www.ingramcontent.com/pod-product-compliance
Lightning Source LLC
Chambersburg PA
CBHW060330220326
41598CB00023B/2666

* 9 7 8 1 0 3 2 3 5 3 0 9 8 *